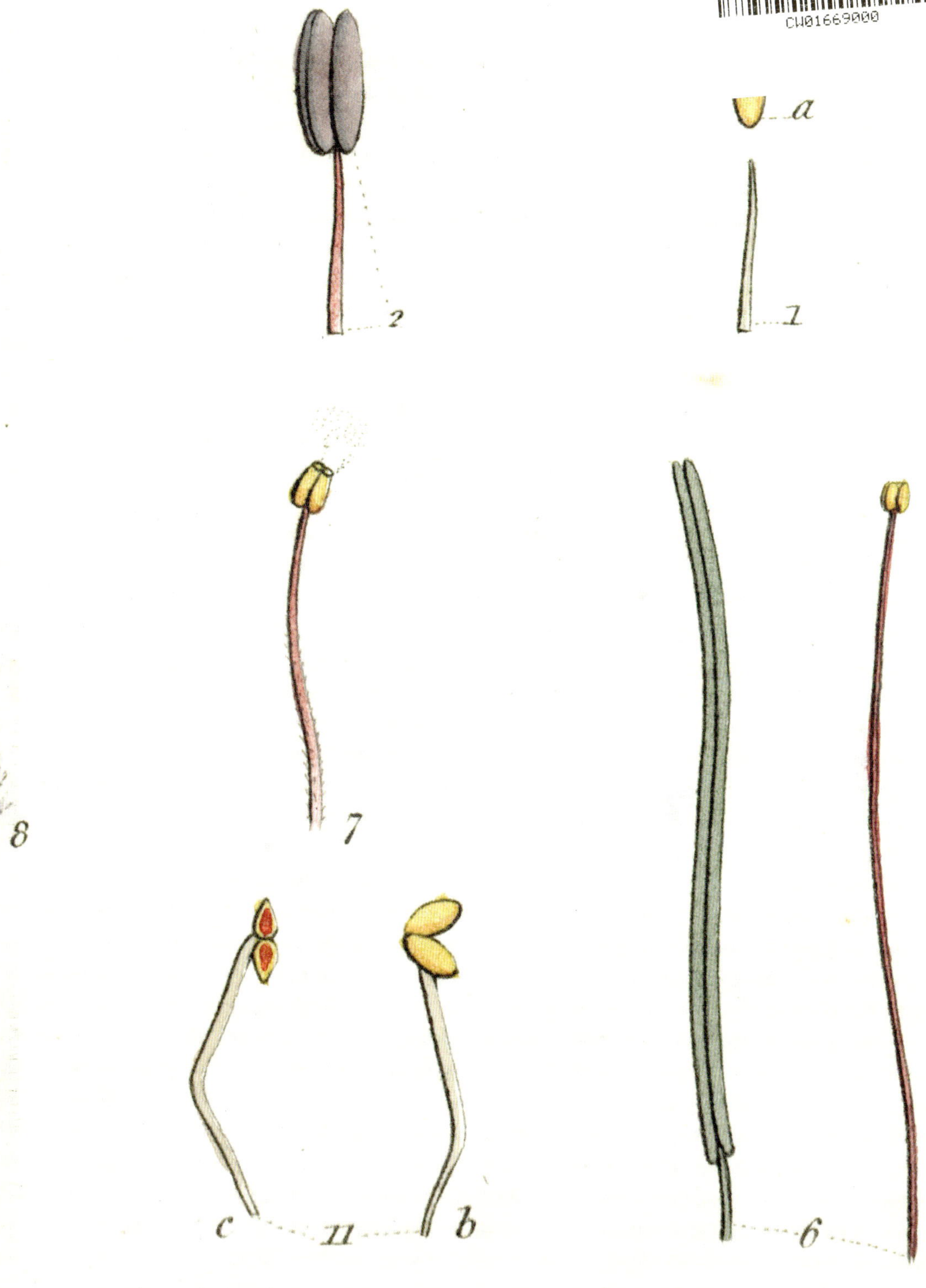

Nov.r 1, 1788.

James Sowerby

THE ENLIGHTENMENT'S
NATURAL HISTORIAN

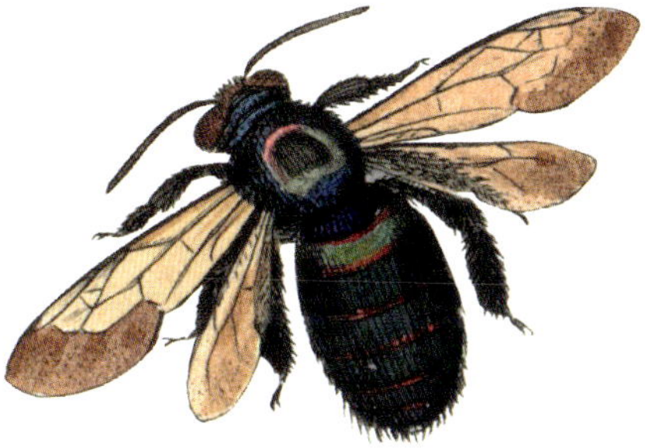

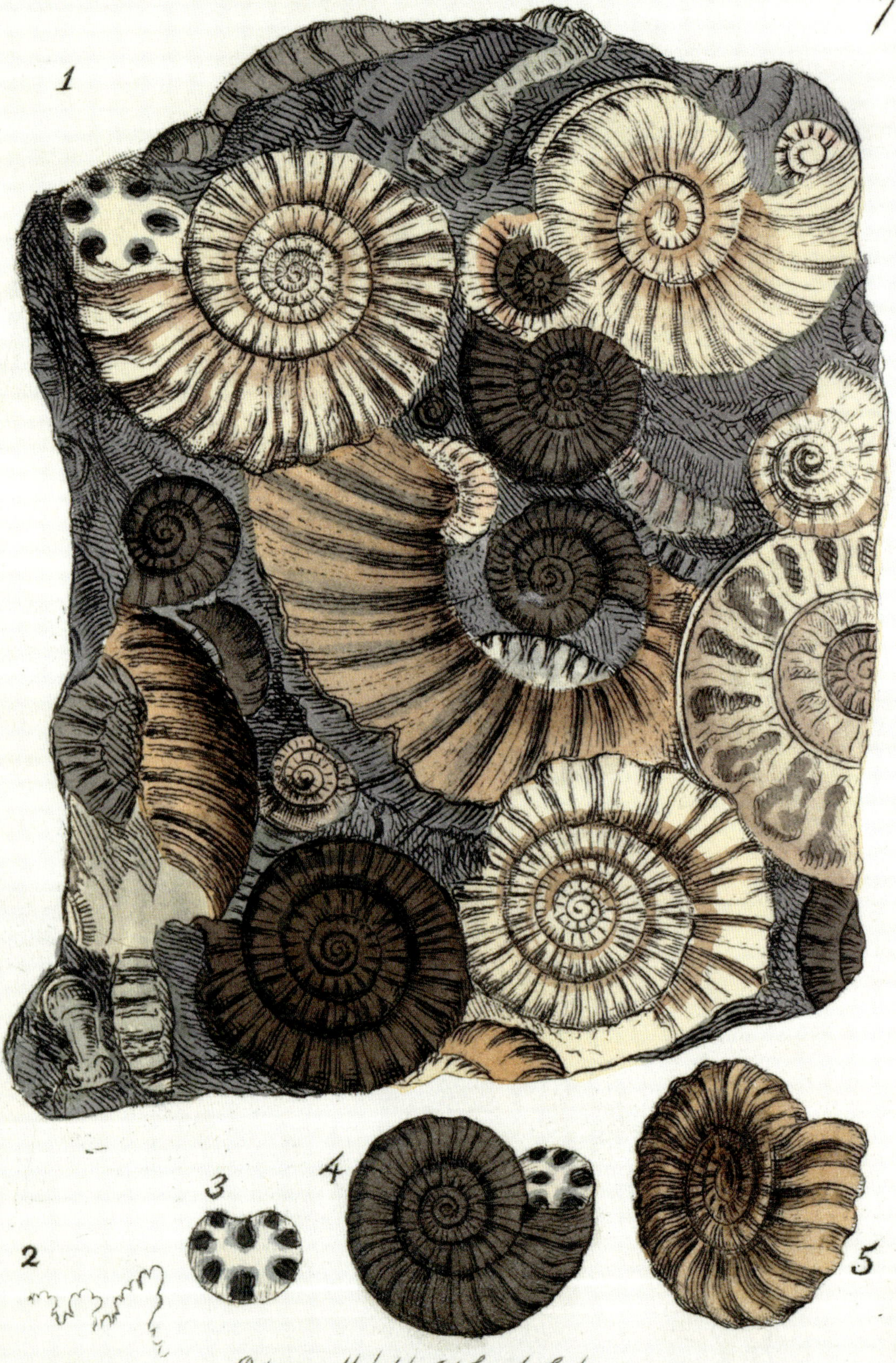

Oct 1 1814 published by Jas Sowerby London.

James Sowerby

THE ENLIGHTENMENT'S NATURAL HISTORIAN

PAUL HENDERSON

To my grandsons Isaac and Jonah

Royal Botanic Gardens Kew

NATURAL HISTORY MUSEUM

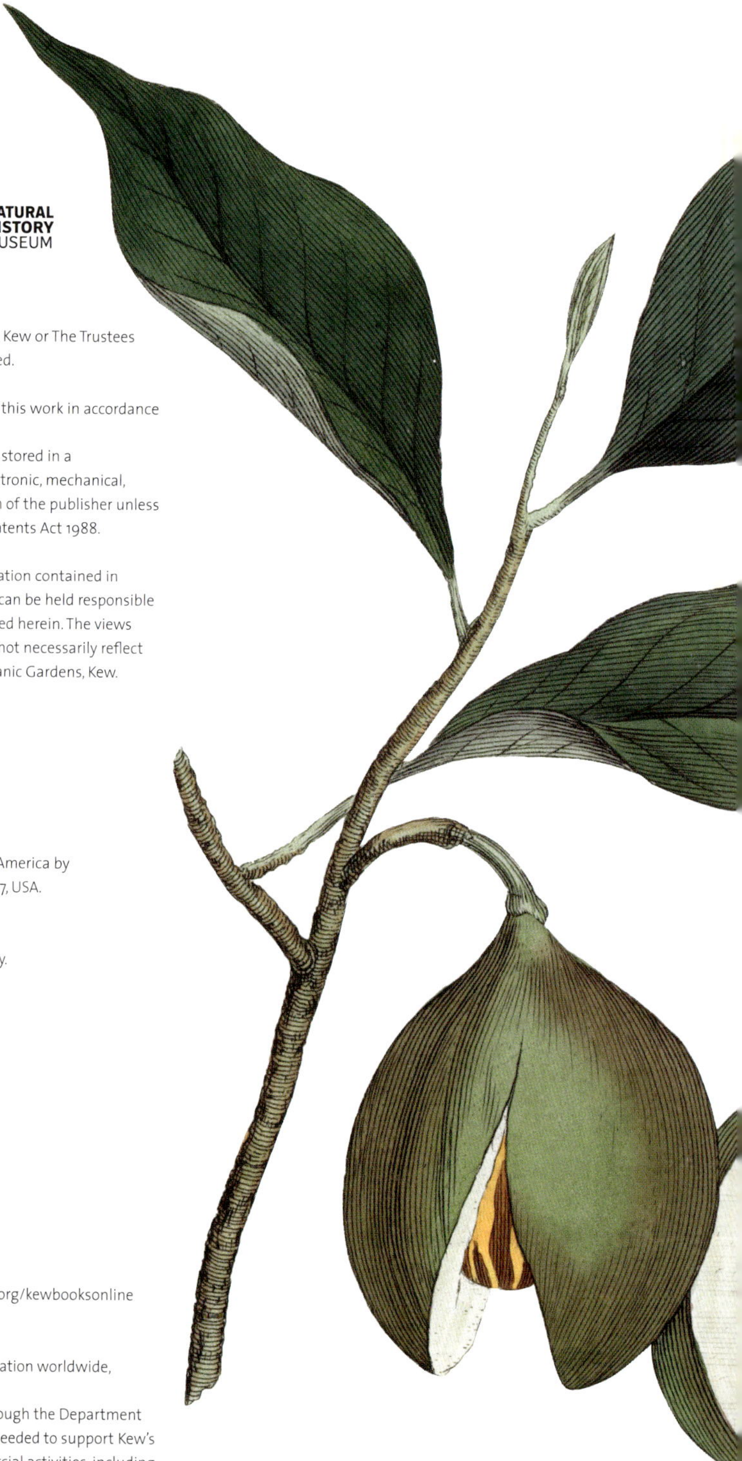

First published in 2015 by the Royal Botanic Gardens, Kew,
Richmond, Surrey, TW9 3AB, UK
www.kew.org
ISBN 978 1 84246 596 7
eISBN 978 1 84246 597 4

Distributed on behalf of the Royal Botanic Gardens, Kew in North America by the University of Chicago Press, 1427 East 60th St, Chicago, IL 60637, USA.

British Library Cataloguing in Publication Data
A catalogue record for this book is available from the British Library.

Copyediting: Michelle Payne
Proofreading: Matthew Seal
Cover, text design and layout: Heather Bowen
Production Management: Georgina Smith

Printed and bound in Italy by Printer Trento

FSC www.fsc.org MIX Paper from responsible sources FSC® C015829

For information or to purchase all Kew titles please visit shop.kew.org/kewbooksonline or email publishing@kew.org

Kew's mission is to inspire and deliver science-based plant conservation worldwide, enhancing the quality of life.
Kew receives about half of its running costs from Government through the Department for Environment, Food and Rural Affairs (Defra). All other funding needed to support Kew's vital work comes from members, foundations, donors and commercial activities, including book sales.

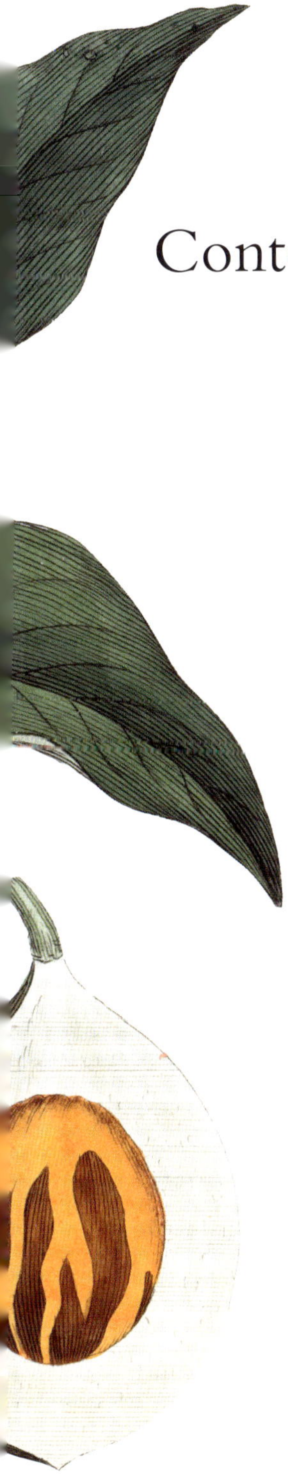

Contents

Acknowledgements

The pleasures of researching and writing about James Sowerby have been greatly enhanced by numerous friends, colleagues and others, who have generously shared their insights, knowledge and opinions. To all I am much indebted. But the start of the project was energised most by Ron Cleevely's significant scholarship; without his early publications and subsequent interest, my task would have been much greater.

The 'Sowerby world' covers so much of natural history that I would have failed without the advice and candid criticisms from those more knowledgeable than me. I thank several colleagues from the Natural History Museum: Charlie Jarvis for botanical help and reviews, especially on matters Linnaean and historical, Dick Vane-Wright on entomological details and history, Chris Stanley on mineralogy and colour, Alan Hart and Peter Tandy on mineral specimens, Sara Russell and Caroline Smith on meteorites, Geoff Boxshall on copepod parasites, Zoe Hughes and Noel Morris on Sowerby's fossil specimens, Mark Spencer on Sowerby's herbarium, and Holger Thues on Sowerby's models of fungi.

I have had stimulating and rewarding discussions with Stephen Harris (Oxford University Plant Sciences) on *Flora Graeca*, Sarah Watkinson (Oxford University Plant Sciences) and Paul Cannon (Royal Botanic Gardens, Kew) on mycology and wood rot, and Tom Kennett (Linnean Society) on James Edward Smith. I thank them also for their constructive reviews on parts of the draft manuscript (any remaining errors being entirely my responsibility). Hugh Torrens (Keele University) helped over the topic of early scientific societies, as did the late Colin Pillinger (Open University) on the history of meteoritics, and Paul Smith (Oxford University) on Sowerby's whale.

My research has taken me to many libraries and places. The fine collection of Sowerby specimens, manuscripts and books at the Natural History Museum is a rich vein of fascinating material; the library staff has been fantastic in aiding my discoveries. It is perhaps invidious to name some when many contributed but I am especially thankful to Andrea Hart for bringing my attention to, and giving advice on, botanical artwork, and to Hellen Pethers for her continuing interest, questions and support. The Royal Botanic Gardens, Kew and the Linnean Society – both rich in Sowerby-related material – provided much assistance and helpful research. My special thanks to Chris Mills and Lynn Parker at Kew, and Elaine Charwat and Gina Douglas at the Linnean Society. Other libraries which helped with important information were: Bristol University, the British Library, the Cornwall Record Office, the Geological Society, the Horniman Museum, Oxford University, the Royal Horticultural Society Linden Library, the Royal Society, and University College London. The State Hermitage Museum, St Petersburg, also gave special access to material. I also thank the Royal Horticultural Society, the Linnean Society and the Oxford University Museum of Natural History for their generous approach towards the provision of publishable images. I have greatly appreciated the splendid support and enthusiasm of the publishing teams at Kew and the Natural History Museum.

Many friends and members of my family have been kind enough to share my passion over the few years of research and writing. Some, including my wife Elizabeth and daughter Laura, gave helpful suggestions, and Michael Venning assisted with photography.

I hope that all those who have contributed in their different ways will find satisfaction in the published outcome.

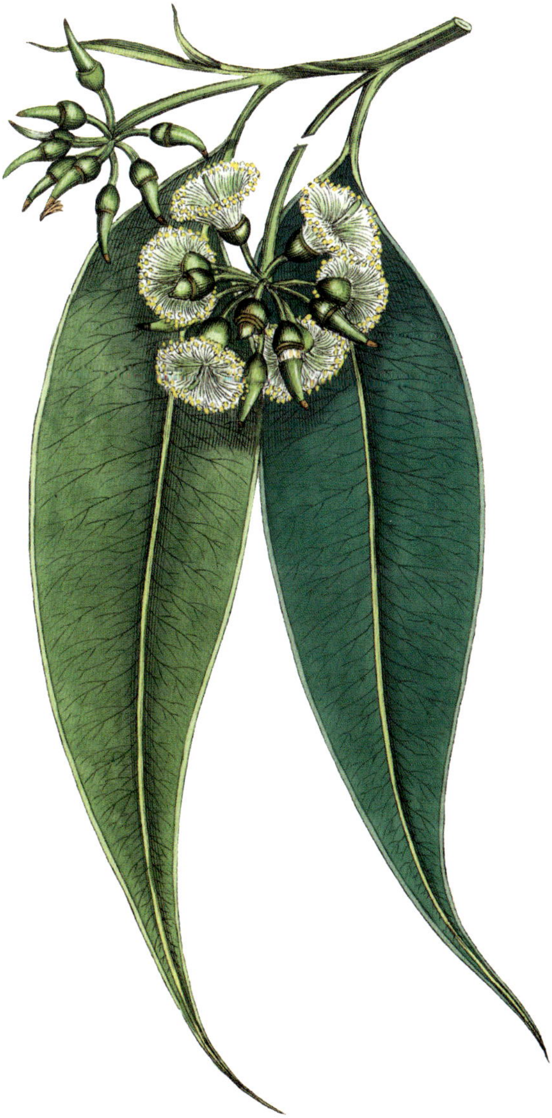

Prologue

It was February 1805 in Dallington, near Northampton, England. Mr John Hawkins, a gentleman of distinction, a past traveller and someone who knew his botany and mineralogy, had just finished a letter to another gentleman, Dr James Smith in Norwich, President of the Linnean Society.[1] Hawkins was agitated and wanted some agreement. He wrote:

> many persons of consequence may go as you say, to Mr Sowerbys but do they go there twice? Whereas these same persons are in the habit of going to Mr White. The one an Engraver lives in a distant part of the suburbs, the other a Bookseller in the most frequented part of the town and acknowledged to have the largest stock of books of any person in the trade besides dealing particularly in works on Natl History. Can we then hesitate in our decision?[1*]

Who was Mr Sowerby and why had he possibly been rejected? Why were Hawkins and Smith possibly in dispute about him and was he worth bothering with? Furthermore, why were 'persons of consequence' going to his house?

James Sowerby was not a 'gentleman' nor was he in paid employment. He was an artisan who had a purpose and a strong dedication to that purpose. He was, as Hawkins observed, an engraver but he was much more than that. He was an artist, a scientist especially in natural history, an author, a father of an already creative family, an active Fellow of the Linnean Society, a publisher and something of a businessman. Hawkins had belittled him in his short paragraph. But Hawkins was a 'gentleman' and not all gentlemen found it easy to deal with artisans, especially if they were not keeping to their station in life.[2] Moreover he had a challenging task to fulfil and he needed clear and certain decisions with which he felt comfortable. Sowerby did not fit into any mould that he was used to or perhaps expecting and, in addition, Sowerby had the aggravating habit of standing up for himself.

Hawkins was a man of integrity and with a strong sense of duty; characteristics that were essential for his role as an executor of the will of Dr Sibthorp, one-time professor of botany

1 Notes to the text begin on page 311.
* Extracts from letters etc. are transcribed as given and not corrected for spelling, grammar or other errors.

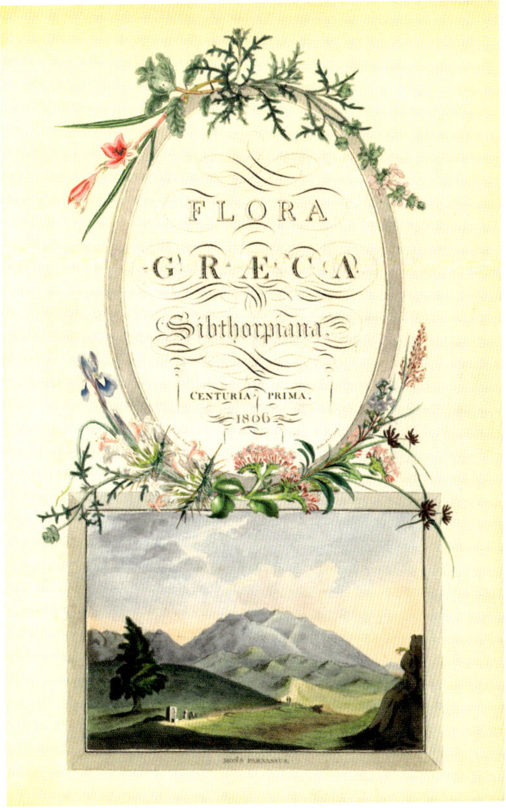

P/1 The frontispiece to the first volume of *Flora Graeca*. The picture shows the Greek mountain *Mount Parnassus*.

at Oxford University, with a particular remit to translate Sibthorp's generous endowment into a finely published Flora of Greece, or *Flora Graeca*.[3] This was to be based on the travels of Sibthorp and Hawkins himself, and the collection of some three-thousand specimens with associated drawings and sketches by the highly regarded botanical artist Ferdinand Bauer who had accompanied them. It was not as easy a task as it might seem – the total project was to take over thirty years, issued in parts and costing more than £30,000. Not surprising therefore that the *Flora* was to be described later as one of the rarest and most beautiful in both plate and typography of all botanical works.[4] It epitomised magnificence over accessibility. Its progress was to throw unusual and interesting light on to the roles of different individuals, the social tensions that arose between them and the workings of natural history at the turn of the eighteenth century. Hawkins's letter to Smith was just one of many he would write in dealing with the fraught administrative matters of production and sale of the first part. His letter showed that he had not wanted to involve Sowerby but increasingly he recognised the need for him as an important contributor to the *Flora Graeca* project. How this came about we will return to later when we learn of Sowerby's struggle to be included in yet another major work.

This book is the story of a remarkable man. Unusual in his breadth of scientific interests which he applied successfully; unusual in his desire to learn throughout his life and to impart his knowledge widely; unusual in going against the current practices by being at one and the same time his own researcher, writer, illustrator, teacher, publisher and bookseller; unusual in his considerable output of innovative, high-quality and influential works; and unusual in becoming the patriarch of a successful line of natural historians. It is also a story of a modest, self-made man, tending to avoid the company of others outside his family, quietly revealing his belief in and respect for God's natural bounties, and with a lasting determination to show his true worth despite his relatively humble status.

Hawkins and Sowerby were moving in a small sector of society enthusiastically committed to the pursuit of natural history and somewhat removed from many of the exigencies that others might be subject to. The eighteenth century saw these activities following different but interwoven strands. Herbals and medical books, written by apothecaries and medical men, sought to show which plants and natural concoctions were beneficial to health and for curing ailments. Local and regional floras often produced by keen botanists, who had other occupations, helped people to appreciate the value of their own environment. People, especially those with sufficient wealth and time, played an active role in disseminating natural history through their cabinets of curiosities and their writings. This was an age of collecting, cataloguing, observing and classifying.[5] Collecting involved touring locally or further afield and sharing the results through a network of fellow enthusiasts, most of whom were amateurs but who knew their specialism. The desire to see the exotic had been stimulated by the cultural pursuit of the gentlemanly Grand Tour to the principal centres or regions of Europe as well as by the much less frequent but highly significant voyages such as James Cook's in 1768–1771 to the Pacific, Tahiti, New Zealand and beyond. The reports, the unusual specimens and the beautiful artwork fired the imagination and stimulated participation in this enlightening process. Involvement was part of the culture and a way of gaining status. It gave a sense of purpose at a time when the range of occupations for the wealthy might otherwise be limited, and it left to posterity numerous finely printed and illustrated books.

Individuals spurred the process through their patronage and ideas or systems. A key player was Joseph Banks, renowned for his successful role in Cook's first voyage and President of the Royal Society, who used his powers and wealth to encourage others, to generate links and establish his own research herbarium. His influence spread to King George III and Queen Charlotte and led to the founding of the Royal Botanic Gardens, Kew (hereafter referred to as Kew Gardens). The sexual system for the classification of plants devised and promulgated by the famous Swedish scientist Carl Linnaeus was another stimulus.[6] Its strength lay in its practical simplicity so that almost anyone could use it to help with their own plant identifications. People wanted accuracy in the artistic depiction of plants and fossils as well as in naming and description. How else would they really know if they had something rare, unusual or particularly interesting? The artist's skill was often crucial.

Mining and the use of ores in feeding industrial development and exports stimulated the collecting of minerals and rocks. It did not have so many adherents as the biological realm but those who were involved, especially when living in or near the mining areas, made significant collections and participated in exchange and writing. The activity was fashionable and respected. Some combined this with interests in other areas of natural history; Sowerby was one of them.[7]

Pursuers of mineralogy were aided by the existence of manuals, keys and guides as well as instructions on analysis, sometimes involving quite risky experiments, written by respected authors.[8] Some collectors were sufficiently keen to set up their own small analytical workshops and to try to become expert in the use of the chemical blowpipe. Sowerby did likewise; he went even further and encouraged chemical collaboration with Humphry Davy and Michael Faraday at the Royal Institution. The classification and analysis of minerals was a topic of active interest and experiment; Sowerby and his sons joined in.

Collectors were also encouraged by the existence of good mineral and fossil dealers. Inevitably, it was often easier to collect, transport, store and display mineral specimens than botanical ones. Dealers could acquire, often from abroad, better and more interesting examples of the same mineral. Mineralogists could aspire to improving the visual wonder of their collection by the acquisition of fine crystallised or coloured specimens, even if at some significant cost. Many were not so interested in the ideas of origin, of geological processes, or even of uses, but they were fascinated by the variety of material and gripped by the passion that any true collector has, whatever it is they are collecting.

The natural history community was tightly knit. Little of it was institutionally based for there were few appropriate institutions beyond some universities and botanical gardens. The members were mostly amateurs including gentlemen and ladies, landowners, clergy, and gardeners. They relied on each other for advice, specimens, information and help with identifications, exchange of specimens or provision of simple equipment. The sending of mail was their main means of communication. To enter this community or network was not necessarily easy. A letter of self-introduction with an explanation of what you had to offer was necessary but not always effective. If you were fortunate enough to have the right connections you might gain access through recommendation or introduction. Joseph Banks, in particular, held strings which he deployed to influence the links and activities within a network. James Sowerby had steadily worked his way into several communities but it was highly advantageous that Banks viewed him very favourably.

The establishment of some Societies helped hugely to improve communication. The founding of the Linnean Society in 1788 for the study and dissemination of knowledge concerning natural history was especially important; so too was that of the Geological Society of London in 1807. Sowerby was elected an early member of both. Members who were able to get to London could meet to discuss their researches and to hear papers read, these later being published and distributed. Networks would be developed and consolidated but a person desirous of involvement would have to have demonstrated a sufficient knowledge and degree of activity in a relevant area to be elected in the first place. Fellowships of these societies, and particularly of the Royal Society, were marks of prestige. An artisan would usually have to be a hybrid artist cum natural historian to be eligible for membership.

The results of the endeavours of the natural historians, beyond their own greater knowledge and understanding, were mostly publications of learned papers, books, or regular printed issues.[8] Authors wrote for the cognoscenti as well as for some reflected glory; if their books were illustrated they were likely to be expensive and would be produced in limited numbers, often by subscription. It is no small wonder then that their accessibility was often seriously restricted. William Curtis, an early employer of Sowerby, and indeed James Sowerby himself were two of those who pioneered a real change in this practice; accessibility was an important challenge for them, even if it was partly based on the need for income. Both were keen to innovate, to teach and to enthuse others in their subjects. Both established their own physical means to help them in this task – a garden for one, a museum for the other. They reached out while most others in the closer community were happy to contribute only in a more traditional way. They are exemplars but their positions were far from easy.

The thirst for knowledge and culture was not of course confined to natural history and other sciences. Tours were undertaken for broad cultural reasons. There was enthusiasm for paintings and prints of many topics, not just portraits, historical events or beautiful flowers. Painting was also a genteel activity for ladies – whether of landscapes, natural history, or family members. England and London were increasing in population; there was a sense of growth. People had caught an air of excitement from the Industrial Revolution and new discoveries of the period. The founding of the Royal Academy in 1769 and of the Royal Institution thirty years later helped slake the thirst for knowledge of the many who attended their exhibitions or lectures. Learned, philosophical and even local scientific or arts societies were started. Culture was booming; enlightenment might have seemed almost tangible.

Let us return to Hawkins's letter to Smith which opened this Prologue and to his rhetorical question as to whether persons of consequence go to Mr Sowerby's house twice. The answer is certainly 'yes' – and many more times than twice. Not just gentlemen and ladies but people of different standings and from various walks of life. Lambeth – the 'distant part of the suburbs' in Hawkins's letter – was little hindrance to them. They went there to bring or to exchange specimens, to discuss natural history, to have lessons in drawing or in the use of colours, to visit his museum, to buy his different publications or artwork, and to get assistance with their own publications. His house in Lambeth was a nerve centre of a remarkable network that spread out to many parts of the British Isles and sometimes beyond, despite the difficulties associated with communication and travel. Its success was partly a result of his willingness to adapt to circumstances, to be helpful, sympathetic and sometimes even subservient. He was prepared to ask remarkable favours of people, but always in the cause of science, and to be an innovator and experimenter. He acquired a good eye for what would succeed in a world where people were thirsty for knowledge about the natural world. His house was also important because it was home to a good family working system. Here he was perhaps lucky because the personalities and strengths of his wife and some of his children proved to be a good match for his objectives. Sowerby constructed a model for science that was certainly unusual but ultimately its success lay in the quality and scope of his work – work that has given us a fine and lasting legacy.

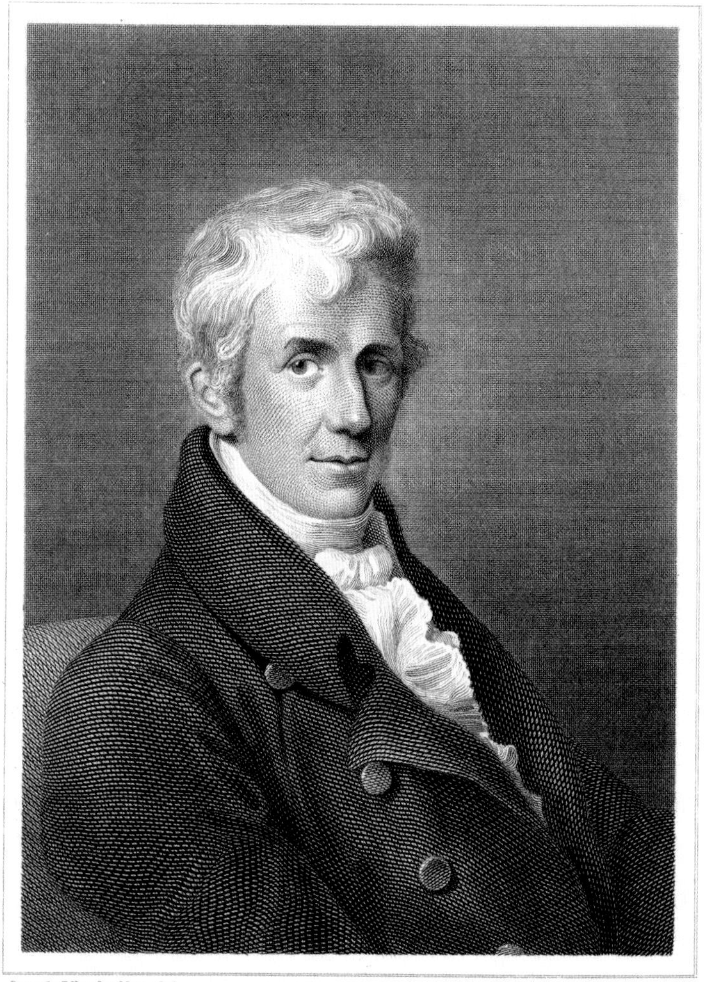

Drawn by T. Heaphy. May. 1816. Engraved by J.C. Edwards.

JAMES SOWERBY, F.L.S: G.S. &c.

Nat: March 21, 1757. _Ob: October 25, 1822.

CHAPTER ONE

Precarious start

1/1 James Sowerby. Engraving by J. C. Edwards after James Heaphy, 1816.

James Sowerby was born on 21 March 1757 at number 2 Bolt-in-Tun Passage, off Fleet Street, London and he tumbled into a world of bustle and change. The family was not very large with four siblings to greet him: his eldest brother John of eleven years of age, a sister and two other brothers. Another brother had died in infancy and two succeeding brothers also died young. James was the son of John Sowerby, a successful lapidary, and his wife Arabella Goodreed. They were not old when James was born – John was thirty-nine and Arabella thirty-two – but his father's untimely death only nine years later was to be a pivotal moment in James's life.[1]

The family business dealing in cut and polished gems, precious stones and beads was still a profitable one – interest in the world of rocks and minerals, as well as in jewellery was at least steady – and John was sufficiently confident to take his son, James's brother John, as a seven-year apprentice when he was fifteen years old. The indenture, dated 1 April 1761, still exists and has the usual contractual constraints of the time. 'He shall not commit Fornication. He shall not play at Cards, Dice, Tables or any other unlawful Games. He shall not haunt Taverns or Playhouses.'[2] This was surely an auspicious moment. The family home was the successful and established family business, John was a hardworking son and all the children were healthy – not something that could be taken for granted in eighteenth-century London.

John, the father, was a respected businessman, a Freeman of the City of London and a liveryman of the Worshipful Company of Haberdashers. For James, a promising and secure future seemed to lie ahead.

If the family home was a propitious environment, so too was its location for business and contacts. Bolt-in-Tun Passage was adjacent to Bolt-in-Tun Court with a busy inn at its southern end from which numerous coaches went to many parts of southern England and Wales – towns such as Hereford, Bath, Dover, Chichester, and Aberystwyth.[3] Here was considerable potential for a passing clientele as well as for the more regular customers or visitors who came to Fleet Street for its trades, shops, residents, eating and drinking establishments and even waxworks. The trades were numerous and resulted in a vibrant mix of publishers and booksellers, banks, goldsmiths, shoemakers and chandelier makers, to name a few, while the inns, taverns and coffee houses were well frequented. Mrs Salmon's waxwork provided a different sort of entertainment: it was renowned for its display of kings and queens, initially with some modelled in gruesome detail, such as King Charles I on the executioner's scaffold, or showing the unusual and improbable, such as Margaret, Countess of Heningbergh, lying on a bed of state with her three hundred and sixty-four children all born at one birth. Several of the models had moving parts.[4] The tradesmen were also renowned: William Parker received the commission for the chandeliers for the Tea Room of the Bath Assembly Rooms opened in 1771, and later also for their Ball Room.[5] Hoare's Bank still survives as a private deposit bank.[6]

Fleet Street was smelly, noisy, crowded and dangerous from the heavy traffic of coaches and wagons – side pavements were not laid down until 1766. At its eastern end was the gloomy, mostly windowless building of Fleet Street prison on the eastern side of Fleet creek, housing mainly debtors who, as inmates, were treated badly even if in other respects they might have been honourable.[7] But this did not stop the street from being a popular residential area with reasonable housing in the adjacent courts and side passages. The people who came preferred the coffee house culture to the sedate and smarter environment of the London Squares. It was also desirable because people of repute lived there – a catalyst for others to try to capture some renown. Samuel Johnson lived in and about Fleet Street for many tens of years, with his last few (1776–1784) passed in Bolt Court. He is mentioned, with inimitable style, by the traveller Thomas Pennant in his *Account of London*:

> I must by no means omit Bolt Court the long residence of Dr. Samuel Johnson, a man of the strongest natural abilities, great learning, a most retentive memory, of the deepest and most unaffected piety and morality, mingled with those numerous weaknesses and

prejudices which his friends have kindly taken care to draw from their dread abode.[8]

Bolt Court is still there on the northern side of Fleet Street, but Bolt-in-Tun Passage which was almost opposite it has disappeared.[9] James Sowerby would almost certainly have seen Johnson on many occasions, possibly accompanied by James Boswell going into the Mitre Tavern – one of Johnson's favourite haunts. James might not have been aware of the presence of James Ferguson, the famous astronomer, who was Johnson's neighbour in Bolt Court in 1776, but he should soon have come to know another: Dr Lettsom, a Quaker and one of the most popular physicians in London as well as being a gentleman of considerable influence. Lettsom was soon to become one of Sowerby's early supporters.

Fleet Street was then near the geographical centre of London. Two kilometres' direct travel to north, east or west would be sufficient to leave much of the city behind, while the Thames to the south created something of a natural barrier but with Lambeth Palace, timber yards and other industrial buildings occupying its southern banks. Development to the south was about to start in earnest. Both Fleet Street with its side alleyways and Lambeth were going to feature strongly in James's life.

James was a lively and innovative young boy who was well able to look after himself and take advantage of his circumstances even when these were difficult – strengths that would serve him well when his father died at the young age of forty-seven. James was only nine while his eldest brother, John, was just coming out of his teens and unable to finish his apprenticeship, having served less than five of the seven years. The family business suddenly lacked necessary skills and experience – John, the father, had no other business partner or associate. The daily bread now had to be earned using the limited expertise that son John had acquired and by relying on the stock of gems and stones that had been built up over the years.

James was sent to day school for his education, partly for a good grounding in Latin and French, which were prerequisites for an educated person.[10] He learned these reasonably well and was able to use his writing skills in French to advantage, especially in later life, even though spelling was not his strong point. The accepted shift from the use of Latin to English in scholarly publications at this time meant that while James could read Latin, he wrote it only very rarely. He was not the most studious of boys but his education and early life experiences gave him a good sense of values, literature and knowledge of the natural world. He had the freedom to express himself through his creative skills by making artistic paper cut-outs, figurines, miniatures, moulds of men in armour and other toys, carving model ship heads, and creating

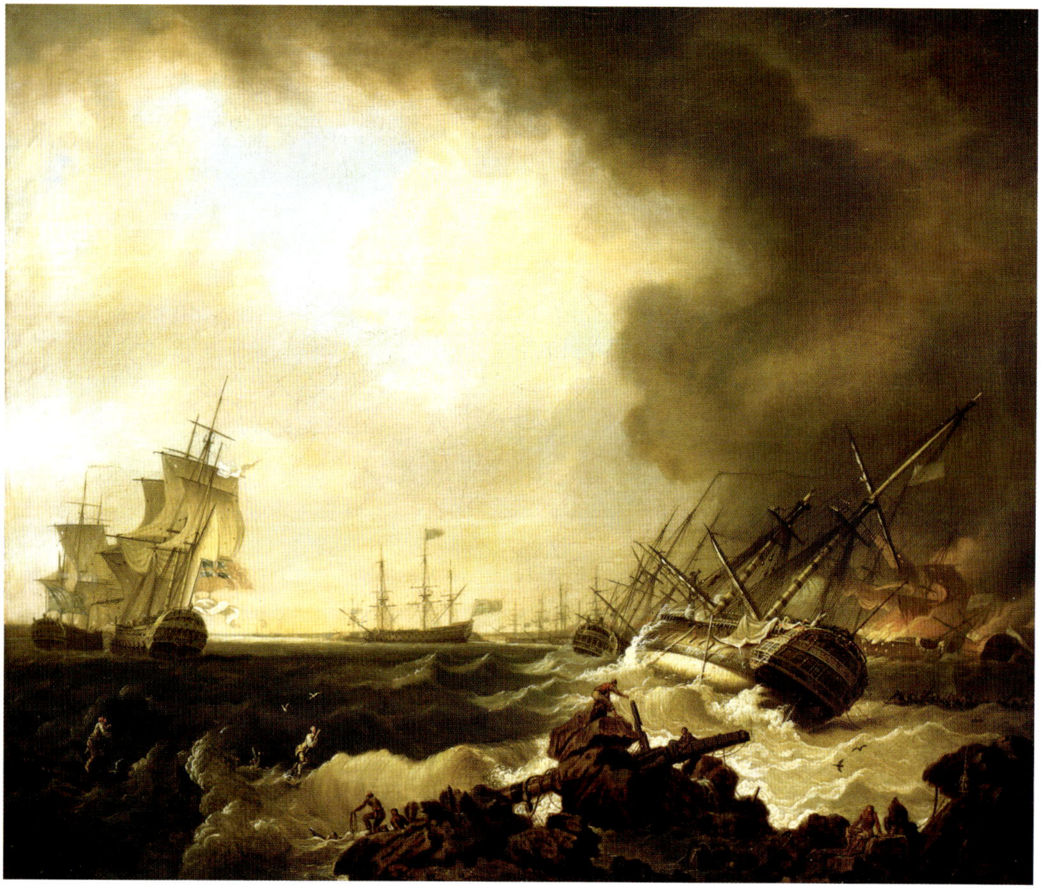

1/2 The Battle of Quiberon Bay, 21 November 1759: The Day After. Oil painting by Richard Wright, 1760. This battle was an important naval encounter in the Seven Years War (1756–1763). Credit: National Maritime Museum.

coloured sketches of his own and his mother's friends. By selling several of these sketches he made enough money to supply his growing artistic interests, and possibly even sufficient to help sustain his own family.[11]

These interests and skills did not go unnoticed. Lapidary was not going to be James's world; the strengths of his friendships and that of his family's network led to the raising of sufficient money to pay for James, at the age of fourteen, to be indentured as apprentice to the artist Richard Wright for seven years. The sum was £100, a fair part of the annual earnings of an ordinary clergyman. This was serious business indeed for not only was the money significant, so was the artist. Wright had arrived in London from Liverpool in about 1759 already with something of a reputation as a marine painter, a reputation which he rapidly enhanced by winning the first prize (or 'premium') of 30 guineas in the first competition for a marine painting held, in 1764, by London's Society for the Encouragement of Arts, Manufactures and Commerce. More prizes came his way from this Society while at the same time he was exhibiting successfully at the 'breakaway' group the Society of Artists of Great Britain (later, in 1768, to become the Royal Academy). Here was a family man of rough manners with a significant professional output, able to make good and no doubt profitable connections with royalty, aristocrats and the print-buying public.[12]

Wright's canvases were largely marine-themed, although he also turned his brush towards moonlit vistas and landscapes. Men of war, swirling black and white clouds, cannon fire, bulging sails and mountains of sea foam formed part of the necessary ingredients for a dramatic, emotive retelling of a historical sea fight or the sea passage of *Queen Charlotte* in a storm: portrayals that would swell the national pride (more especially with the successful termination of the Seven Years War), enhance loyalty to the King and remind the city dwellers of the stark forces of nature. If marine painting was not deemed to be high in the rankings of fine art it certainly had its popular place.[13]

So in 1771 James was going to a worthy Master but one whose talents and fame were about matters very different from those to which James had inclined. He was about detail, finesse and the relatively small scale; Wright about large canvases and action. Wright was a family man with a well-established business, and a wife and children who were developing, with the father's help, as artists in their own right. Was Wright a good choice – if choice was indeed available? He certainly had wide connections and the right sort of friends including George Stubbs, the famous painter of horses and other animals. James could reasonably expect an entrée into this circle. He might also expect to learn new techniques and to broaden his nascent oeuvre. But the match seems strange in a world where appreciation of, and demand for, works of detail such as botanical drawings, portraiture and miniatures was high and, if the quality was right, possibly insatiable. James's supporters must have wondered whether or not it would be a success.

In one sense they need not have worried; James's initiative shone through and he clearly understood the value of money. He was now in a good working environment, among a pleasant family and with the sort of work that he liked and, moreover, was able to adapt to his own purposes. His skills were used and developed to add the detail, of figures and animals, to Wright's original work and also to copy, with sufficient accuracy and quality, the works of other painters and those of his own master – all of which had a ready market. In his spare time Wright allowed him to paint his own pictures and develop his personal collection. Only three years into the apprenticeship, in 1774, James exhibited his painting *A Deceptio Visus* ('a deceptive sight') at the Society of Artists; he had some artistic independence but this title gives little clue as to how he was using it.[14]

The time passed, the job was being done, and Wright was earning a living and keeping his family but not becoming rich. To maintain a reasonable standard of living the artist had to continue with a good saleable output, yet fashions could change and commissions might dry up. Then there was the issue of bad health or even death, a factor which would unexpectedly change or give a harsh premature end to an apprenticeship.

James now experienced just such an event, similar to that suffered by his brother, John, some nine years earlier. Wright suffered a stroke and although he was able to do some artwork by instructing and guiding his pupils it was not to be for long. Wright's last exhibited work was shown in 1773 and he was dead by the summer of 1775.[15] The family was not wealthy, the business was not viable without the master, and while his children continued to paint and indeed exhibit their work, this was not sufficient to provide any support for James. He had to leave the household in Pimlico and ultimately return home to Fleet Street.

Frustration must have welled up within James at his circumstances. He knew he had the skills; he was able to exhibit the occasional picture and sell his work, but he was not yet ready to be a sufficiently competent professional able to earn a sustained, independent living. The frustration was aggravated by another failure to fulfil an apprenticeship, this time to William Hodges, a London painter of some renown who was prepared to take James on trial. Hodges, as yet unmarried, had recently returned from James Cook's second voyage (1772–1775) to the South Pacific where he had been the official artist, making sketches and paintings of people and landscapes. He was immediately involved in checking and organising the engraving work of the illustrations for Cook's official account of the voyage. He had not really been able to settle into a steady professional routine, something that was to elude him for some time, so an untrained assistant or apprentice might have been more of an irritant than a benefit. James had previously been working in a harmonious and productive family household; now he was effectively alone with the master and the visiting engravers in an environment far from his expectations or standards, being a dirty painting room and with mean meals to boot.[16] He soon decided to leave.

Now firmly back in the family home in Bolt-in-Tun Passage there was an air of freedom but also of new responsibilities. The family would not be able to support him; he must do this for himself and do it with energy. He exhibited three paintings at the Society of Artists in 1776 – *A Landscape, A Ship on Fire*, and *Fish* – titles that resonate with his innate interests or training. He was soon painting portraits and miniatures, copying paintings and earning a reasonable living while developing his skills. The daily press had noted the quality of his exhibited work, his reputation was building through his commissions and there was a demand for his particular art.[17] But what might be the best next steps for a patronless young man such as James on the brink of his professional career? What opportunities were actually available in Hanoverian London, or for that matter elsewhere in the country?

Writers of social and working conditions were able to make a reasonably justifiable categorisation of the social levels of people. The 'lower orders', beneath those classes with members including the gentry, the clergy and the merchants, ranged from the master craftsmen and small shopkeepers down to the vagrants, beggars and the destitute.[18] While Sowerby's family could rightly be said to be within the top of the 'lower orders' ranking, this does not mean that life would necessarily be comfortable and certain, even while good health was sustained. As a skilled artisan, if he was to have some stability in his artistic life, James would most probably have needed to be either regularly employed, perhaps as a journeyman, or to be sufficiently renowned to receive commissions from a changing body of clients. His temperament and the circumstances of the time led to a rather different outcome.

A certain ruddy faced gentleman, by the name of William Curtis, was already well known to many people. Short and energetic, he emanated kindness and warmth as well as his ardent love of natural history, especially botany. He had been an apothecary, practising in London, but was now *Praefectus Horti*, or horticultural director, and Demonstrator to the Society of Apothecaries at the internationally renowned Chelsea Physic Garden – a post he had held since 1772. He had published a little on natural history but now planned a major work on the plants that were to be found growing wild in the environs of London; this to be illustrated with fine coloured

1/3 William Curtis. Watercolour miniature by Peter Paillou, 1798.

1/4 Deadly nightshade (*Atropa belladonna*), by Sowerby for *Flora Londinensis.* Curtis wrote that the reduction in the number of such plants was not just because of its toxicity but also because of 'the rage for building, joined to the numerous alterations perpetually making in the environs of London'. The plate is also interesting because of the detail given to the different parts.

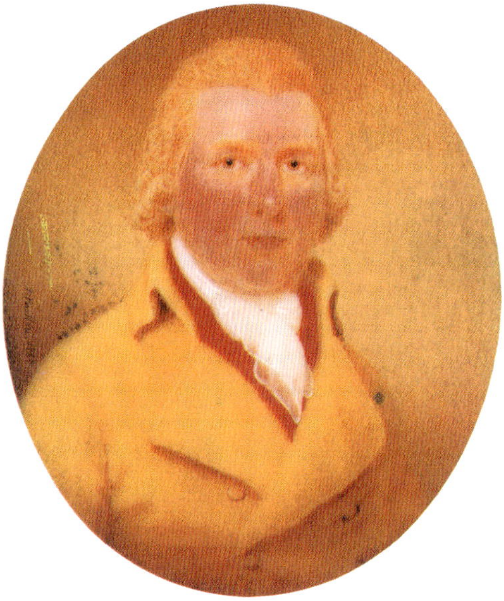

plates and to be a monthly serial until complete. He needed artists and they had to be good ones.[19]

Exactly how and when James Sowerby came to be known by Curtis is uncertain but he probably knew of Curtis's new project and applied to him for a role within it. Or he might have been recommended to Curtis, who was on the lookout for talent. Either way he was in Curtis's employment no later than 1776, not long after he had left Hodge's studio. He was also drawing illustrations, possibly on Curtis's recommendation, for John Lightfoot, who was producing a book on the flora of Scotland as a result of his five-month travels with Thomas Pennant in 1772.[20] Indeed, Pennant was subsidising the work and his personal artist, Moses Griffith, was providing most of the drawings. Such involvement with keen and active botanists who numbered several influential people among their friends was a defining step for Sowerby. Another was realising the opportunity for more guidance and training in art, not this time as an apprentice but through the more formal method as a student at the Royal Academy Schools.

Being commissioned by John Lightfoot was a timely opportunity. Lightfoot was a Church of England clergyman and another ardent natural historian. He was in just the right circumstances to pursue his scientific inclinations; he had been appointed curate and lecturer at Colnbrook in Middlesex and then additionally in 1767 as chaplain to the wealthy Dowager Duchess of Portland in Beaconsfield.[21] Lucky was the man who had a good income, job security and time. Lightfoot's Flora of Scotland was just the sort of work that the many naturalists of the day wanted. Interest in natural history was burgeoning; gentlemen and ladies not only enhanced their cabinets with fine or unusual specimens, they also sought to develop some scientific interest and make their collections a means for social exchange and purpose. Accounts of regional flora, and to a lesser extent fauna, were therefore desirable.

Lightfoot had written to Curtis in March 1777 asking for help with producing some of the illustrations for his work.[22] Moses Griffith, the

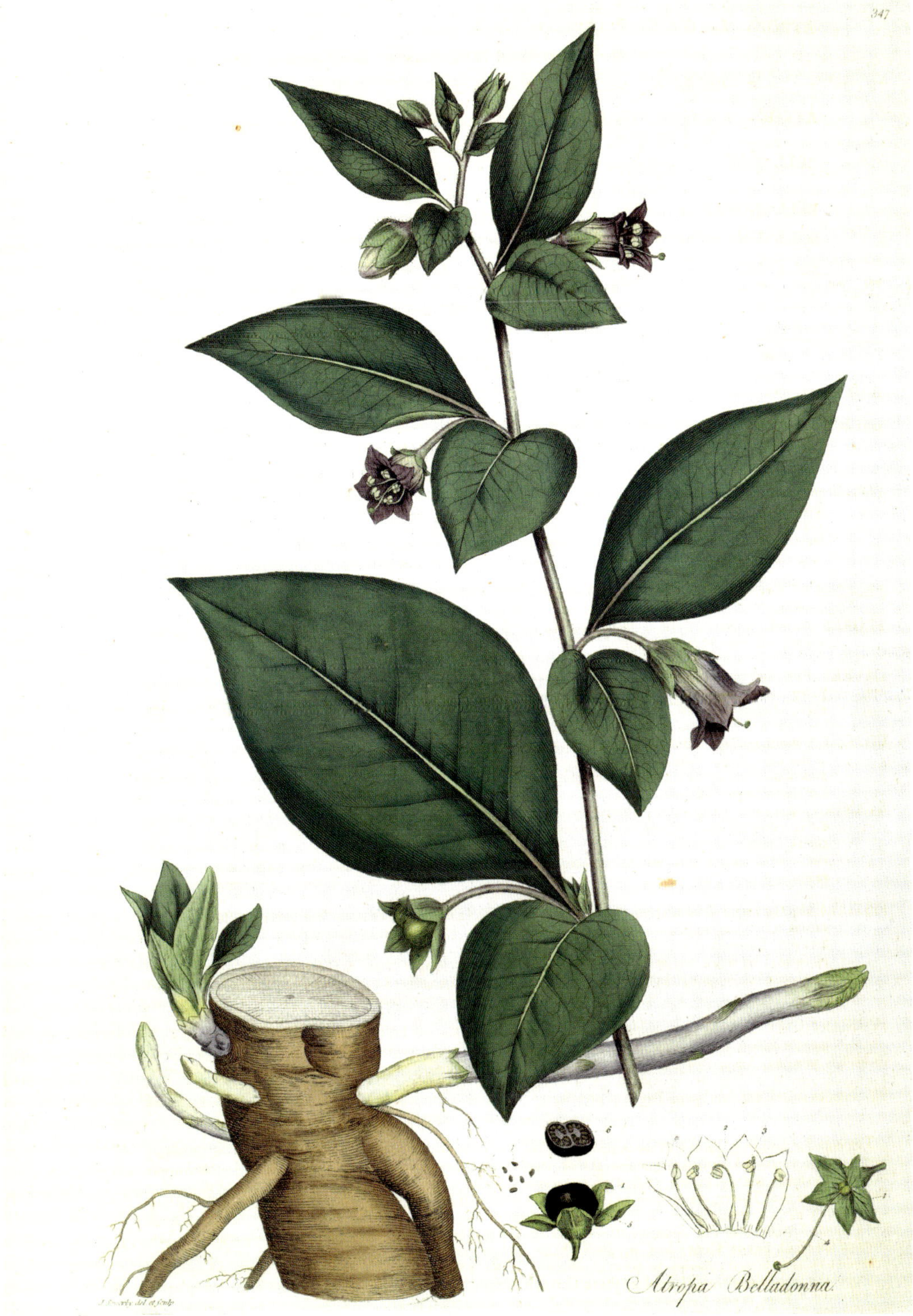
347
Atropa Belladonna

1/5 Verdigris mushroom. (*Agaricus aeruginosus*, synonym *Stropharia aeruginosa)* by Sowerby. Curtis wrote that he wished that every fungus was as distinct in its characters as this one and that 'to know the fungi well we must watch them daily and yearly; in short *we must live with them.*' One of several Sowerby drawings of fungi for *Flora Londinensis*.

artist, was unable to produce them fast enough and so Sowerby filled the gap by drawing the seaweeds. *Flora Scotia*, in two volumes, emerging later that year was immediately well regarded for its scope and rigour, dealing with some 1,250 species, including the non-flowering plants (mosses, fungi, ferns and algae), all as a result of direct scientific fieldwork. Lightfoot moved in a small but important circle of gentlemen of influence; a circle which was enhanced by his association with the Duchess of Portland. It included people such as Joseph Banks, Daniel Solander and Samuel Goodenough, all of whom were active and well regarded in their scientific pursuits. If Sowerby was to be set on a path involving natural history rather than portraiture, then this group could strongly influence his journey. *Flora Scotica* would gain him further recognition. His work for Curtis could be even more powerful.

Energetic as ever, Curtis had already produced content for his London project. Some of the first folio-sized plates had been in the hands of a few people, including the Professor of Botany at Edinburgh, John Hope, as early as 1774 but the work started to appear in unbound issues or 'numbers' of six plates, each with their associated letterpress, in 1775.[23] *Flora Londinensis, or, plates and descriptions of such plants as grow wild in the environs of London,* aimed to give a botanical account of the plants, in Latin and English, with information on their uses in medicine, agriculture and the rural economy, together with large, finely crafted plates. So as not to tread on the toes of others who were working on county floras, Curtis restricted himself to those plants growing within ten miles of London but declared that once the work for London was finished he would embark on another for the whole of Great Britain. The plates were of exceptional quality and very suitable for framing and display. Each number of six plates cost 2s.6d. if uncoloured or 5s. coloured.[24]

Curtis's *Flora Londinensis* was a monumental and expensive project which needed resourcing and a sustainable list of subscribers. For the latter, quality of the product was paramount; the artist or artists had to be some of the best and they would expect suitable recompense. One was William Kilburn, employed to do the early illustrations; he conveniently lived in Bermondsey, near to the small plot of land just on the southern side of the River Thames where Curtis cultivated indigenous plants.[25] But could Sowerby also have made some of the first drawings? Unfortunately, few of the early plates are signed by the draughtsman or engraver; this was not unusual for artists were not consistent about signing their work, so it is not straightforward to assign artists to particular plates. In any event Sowerby was working for Curtis and joined him in botanical excursions around London. By 1777, Curtis had published the fasciculus of the first twelve

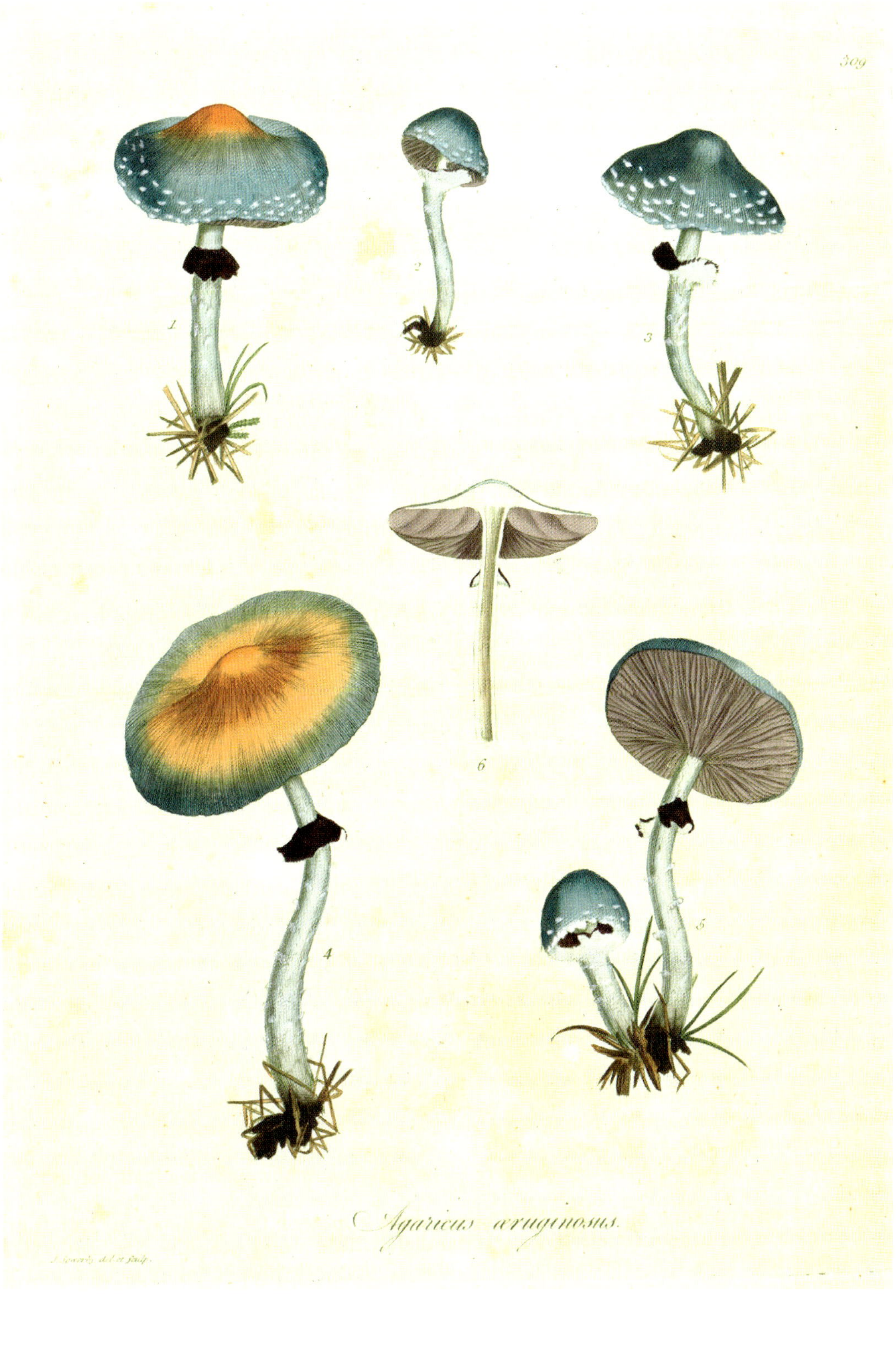
309
1.
2
3
6
4
5
Agaricus æruginosus.

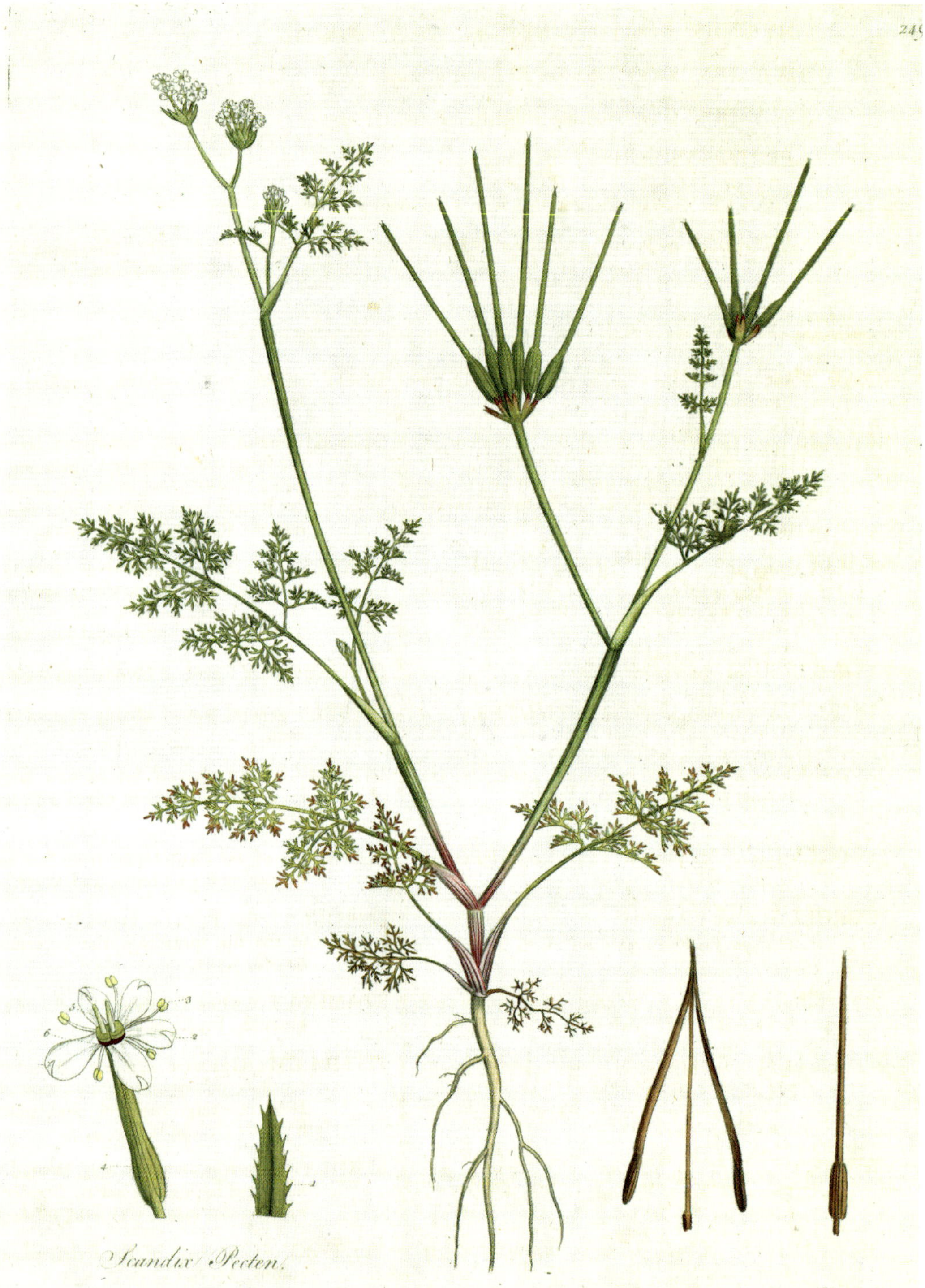
Scandix Pecten

1/6 *Scandix pecten*. Shepherd's needle or Venus's comb. Recorded as being common in cornfields. Named for its fruits with long beaks which can resemble teeth of a comb. Sowerby's plate for *Flora Londinensis*, drawn with fine detail.

numbers of his *Flora Londinensis*, comprising 72 plates, at a cost of £3.4s. in boards, coloured, or £1.4s. plain. His preface records that the publication rate had been slowed because of the loss of one of his artists but this had been rectified by employing two more, both eminent, 'so that the drawing and engraving, which before fell to the share of one person, being now divided betwixt two... ' One of these must have been Sowerby. The work has been said to contain 'the finest illustrations of British plants ever published'.[26]

In that same year Curtis also published his *Linnaeus's System of Botany*, an accessible and helpful guide to how Linnaeus's sexual system of plant classification worked; a system which was rapidly adopted in the botanical world. Curtis wrote it for the use of his students but it was probably widely consulted not least because of its four plates, each one with several drawings by Sowerby of the components of plants that are used in the classification. The small drawings are not in the style of those in the larger *Flora Londinensis* but they exemplify Sowerby's ability at this early stage to depict accurately and clearly the necessary detail while retaining an attractive lightness. Curtis, in his preface, had this to say about Linnaeus's own description of his classification:

> and had the drawings which have been given to illustrate it, been equal to the Author's descriptions, it would doubtless have made a more rapid progress, and been more generally and perfectly understood than it is at present. Those who in this country have undertaken to explain, and make a knowledge of his system easy, (and which of late have been sufficiently numerous) have given us over and over again the same hackneyed plate of the Classes, retaining all its imperfections, and gradually accumulating new ones; so that the designs which were at first scarcely intelligible, are now rendered useless, nay worse, as they tend to mislead.

Curtis wanted strongly to avoid such a repetition; for this he had full confidence in Sowerby's art and its botanical accuracy.

It is right that Linnaeus's classification of 1753 is well served by good illustration, for the system is based on the visual characteristics of part of the flower and its purpose was to facilitate identification. For the flowering plants he proposed twenty-three Classes, each based on the number and position of the (male) stamens. So, for example, the Monandria class has a single stamen, the Pentandria five. All the non-flowering plants such as the mosses, ferns and lichens he grouped together in his twenty-fourth Class, the cryptogams. The twenty-three Classes are subdivided into Orders on the basis of the number of the flower's (female) pistils.

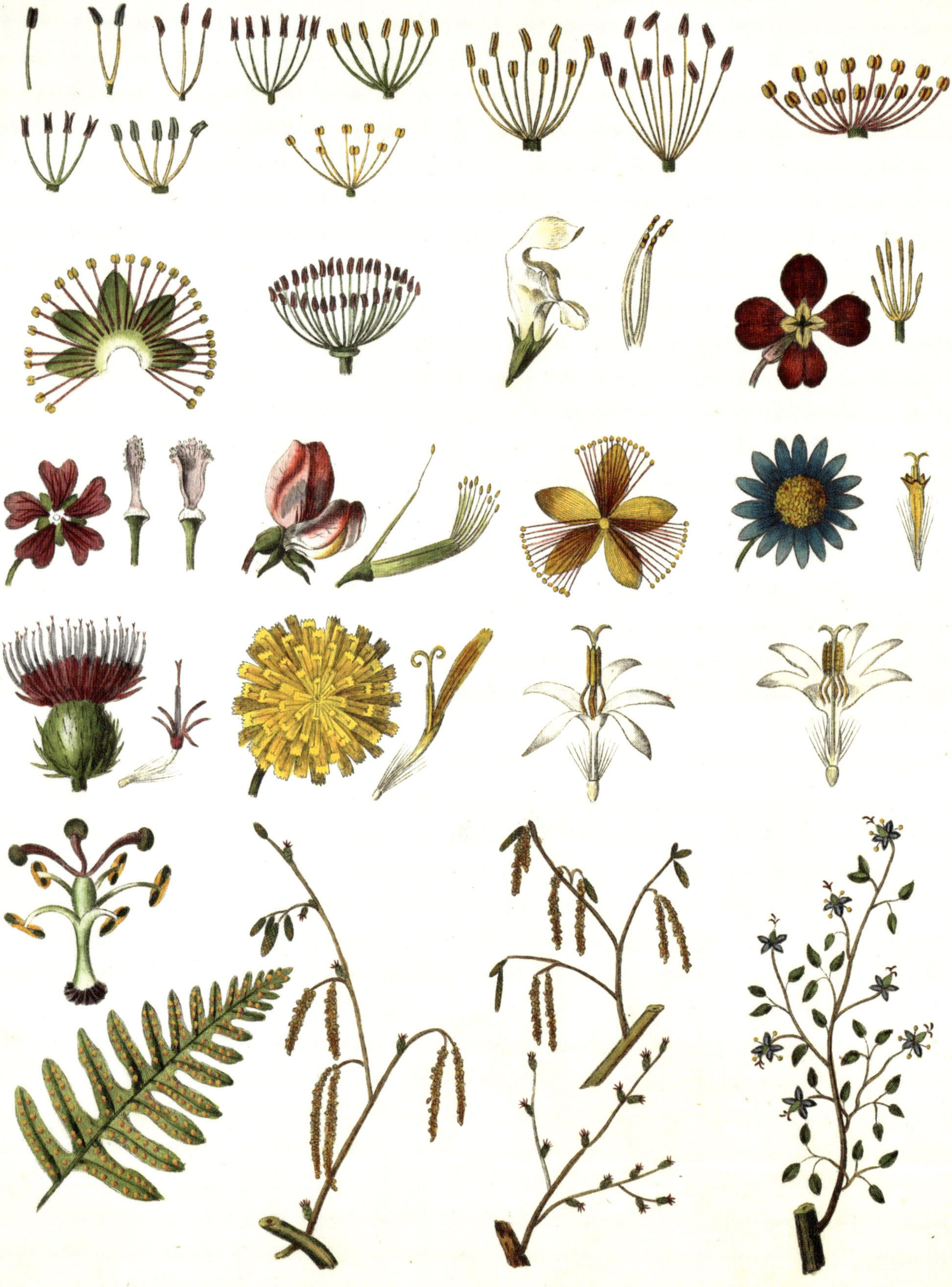

1/7 'Classes explained and illustrated by figures'. The top illustrations show the stamens for the classes 1 to 11, Monandria to Dodecandria: the others depict the other classes, for example that on the bottom right-hand side for class 23, the Polygamia, showing 'hermaphrodite and male or female situated on the same plant'; and that on the bottom left depicting class 24, Cryptogamia, with no stamens. The first plate, by Sowerby, in Curtis's *Linnaeus's System of Botany*.

If Sowerby was gaining credit for his work and there was a niche to fill, why should he now choose to join the Royal Academy Schools? Among the several reasons was his own diffidence about his abilities and the quality of his work. He had sometimes been asked to do work but declined because of lack of confidence; the Schools offered a chance to redress some weaknesses resulting from a shortened apprenticeship. The opportunity was also rather new, for the Schools had only been in existence for a few years (since 1769) and, provided that an applicant was sufficiently promising to gain entry, the tuition for the six-year studentship was free. Professors and Royal Academician 'Visitors' delivered the lectures and other tuition, serving in rotation for nine months of the year. Here was a real chance to learn technique and gain confidence in expression, and without any financial burden on family or the need for sponsors. It was not to be missed.

With an appropriate recommendation, Sowerby was enrolled as a painting student on 31 December 1777. (The Keeper of the Schools was George Moser, Edward Penny his Professor of Painting, Samuel Wale the Professor of Perspective, William Hunter, the Professor of Anatomy, and Frederick Barff, the Professor of Chemistry.[27]) Artistic competitions were set at the start of each year; students would compete for the prizes and hope to raise their reputations as well as benefiting from some monetary gain. Sowerby was now aged twenty and, with the support of Curtis and the Schools, was at last set on a clear course to start his career. But one question remained. Would this course lead to a life as an artist, even a celebrated one, who worked to help fulfil other people's needs and ambitions, or might it lead to one where his artistic skills would form an important part of his own grander vision or broader desires?

341

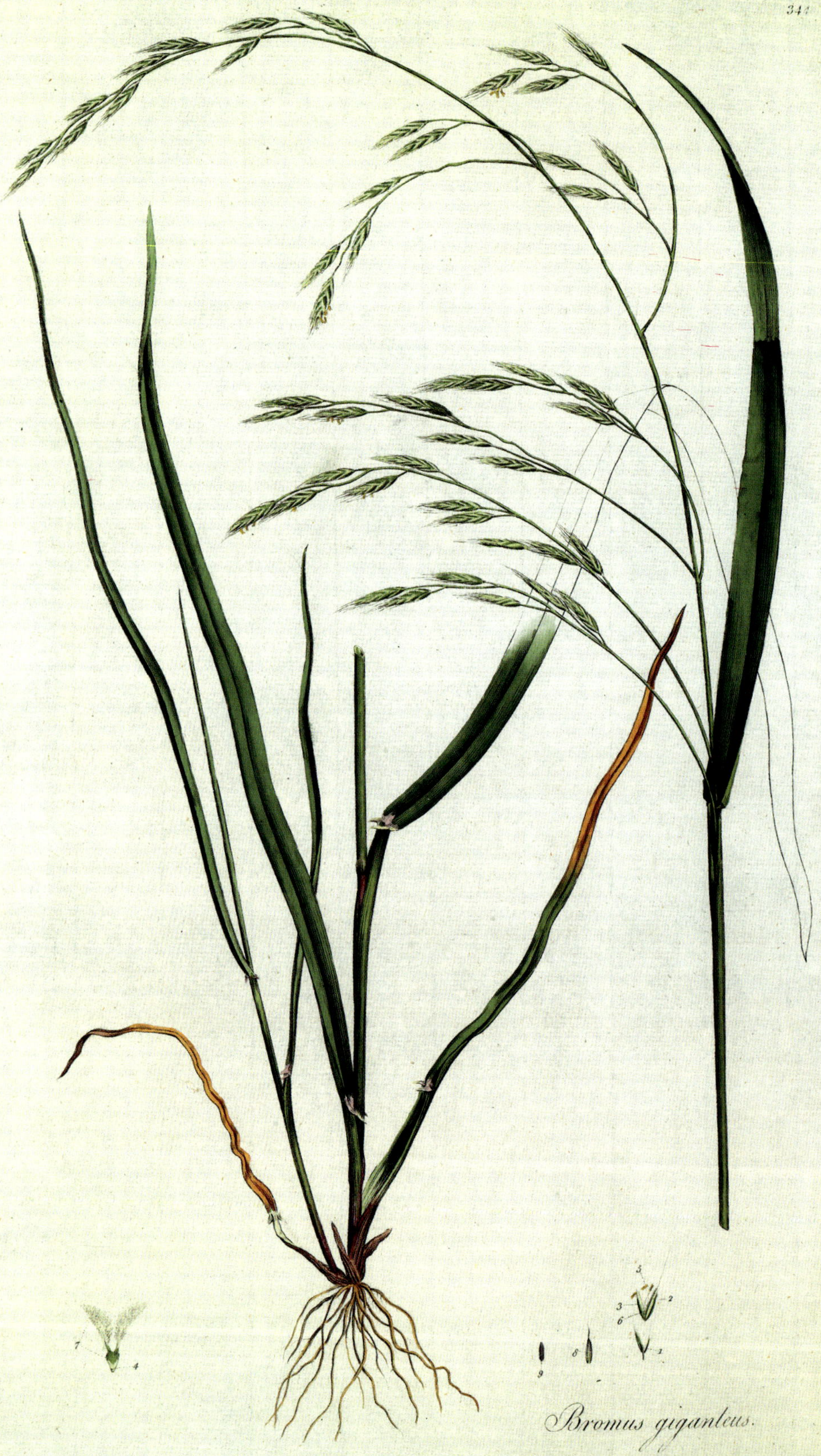

Bromus giganteus.

J. Sowerby del. et sculp.

CHAPTER TWO

Rising reputation

2/1 Tall brome grass (*Bromus giganteus*). Engraving of Sowerby's drawing for *Flora Londinensis*. Curtis noted that it was to be found by the sides of the River Thames and in ditches. Sowerby was able to capture the real nature of grasses in his drawings.

John Stuart, the third Earl of Bute, was a very wealthy man. He was also a patron of the arts and sciences, a lover of botany and mineralogy, and a man of some power. Lord Bute succeeded to his father's estates in 1723 and later was one of the Lords of the Bedchamber of Frederick, Prince of Wales. The Prince died prematurely in 1751, causing Bute to become tutor to Frederick's son, the future King George III – an event which was to give him considerable influence over several years. Another death, this time of his wife's wealthy father Edward Wortley Montagu in 1761, led to more money flowing into the Bute family's coffers. His wealth was to provide significant benefits to the pursuit of natural history.

Bute's connections and his interest in botany came together with his assistance to Princess Augusta, the widow of Prince Frederick, in the further development of the late Prince's private gardens at Kew, especially to make them more of a botanical centre. He energetically entered into planning and supervising the alterations whether they were architectural or horticultural. He appointed the Head Gardener, William Aiton, in 1759 and saw to it that the Linnaean system of classification was used. He was also active in procuring numerous plants and trees, especially exotic ones.[1]

The accession to the throne of George III in 1760 brought Bute rapidly to the centre of political life and in 1762 he finally became prime minister. But he was deeply unpopular with his colleagues and he did not have the stomach for conflict. He resigned in 1763 and, not long after, lost the confidence of the King. His later life (he died in 1792) was spent in some travel, scholarship and also significant patronage. William Curtis was one beneficiary, and where Curtis benefited, Sowerby would also.

Curtis seemed to be in financial difficulty from the start of the publication of *Flora Londinensis*. He approached Bute through a letter asking for help in the form of a loan of £300 at 5% to be repaid in three years – a rather optimistic time span. Fortunately he was advised that Bute was not in the habit of making loans – booksellers would be more appropriate for that – and that he preferred to support works without expecting interest. It would be a good idea, Curtis was told, if the work was dedicated to Bute.[2] Curtis complied; the first bound volume of fascicles had a fine dedication to the Earl whom he called 'the Maecenas of the present age'. Curtis got his money.

Curtis had resigned in 1777 from his post at the Chelsea Physic Garden because of the shortage of time to pursue his own interests of publishing, gardening and a little teaching. He had also been made aware that he was not satisfactorily fulfilling the terms of his appointment, for example by failing to give some lectures. His energies were now additionally directed towards starting a new botanic garden worthy of note and for which people would be prepared to pay a subscription to visit. He desired that people understood the value and importance of botany and regretted the current state of ignorance, a desire encouraged by the generous patronage of several including the Honourable Daines Barrington and Thomas White Esq.; relatively rich people who were botanical enthusiasts. His printed publicity (1778) was entitled 'Proposal for opening by subscription a Botanic Garden to be called the London Botanic Garden: designed for the use of the physician, the apothecary, the student in physic, the scientific farmer, the botanist (particularly the English botanist), the lover of flowers, and the public in general.' He was assured of the contribution of plants from many people including Lord Bute and the Dowager Duchess of Portland, as well as from some other gardens including Kew. He continues:

> the Botanic Garden proposed to be opened and as yet in its infancy, is situated on the fourth-side of Blackfriars-Bridge in Higler's-Lane, Lambeth Marsh, near the Magdalen Hospital. It is surrounded on two sides of it by water, which ebbs and flows with the tide, and which being let in, renders it very convenient for the growth of aquatics.

The garden was to be arranged according to the Linnaean system, with study rooms and a library, and would be open from Tuesday to Friday each week from six in the morning till eight at night. It opened its gates on 1 January 1779. In addition to all that had been described, Curtis regularly gave botanical and horticultural talks and lectures, including such topics as 'on the sexes of plants' and 'on plants compared with animals'.[3]

It may be wondered why Curtis should develop such an enterprise when a major botanic garden existed at Kew. But the latter was a royal residence and access was, of course, very limited. Joseph Banks had gained a foothold early on by meeting William Aiton in 1764 and commissioning the artist Sydney Parkinson in 1767 to draw some of Kew's flowers. By 1773, soon after his return to England, he had gained the confidence of the King and established an unofficial directorial role for himself at the Gardens. Indeed, he would often meet the King at Kew on Saturdays.[4] But Banks also knew Curtis well for the two of them had been members of a small group of botanists on a study visit to Wales in 1773.[5] He was also impressed with Curtis's publications and was no doubt generally supportive. In some ways there would be no competition, except that Banks encouraged as many people as he could to send plants and tree specimens to Kew from as many parts of the world as possible. On that account Curtis might not have always had first pickings.

Sowerby was strongly influenced by his association with Curtis. He learned much about botany, painting and engraving, and about ways of either making or not making a living. Curtis was not good at paying his artists on time, probably because of his problems with the flow of money and the expensive nature of his endeavours. Sowerby could not and did not rely on Curtis even though he was his main means of employment. He was busy at the Royal Academy Schools and exhibited reasonably frequently at the Royal Academy and the Society of Artists, displaying portraits and other works depicting a landscape, flowers and fruit.[6] His last exhibited painting was at the Royal Academy in 1790, by which time he had shown eleven works over seventeen years. Exhibiting helped him to gain recognition and commissions. In 1780, the year he exhibited his *Portrait of a family* at the Royal Academy, more than 60,000 people viewed the exhibition over its thirty-four days.[7]

His reputation was sufficient for him to be independent when his mother died on 1 August 1782, causing him to leave 2 Bolt-in-Tun Passage. Over the next few years he lived in different places – in Fetter's Lane, at Crune Court, and at Brandon Road, Newington – before he settled into his married home in Lambeth for the rest of his life. But the loss of his mother was balanced to some degree by new connections and new friends. There were those who would advance his career, one of whom would help his personal and private life in a significant way. Robert Brettingham De Carle became friends with James through the world of art and the Royal Academy. He was a son of a mason and bricklayer, Robert De Carle, who was continuing the successful

2/2 The Coade Stone Lion, atop the Lion Gate entrance into the Royal Botanic Gardens, Kew. Designed by Thomas Hardwicke. Photo: Jeff Eden.

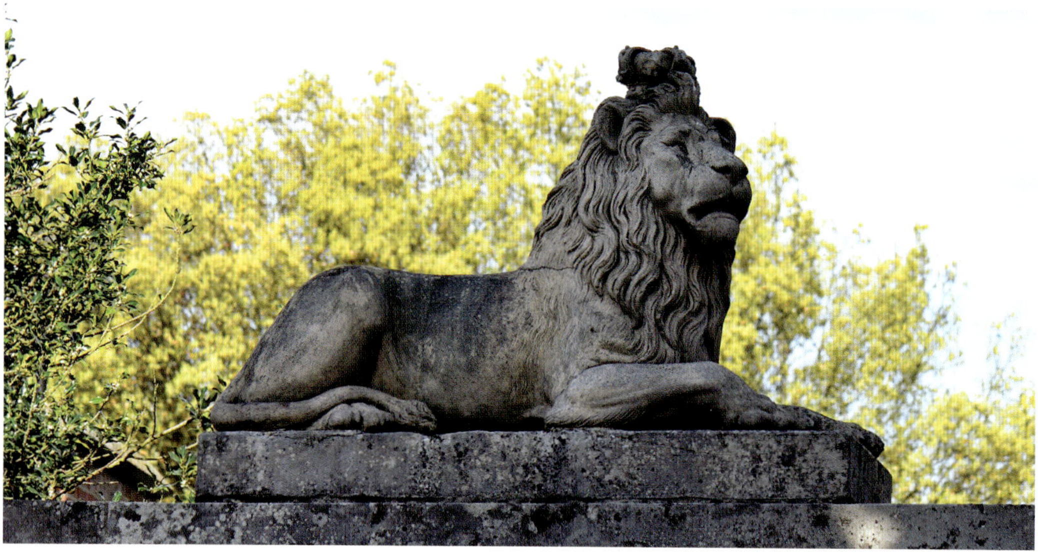

family business in Norwich. Robert's interests were aligned, at least in part, with those of his family but he left Norwich when he was still a young man, came to London where he lived in Lambeth, married Ann Davy in 1774, and took employment at the Coade's Stone Works, also in Lambeth, where he made something of a name for himself and produced several important Coade stone sculptures.[8] Although not a student at the Royal Academy Schools, he exhibited there in 1785.[9] James and he were to be very close friends for the rest of Robert's rather short life – he died in February 1791 at the early age of forty-two.

They were not just friends; they also became part of the same family. The story is relayed by Sowerby's eldest son that his father, on hearing De Carle's lively description of his youngest sister, Anne, painted a miniature of her and then was soon able to find it remarkably true to life when he visited De Carle's family home.[10] This was the beginning of a relationship that led Anne and James to becoming man and wife. It is difficult to judge the truth of this rather romantic tale but it is likely that he came to know his future wife over two or more years. She had the appearance of a gentle attractive woman, with a slightly elongated face, a well-proportioned nose and neat lips. Her eyes lent her a sad or thoughtful look.[11] After several visits by Sowerby to Robert's family home in Norwich during 1784 and 1785, the couple were married on 9 February 1786

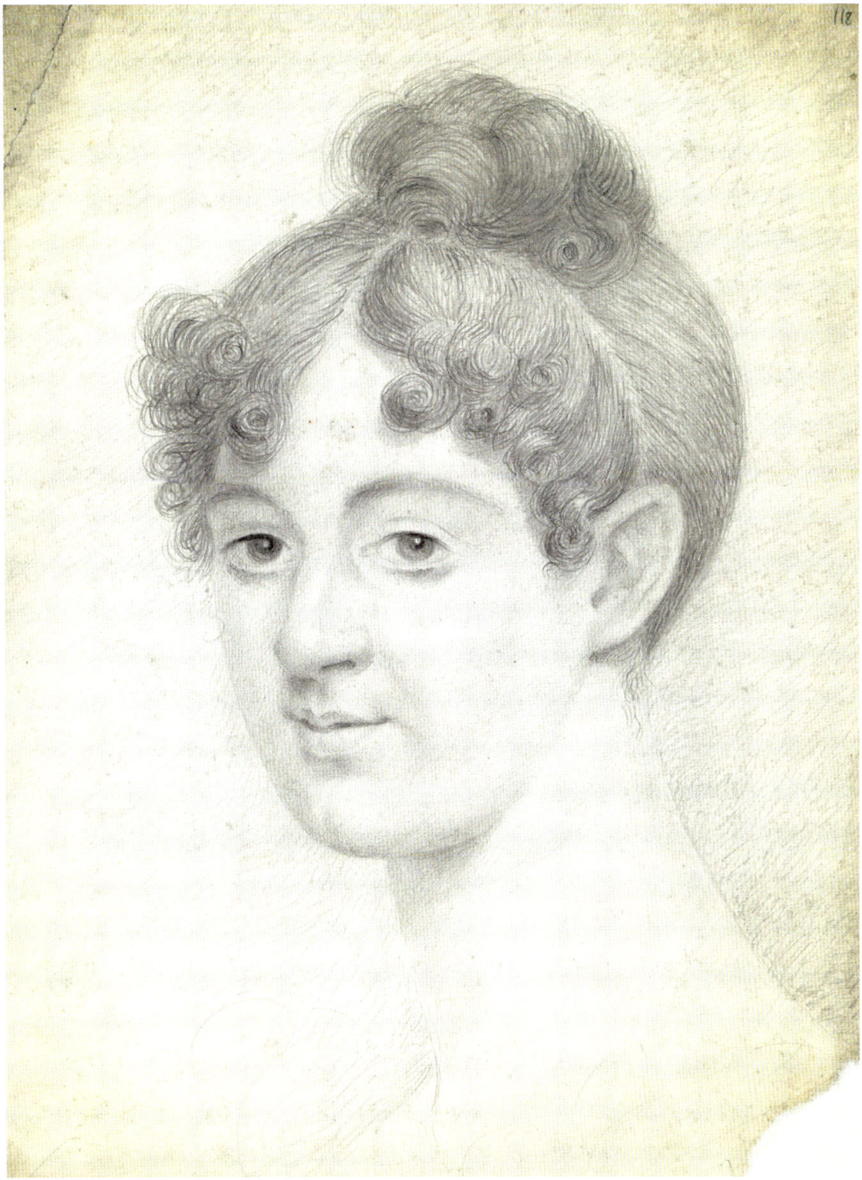

2/3 Anne Sowerby. A pencil portrait by her husband James. About 1797 when Anne was thirty-three years old. Credit: The Natural History Museum.

in Christchurch, Surrey. With James's visits to Norwich he was to come to know and to be known by a circle of very active botanists, several of whom lived in Norfolk and Suffolk. Some, such as Thomas Woodward, were rich enough to devote much of their life to botany.[12]

Woodward lived in the small market town of Bungay, Suffolk. He was particularly interested in seaweeds and fungi and knew that Sowerby's skills and interests could be of use to him. In December 1784 he sent him specimens with strict instructions to draw them. Later in the same month he asked for drawings to be engraved but knowing that as Sowerby was staying in Norwich this might be difficult for him. He wrote of other botanists within his circle: James Smith, James Dickson, William Curtis and a Mr Crowe, and within nine months was suggesting working together on a publication after receiving some fine drawings on fungi:

> the plates of Fungi might be as good a publication as any, & would not interfere either with Dickson or Curtis, the former gives only uncoloured figures which of fungi are of hardly any use ..... If Mr Crowe & you think this place would answer & you could contrive to spend about a fortnight in the Spring and a Month in Autumn at Norwich and here you might have subjects sufficient procured by Mr Crowe and myself for drawings sufficient to supply a fasciculus of 6 plates every 2 months.[13]

The project was never realised but Sowerby now knew these people personally and learned of the high regard in which he was held. Indeed, James Smith was to become a friend and working colleague for the rest of Sowerby's life. James Dickson would provide an early platform for some of Sowerby's most sensitive drawings and Mr Crowe would be a continuing source of knowledge and a provider of specimens. These men, of somewhat contrasting personalities and circumstances, were in regular contact and became some of the most important contributors to botany in England at the time. They had a marked influence on Sowerby in his early days, revealing, in their separate ways, something of the nature of their botanical network.

James Smith particularly, by a remarkable concurrence of circumstances and events, helped to shape the character of British science. Born in Norwich, he was the son of a Protestant dissenter and wealthy draper, educated at home and in Edinburgh where he studied medicine and botany, the latter under John Hope the Professor of Botany. In the autumn 1783 he came to London to study under the surgeon Dr John Hunter and the physician Dr William Pitcairn. He rapidly established contact with Sir Joseph Banks, and was soon devoting much of his time to botany rather than medicine.[14]

Some two months after his arrival, Smith was breakfasting with Sir Joseph when news came of an invitation to Banks to purchase the

2/4 Joseph Banks as a young man (about 1771–1773). Oil painting by Sir Joshua Reynolds. Credit: National Portrait Gallery, London.

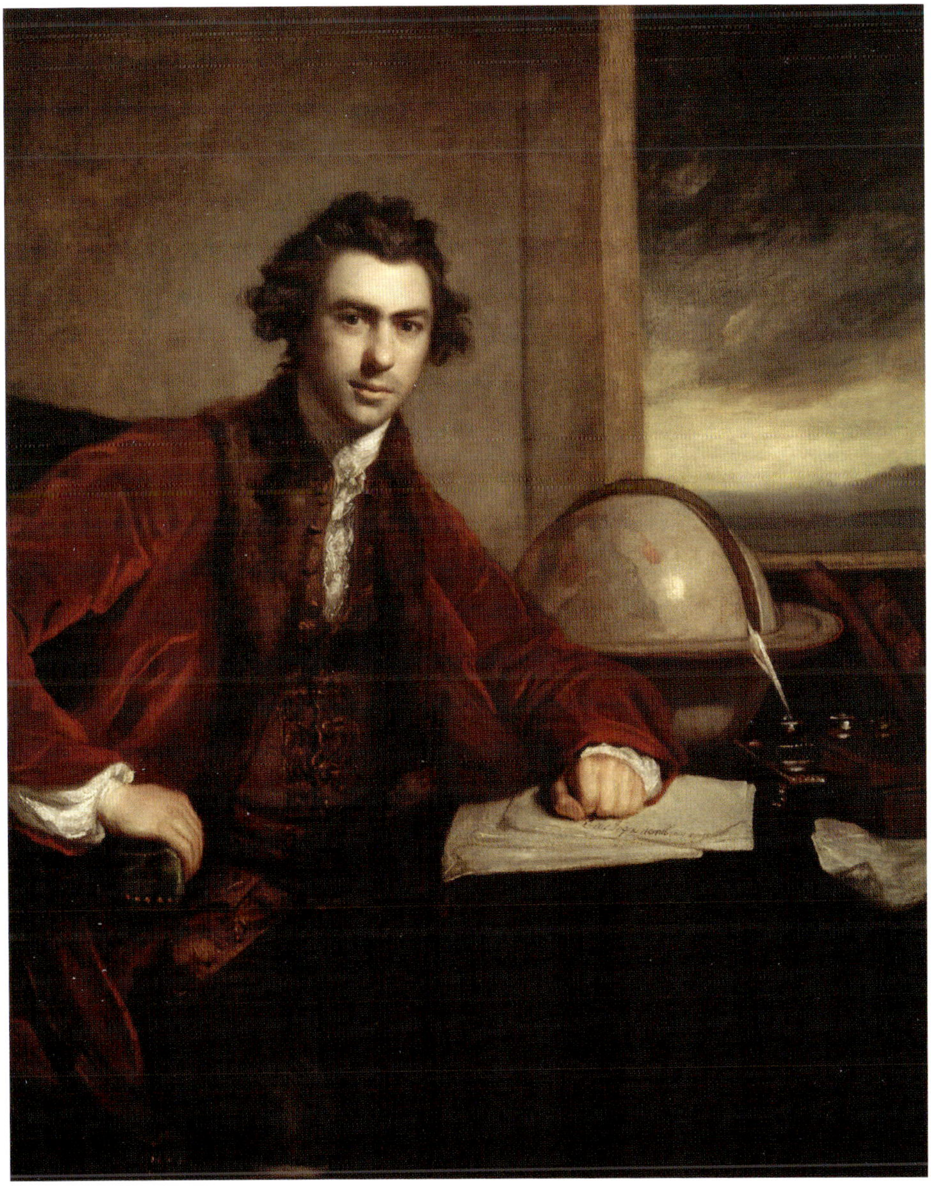

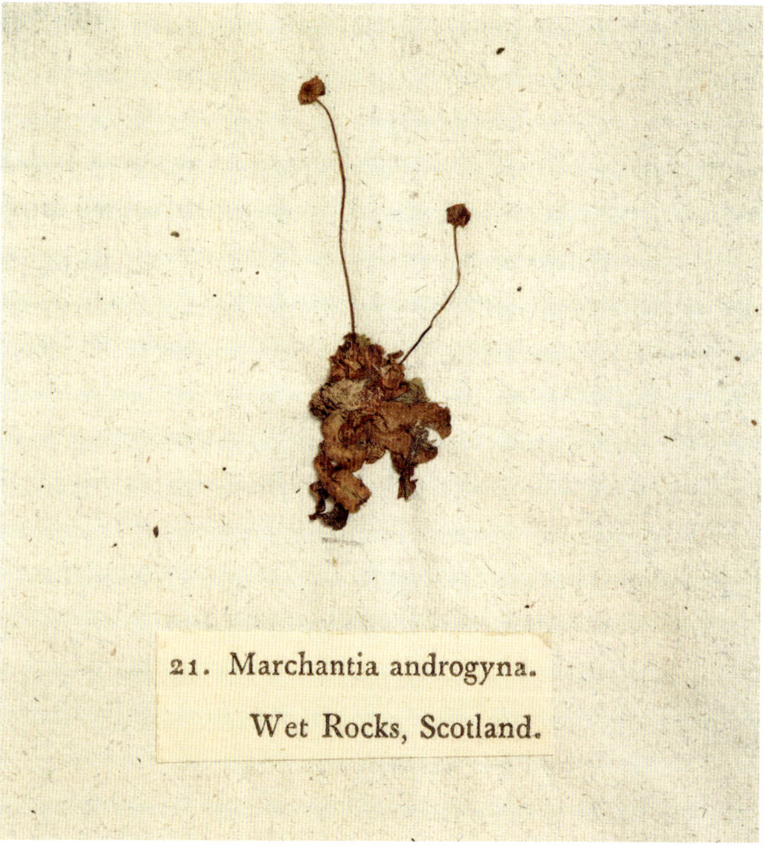

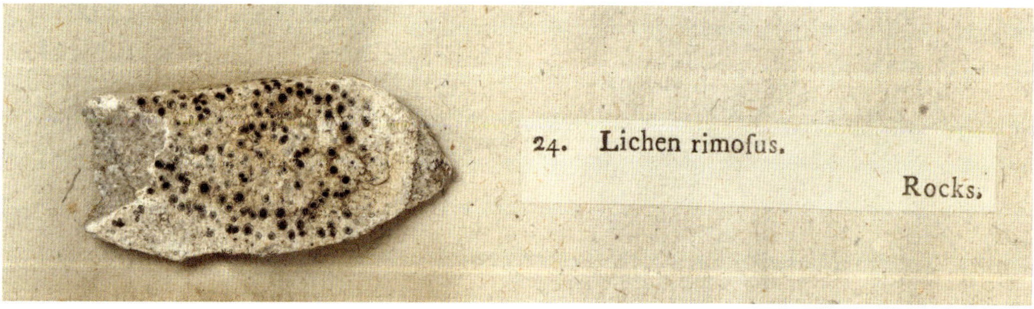

2/5a and b Specimens in Dickson's *Hortus Siccus Britannicus.* (a, left). The liverwort *Marchantia androgyna* (approximately 5x2.5 cm in size). (b, below). A lichen (*Lichen rimosus)* on a piece of crystalline rock which, in turn, is stuck onto a large folio page, approximately 2.5x5.5 cm in size. Dickson produced some fifty copies of this work.

herbarium and other collections of the late Carl Linnaeus, whose younger son had recently died. Linnaeus's widow now sought to sell the collection. Banks, with a significant herbarium of his own, was either not interested or unable to accept the invitation, but he saw in this earnest, young botanist and breakfasting companion a potential purchaser and also a chance for England to have a highly important prize. By the spring of 1784, Smith had persuaded his indulgent father to provide the money; he then purchased the collections, and housed them in his Chelsea apartments on their arrival in October. For a cost of £1,088 (approximately equivalent to £85,000 today[15]) he had obtained a collection of some 14,000 plants, over 3,000 insects, numerous shells and a substantial quantity of manuscripts and valuable books.[16] The Linnaean collection, Smith and Sowerby were soon to be inextricably linked.

James Dickson was in stark contrast to Smith, especially in that his family was a humble one and that he was poorly educated.[17] He worked as a gardener and nurseryman before starting his own business as a seed supplier and nurseryman at Covent Garden in 1772, through which he became friends with many botanists including Sir Joseph Banks. Dickson had a passionate interest in mosses, as well as grasses and fungi. He collected numerous specimens and produced several publications, one started in 1785 and illustrated by Sowerby, to which we will return, and another which not only gives a splendid insight into his enthusiasm but also shows how botanists used different techniques to help people in specimen identification. This remarkable work, A *Collection of Dried Plants, Named on the Authority of the Linnaean Herbarium and other original collections* (1789–1792) contains dried plant specimens, one species only to each good-quality and folio-sized page, with its accompanying printed labels. Some of the specimens were collected by James Smith on his European tour; a few came from British localities or from the Chelsea Physic Garden. The aim was to promote the knowledge of rare and obscure plants and to provide specimens of them for surety of comparison and identification. The method of doing so was, of course, problematic and counter to Dickson's aim for, as Dickson said in his own advertisement for the work: 'it is to be regretted that a work of this kind must be of a very limited extent, on account of the difficulty of collecting so great a number of specimens as it requires. There are, therefore, only fifty copies printed.' Small though the publication number was, it probably still managed to meet the desires of the relatively small band of wealthy botanists in south-east England. Nor did the limited publication number diminish his enthusiasm or persistence. After four fascicles, each with twenty-five species, had been produced he turned his attention to a new series devoted to native British plants, as requested by the purchasers of his earlier work. *Hortus Siccus Britannicus* [Dried British Plants] in nineteen fascicles appeared

2/6 The willow *Salix croweana*, named after the surgeon and botanist James Crowe. Coloured engraving from Sowerby's *English Botany*.

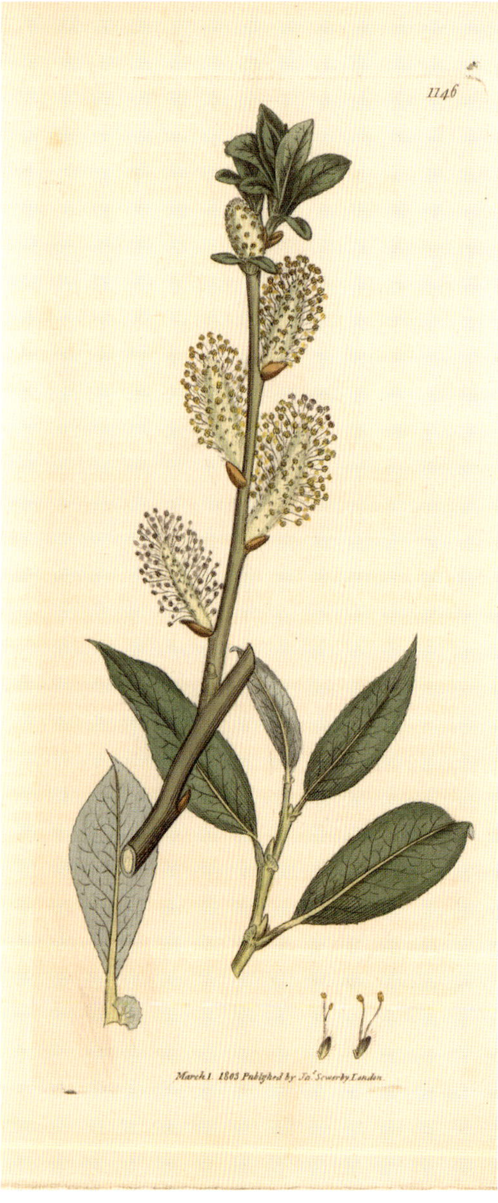

over the following nine years to 1802, with 475 flowering and non-flowering species including lichens, seaweeds and mosses. In one case a species of lichen on a thin piece of rock is stuck onto the page.[18]

Dickson dedicated the various parts of these works to those within his circle – James Crowe, Thomas Woodward and William Curtis, among others. Few copies probably are seen in the world today, not only because of the fragile nature of many of the plants but because many owners removed the specimens to put into their own herbaria. But the work was a labour of love and dedication, with an approach that was unusual in its scale and duration. In contrast, using an artist and an engraver such as Sowerby is a relatively easy and durable approach. As we will see, Dickson turned to Sowerby for his other publication.

Mr James Crowe, a resident of Lakenham, Norwich, was a successful surgeon and landowner with a keen interest in gardening and botany and actively encouraged his friends by supplying them with specimens or aiding their researches.[19] He sent specimens to Curtis and travelled some distance in search of unusual species. He did not publish any botanical works but became an expert in certain plant groups especially willows, establishing his own salicetum, and providing Sowerby with many examples of his species. He did not seek the limelight beyond Norwich but his skill as a

2/7 John Coakley Lettsom with his family. Painting by an English artist, about 1786. Credit: Wellcome Trust. Wellcome Library, London

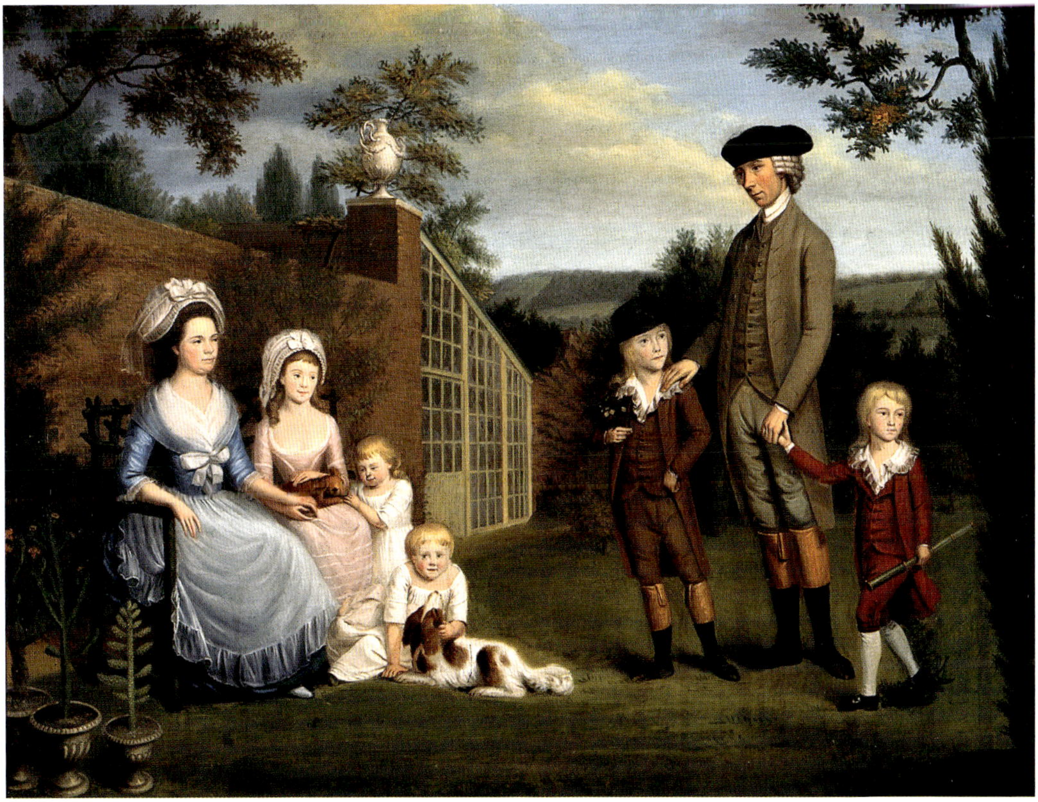

botanist was well recognised, especially by his long-standing friend, James Smith, who named the shrubby plant genus *Crowea,* and also the *Salix* species *S. croweana* in his honour Smith,who had only a small garden at his house in Norwich, sent seeds and cuttings for planting in Crowe's gardens, where there was also a hothouse. Crowe, in return, provided specimens and information for Smith's and Sowerby's botanical works, especially *English Botany.* Mr Crowe, for that is how he was always referred to, often features in the correspondence of the botanical fraternity and was a Founding Fellow of the Linnean Society.

Norfolk was not the only location for Sowerby's network. In June 1785 the physician Dr John Coakley Lettsom wrote, from the house of his medical practice in Basinghall Street, London, just one sentence to Sowerby: *When Mr Sowerby*

2/8 The beautiful *Sprekelia formosissima* (previously known as an amaryllis). One of many watercolours by James Sowerby of plants in John Lettsom's garden at Grove Hill. This plant is commonly cultivated today.

can make it convenient to call upon me, any Thursday or Tuesday from 4 to seven afternoon – or Monday & Friday morning, I shall be glad to see him.[20] A commission was likely to follow.

Lettsom was a Quaker, wealthy through both marriage and his successful profession. He was highly energetic, pursued several interests including natural history, corresponded with renowned people in Britain and abroad, wrote on many subjects but especially on tea, and was very sociable. His wealth had allowed him to build a fine house, *Grove Hill*, in Camberwell, south of the Thames. Adjacent to the house were his library and museum, the former thirty-nine feet long and twenty feet wide, divided into sixteen compartments with a bust over each of some distinguished and appropriate person such as Newton over the arts and sciences compartment, Locke over divinity and law, Hogarth over prints and maps and John Wesley over tracts and pamphlets. Beyond these buildings were the greenhouse, the hothouse and laundry, stables, pleasure garden, kitchen garden, paddock, orchard, succession garden, 'arbustrum' or grove, pond and statue of cupid, Sybil temple, cold bath and basin, rotunda, and Shakespeare walk with a statue.[21] All this was very fine. Lettsom entertained frequently and quite lavishly. His garden contained numerous species – it was a kind of botanic garden and would have been of interest to his guests, horticulturists and others, and undoubtedly helped raise his status and influence. What better person could there be to illustrate some of the special plants and shrubs growing there than James Sowerby? Notes, scribbled by Sowerby on the Lettsom letter of invitation, show that he was soon receiving money for work done and for the cost of some materials. The result was a fine series of watercoloured drawings of plants growing in Lettsom's gardens at Grove Hill.[22] Lettsom was also known to support Curtis in his *Flora Londinensis* endeavour, the second volume (1798) being dedicated to him.

Others were starting to send Sowerby commissions for natural history subjects. A Mr Henry Harrison of Wearde in Cornwall wrote on several occasions in 1785 asking Sowerby to paint a 'Sparrow Hawke' and 'to show the ferocity in its eye with its brilliant yellow'.[23] He would, he said, be happy to have it half life-size. Later he asked Sowerby to paint a Polyanthus for 10s.6d., Sowerby's charges being from half to one guinea depending on the amount of work. But four months later Harrison was forced to write 'I believe you forgot to Execute the Polianthus you promised should be done for me some months since.' Sowerby replied that he needed a sketch or the original material: 'it matters not what the subject if there is an original to be had from Man the Lord of the Creation to the lowest reptile.' He apologised for the delay and then continued, 'but have been engaged in portrait painting I mean

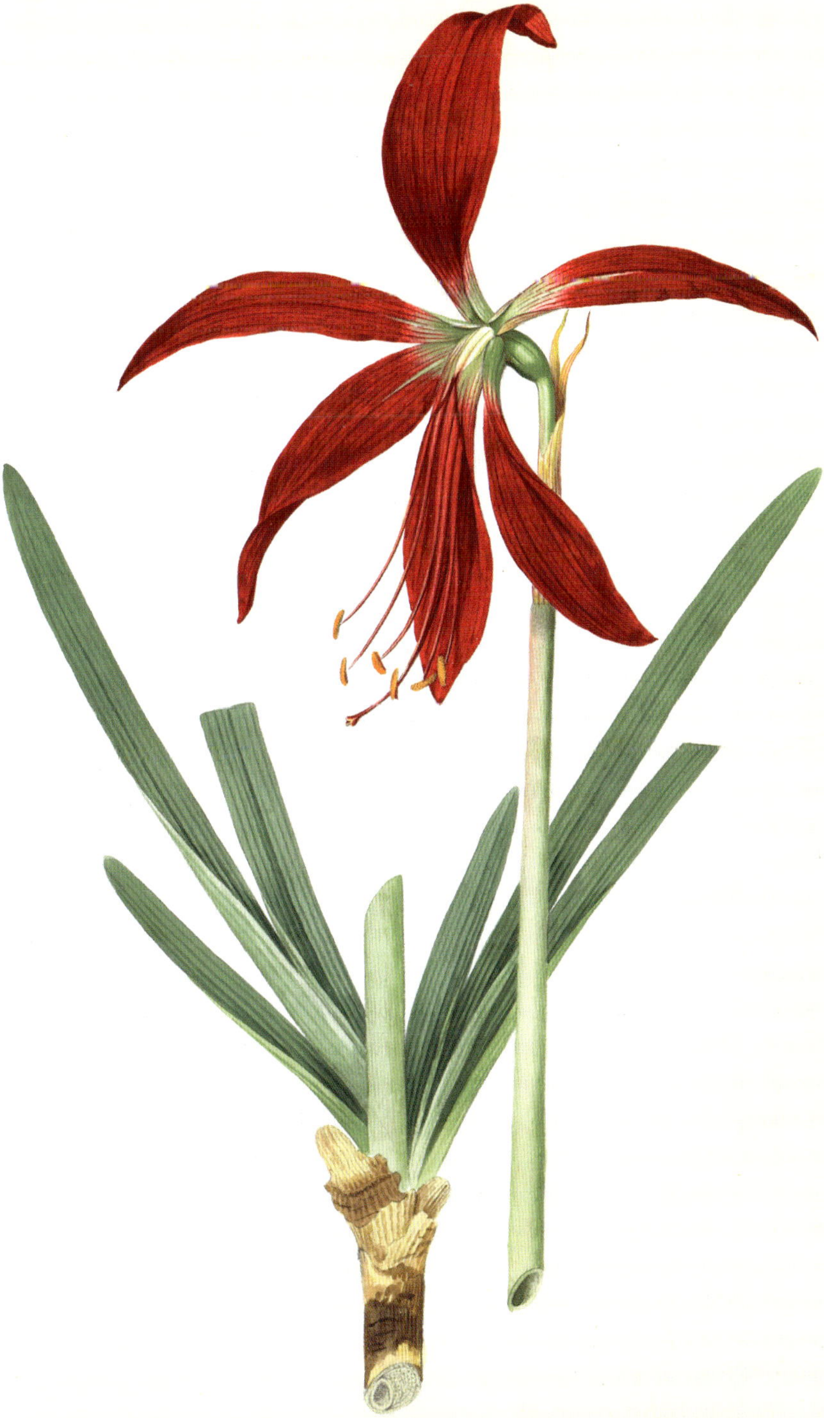

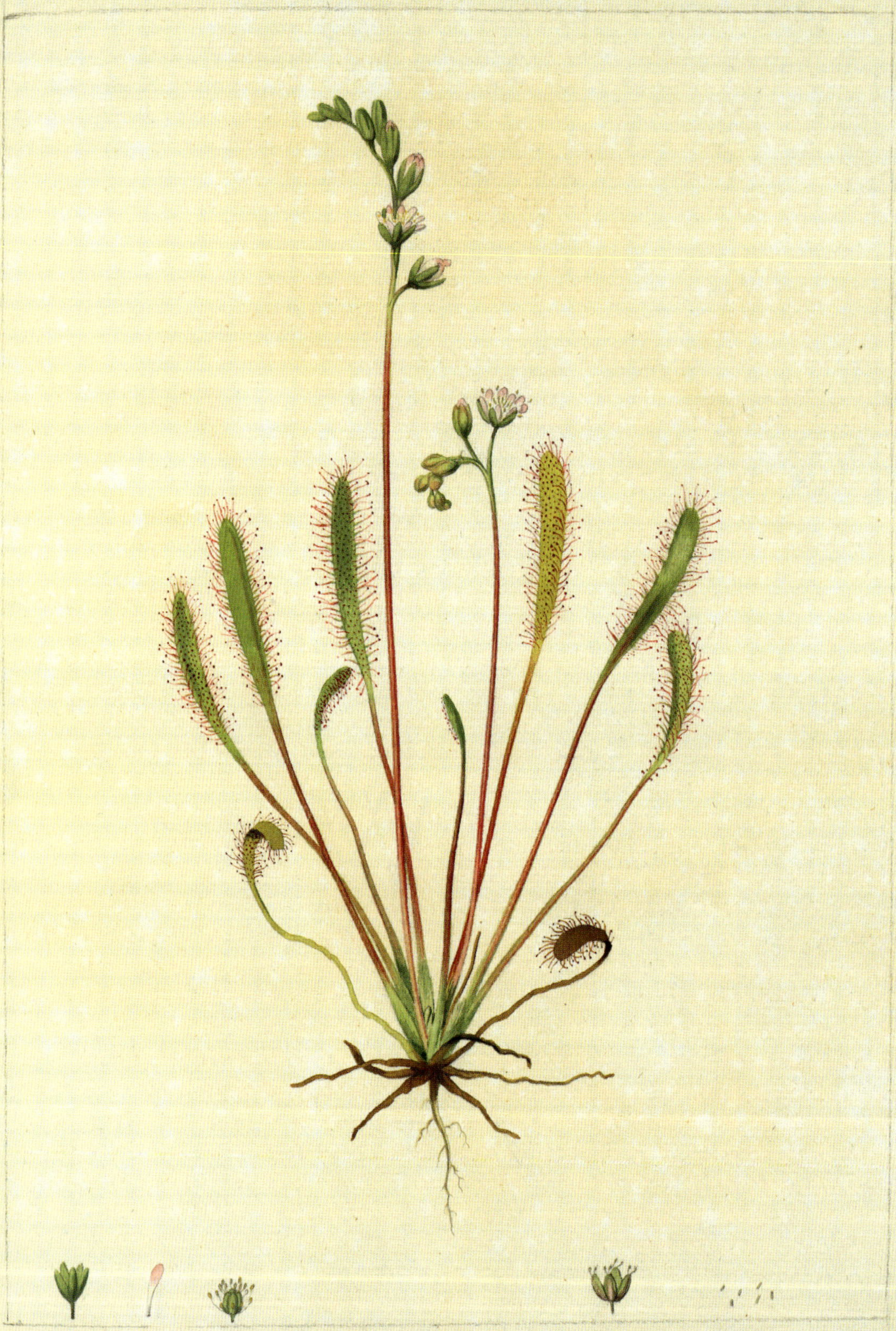

Drosera anglica

2/9 English sundew (*Drosera anglica*). A watercolour of this carnivorous plant in Lettsom's garden, by James Sowerby. The plant traps and digests small insects.

2/10 *Massonia depressa* or hedgehog lily. Native to South Africa and, unusually, pollinated by small rodents. A watercolour of the plant in Lettsom's garden, by James Sowerby.

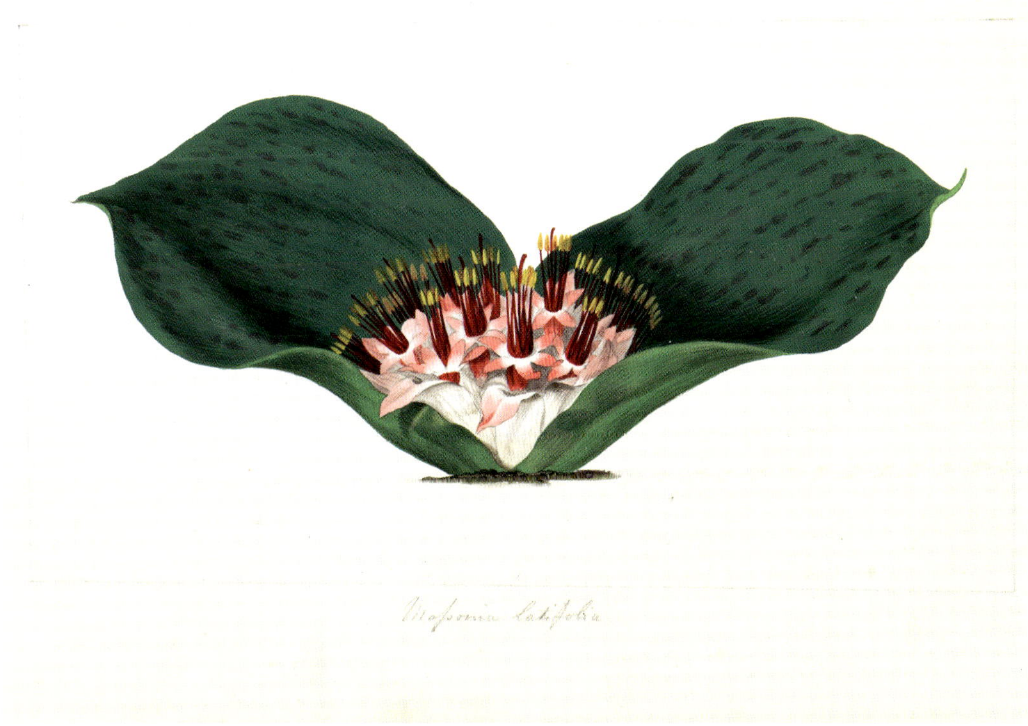

painting likenesses of Men Women & children though I have also painted a striking likeness of a toad for a German naturalist, am pleased with natural curiosities'. The work was done a month later and Harrison was well pleased with much of it such that he asked Sowerby to now do a '... colar'd goldfinch in a pleasing attitude & a Bull finch'. People were prepared to cope with delays so as to have the artist they wanted most.

Sowerby was busy with the many small and larger commissions now coming his way and with teaching drawing to several students. He was content that for him natural history was displacing portraiture, and there was no shortage of work whether painting or engraving. Richard Relhan, chaplain of King's College Cambridge and a botanist, produced his Cambridge Flora, or *Flora Cantabrigiensis,* in 1785, with its seven plates engraved by Sowerby from the original drawings by James Bolton.[24] This was a work typical of its time, having some several hundred pages of text describing, in Latin, the plants according to the

Linnaean classification, and clearly intended for the serious botanist. Sowerby was moving into the realms of specialism and yet while these commissions satisfied his desire to pursue natural history he also felt inclined to address a wider audience which included those who would enjoy being instructed or guided about the natural world around them.

William Curtis, however, was not going to let Sowerby slip away entirely to the employ of others. Not content with the sales of his *Flora Londinensis,* which never exceeded 300 copies and was loss making, he started, in 1786, preparing a new publication, *The Botanical Magazine, or, Flower-Garden Displayed: in which the most ornamental foreign plants, cultivated in the open ground, the green-house, and the stove, will be accurately represented in their natural colours. Intended for the use of such ladies, gentlemen, and gardeners, as wish to become scientifically acquainted with the plants they cultivate.*[25] Sowerby drew the first plate, *Iris persica,* or Persian iris, dated March 1786, and more than sixty subsequent ones. The other main artist was Sydenham Edwards. Both drew the illustrations from living plants. The accuracy and skill with which Sowerby fulfilled his task allowed Curtis on one occasion (for *Limodorum tuberosum,* 116 in volume 4) to simply write 'our figure will make a description of the plant unnecessary'. For this, on one small page, Sowerby has depicted, with remarkable elegance and in life-size, the tall plant with its bulbous root, long stem, curved leaf and flowers. The first part appeared in February 1787, contained three coloured plates and cost 1s. The *Botanical Magazine* was an immediate success, it was soon to have a circulation of around 3,000 copies, and was financially secure. Curtis was to claim that whereas *Flora Londinensis* had gained him 'praise', the *Botanical Magazine* had 'brought him pudding'.

Why did the *Botanical Magazine* have this contrasting success? Each issue was much cheaper, at 1s. than the coloured versions of *Flora Londinensis* at 5s. Of course, the *Flora* had twice as many plates (six compared to three) but the outlay each time was an important factor. The *Botanical Magazine* was also smaller, with a page area a quarter that of the *Flora.* It was less grand and more accessible. It appealed to a wider audience with its easy English text, yet still purporting to be scientific and containing clear guidance on plant cultivation. It was published only in colour, whereas the *Flora* posed the dilemma of subscribing at 2s.6d. for the uncoloured issue or 5s. for the coloured. Finally, it appeared rather regularly, unlike the *Flora,* at about twelve issues a year. And, of course, the illustrations were of a high standard.

The quality of Sowerby's work for Curtis's publications was very well recognised; many authors were seeking his skill for the illustrations to their publications and some were prepared to extol him in their advertisements or

2/11 *Iris persica* or Persian iris. The first plate, drawn by James Sowerby, in Curtis's *Botanical Magazine*, 1787. It was probably chosen because 'its beauty, early appearance, and fragrant blossoms, make it highly esteemed by all lovers of flowers'.

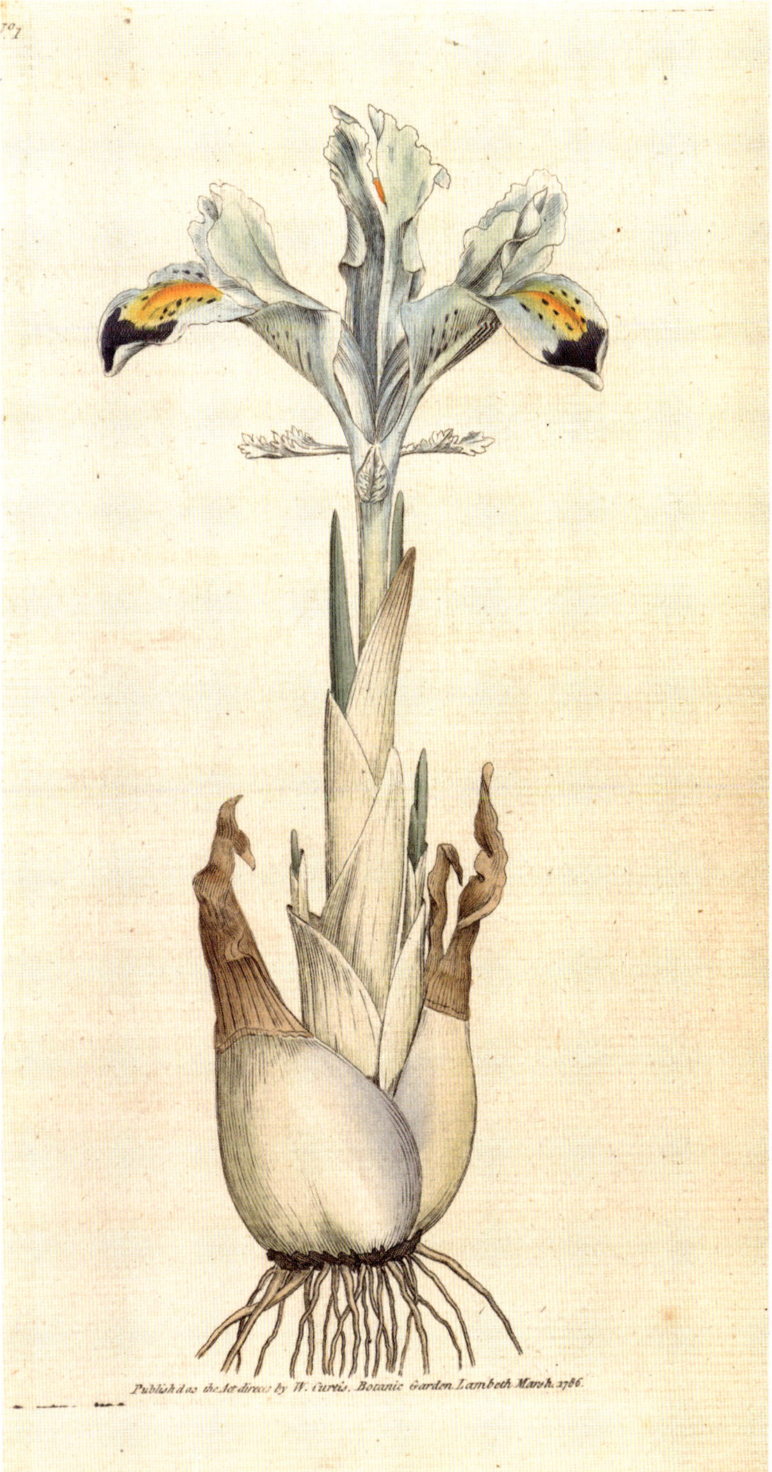

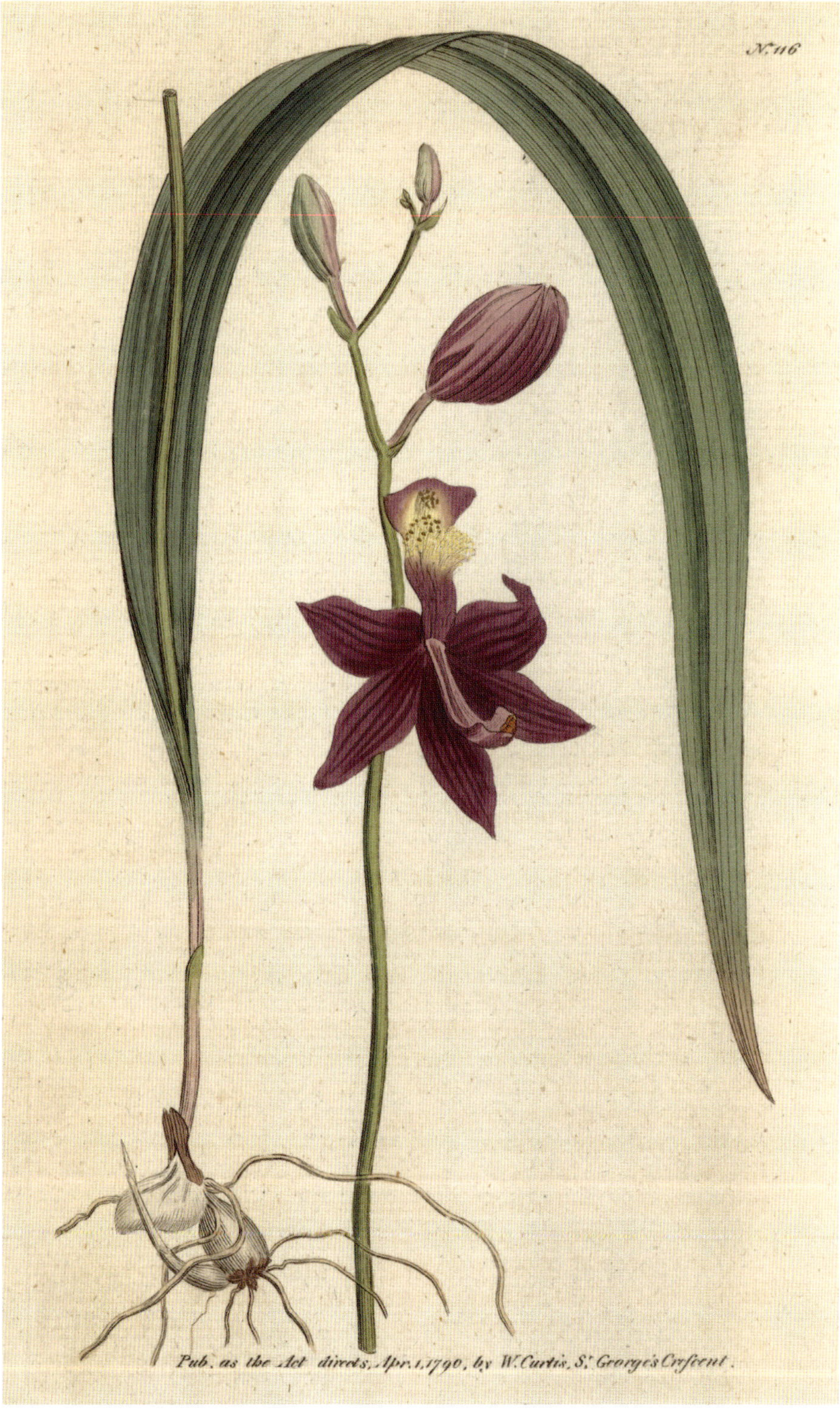
N.º 116
Pub. as the Act directs, Apr. 1, 1790, by W. Curtis, S.t George's Crescent.

2/12 (opposite) *Limodorum tuberosum* (synonym of *Calopogon tuberosus*.) The plate in *Botanical Magazine*, drawn by Sowerby and about which Curtis wrote that 'our figure will make a description of the plant unnecessary'. This orchid is native to North America.

2/13 (right) *Fuchsia coccinea* or scarlet fuchsia. A bold illustration, drawn by Sowerby for *Botanical Magazine*, of a plant which at that time (about 1790) was expensive to purchase but which Curtis claimed that 'as it is readily propagated, both by layers, cuttings, and feeds, it will soon be within the purchase of every lover of plants'.

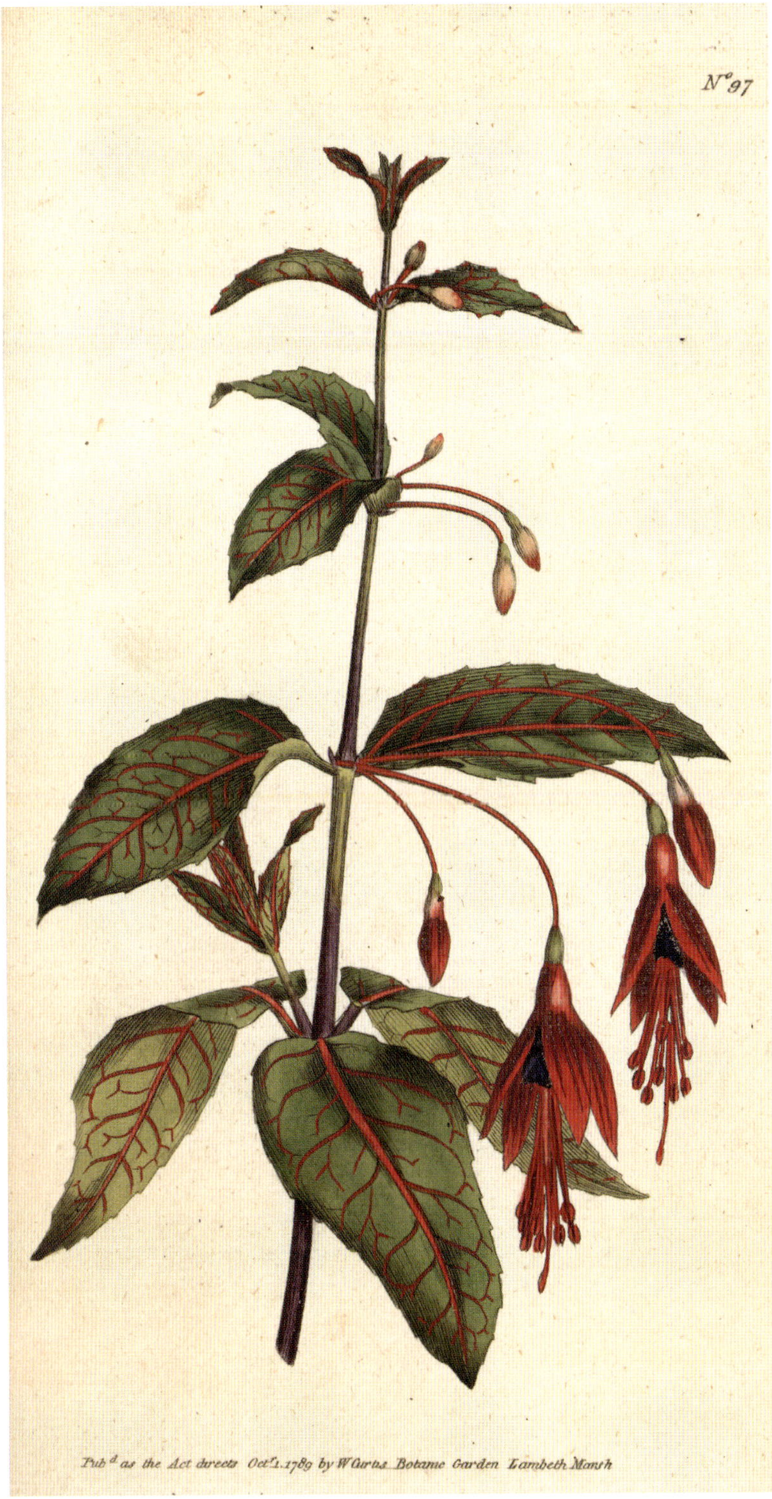

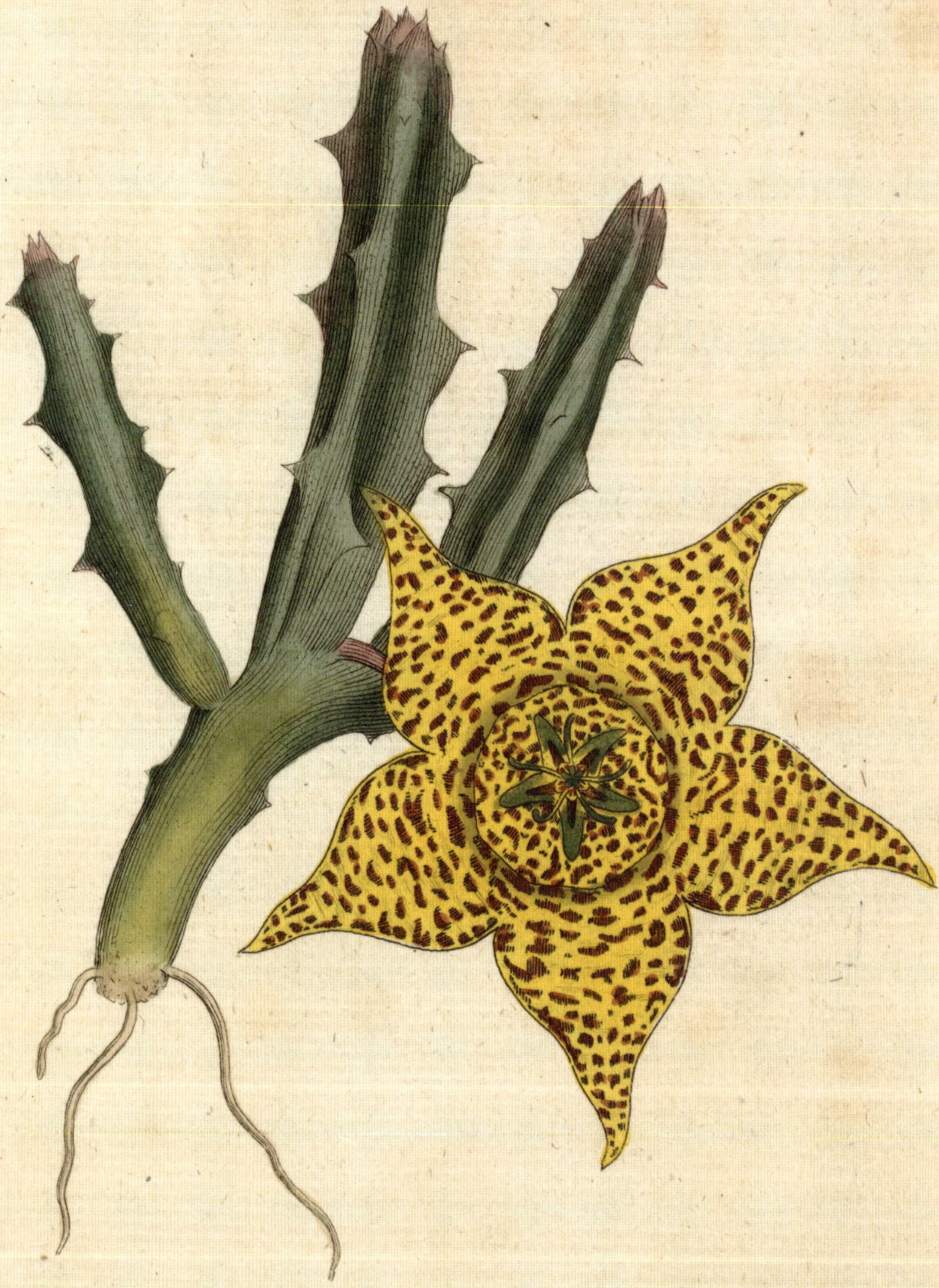

2/14 *Stapelia variegata* or variegated stapelia. Shows one of the more unusual plants illustrated early on in the *Botanical Magazine*. A native of the Cape of Good Hope, which in Britain can be kept in a heated greenhouse.

introductions. One such was in *The Pomona Britanica, or, Fruit-Garden Displayed containing scientific descriptions of all such fruit-trees and fruit-bearing plants as are cultivated in Great Britain*, published in 1788 and written by 'several hands' with revisions by A. and W. Driver. The two-page advertisement said 'the Specimens, will in general, be engraved from drawings executed by Mr. SOWERBY, whose abilities several recent botanical works will enable the public to form a judgement', with the footnote 'Vide Flora Londinensis, Botanical Magazine etc.' Its title indicated that it would be an extended publication but that was never realised, only a small (first?) portion appearing.

The eight coloured plates are of fine drawings showing wood, leaves and fruit, of types of cherry, raspberry, bergamot, apricot, apple and plum. Some of the original drawings are in the Natural History Museum, London; they have a clean sensitivity with the portrayal of lifelike textures and colours. On one drawing of a cherry a pencilled note, in Sowerby's hand, is about the taste of the fruit in comparison to others, and it is likely that Sowerby contributed something towards the text.[26] The text also contains hints of the current state of knowledge of plant growth and horticulture, as for example in its account of 'air':

> Air has been considered by divers authors, as one of the principal agents of vegetation; it is an elastic fluid, with numerous particles of different qualities or nature floating in it, which are properly adapted for the existence of vegetables and without which they cannot live, much less come to perfection. It operates not only upon their surfaces, but within the bowels of the earth, rarefying its crudities and collecting a variety of salts and juices, either peculiar to itself, or such as are gathered from their various lodgments in the earth, insinuating its substances into the liquid sap of vegetables, entering the roots and by the cooperating powers of the sun rarefying the juices in the upper part of the tree.

Sowerby would not have been the author of this kind of information. He knew his limitations and his abilities. Now he was attempting to have better control of his own destiny. For every year after the appearance of the Drivers' *Pomona Britanica* until his death, Sowerby was publishing works of which he was an author. For many of these he was the sole author and for most he was in a controlling position.

2/15 a and b. Two engraved plates in the small book *The Pomona Britanica.* (a, left)The Swiss bergamot pear. The fruit was described as: 'a native of France, much esteemed at the table for its beautiful appearance, and the sweetness of its abounding juices'. There were many varieties of pears and the naming of them was not always consistent. (b, right) An unnamed plum featured at the end of the book in some copies, without any letterpress.

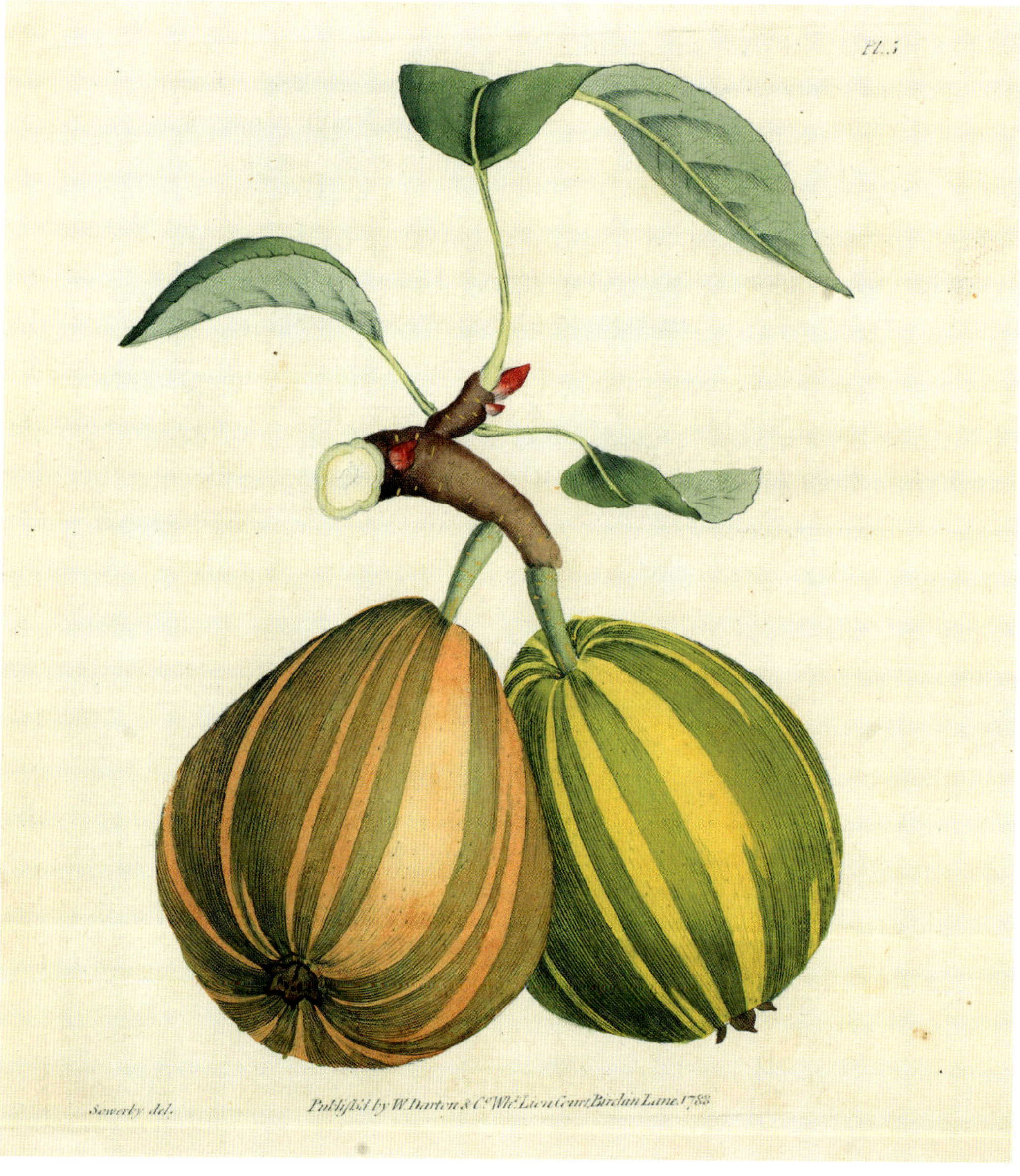

2/16 a and b (overleaf) Sowerby's original coloured sketches of (a, left) plums with bloom on their surface - see also 2/15 (b) - and (right) apples, with his pencilled notes on the nature of the fruit (such as 'a soft and mellow eating apple'). Intended for Driver's *Pomona Britanica* (1788) but not generally published.

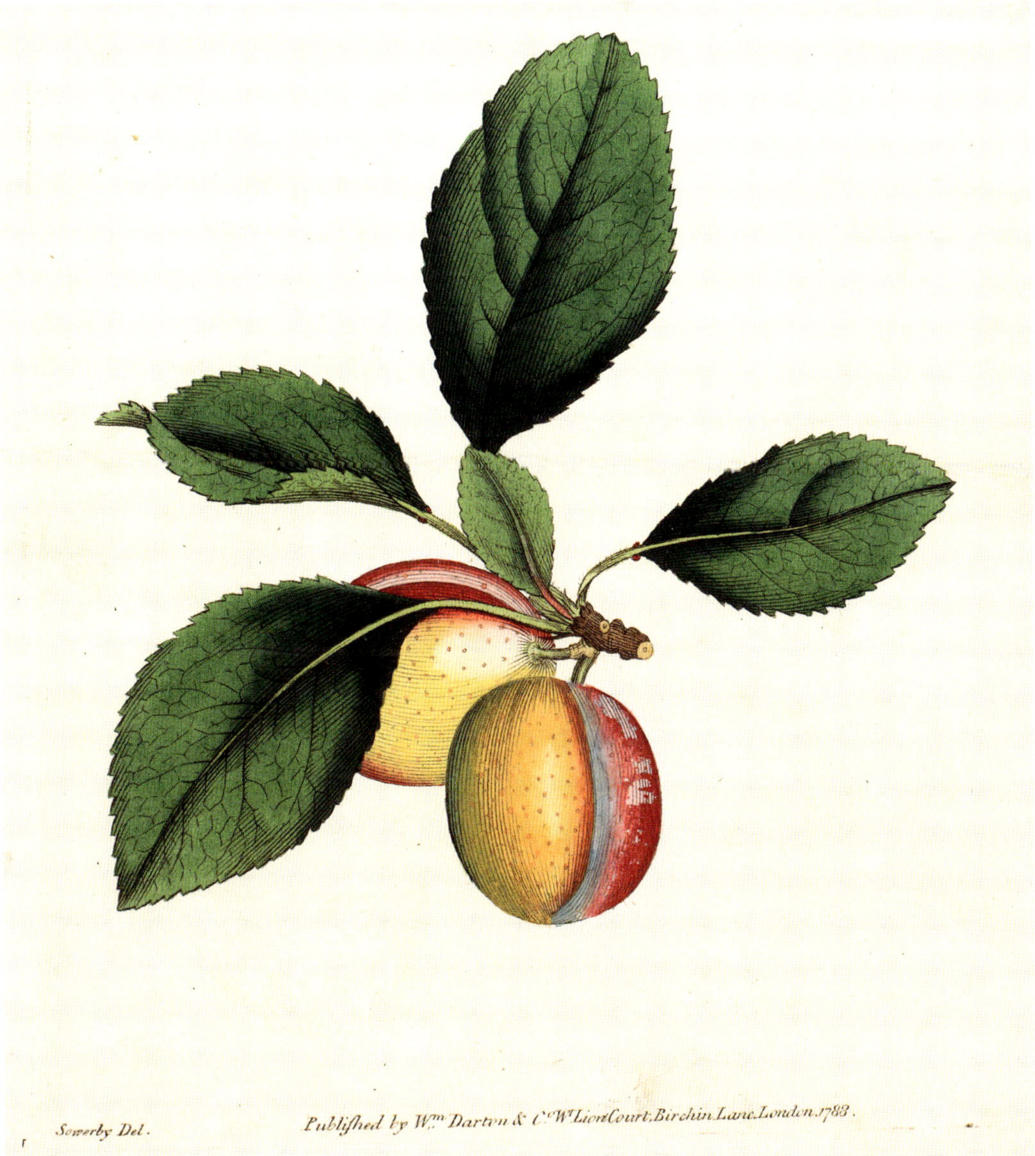

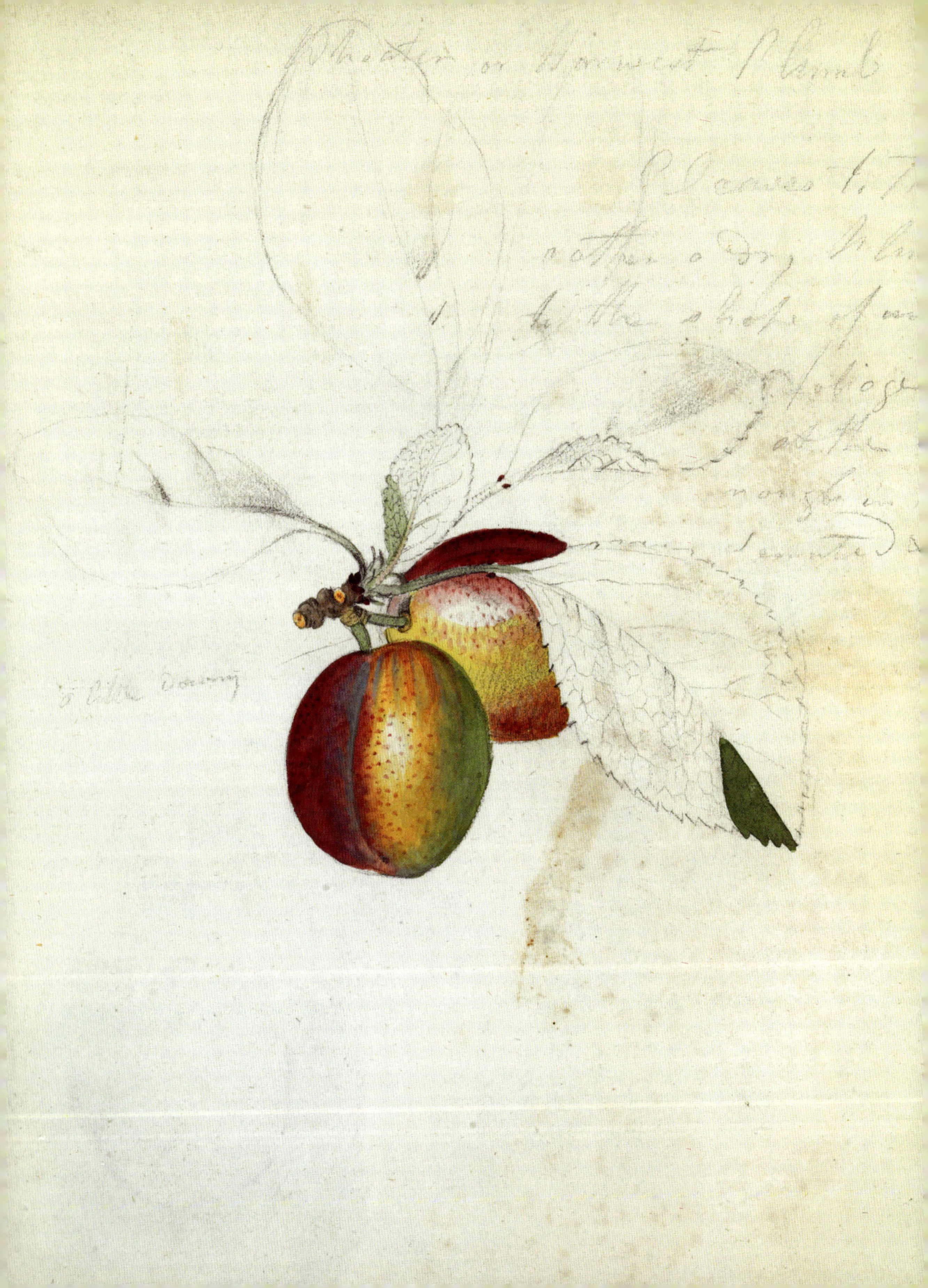

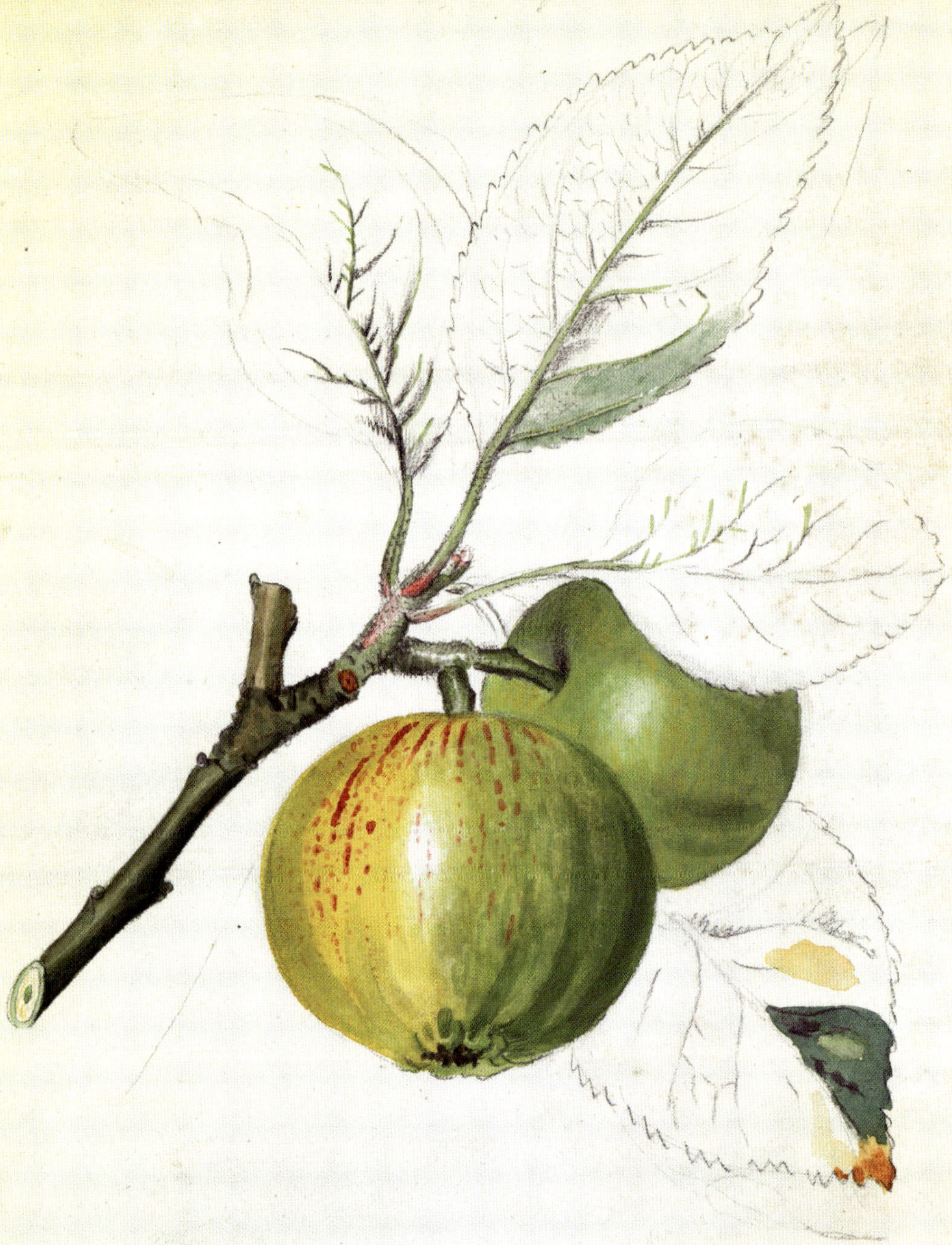

This sort is generally a little streaked with a brigh
d but seldom stays on the tree so ripe as the, near
ere represented a soft and mellow eating apple
ther a large and open core, front of foliage a bri
hining green a little crumpled, back ~~hoary~~ rather a strong
owney pubesence ~~at~~ the end of the branches term
[illegible] the leaf and the back of it

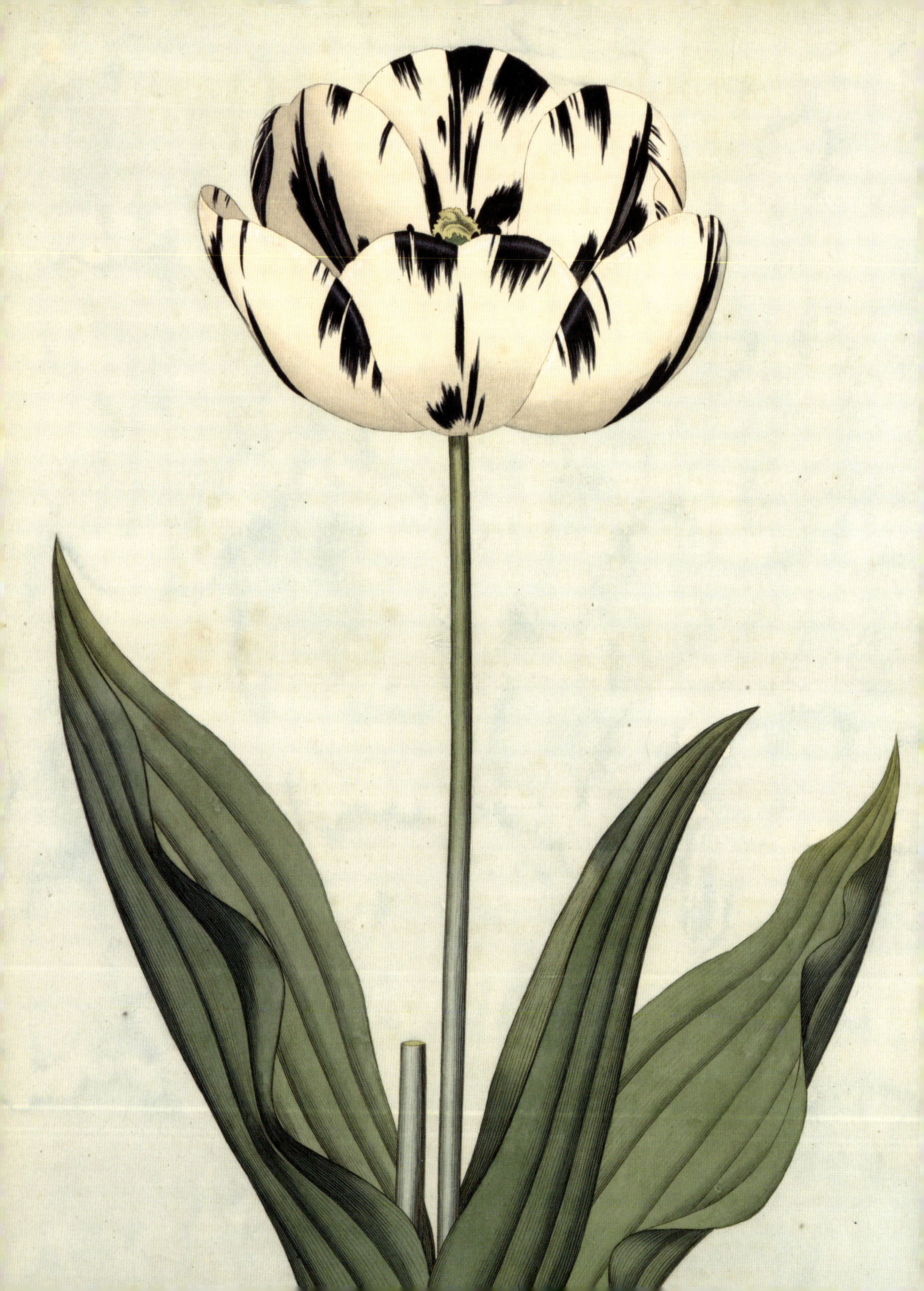

CHAPTER THREE

The distant suburb

3/1 The Rodney, a Bybloemen tulip. One of seventeen hand-coloured engravings in Sowerby's *Flora Luxurians*. A Bybloemen 'signifies a tulip with a white bottom, which does not belong to any regular family or tribe of tulips'. This one cost ten guineas.

Marriage brought more than a wife; it brought also a house and home. Anne's father, Robert De Carle, gave the newlyweds a house in the rapidly developing area of Lambeth south of the River Thames, near to Westminster Bridge and on the southern side of Bridge Road. The house, 2 Mead Place, was one of a plain-fronted terrace, each about eighteen feet wide, with four floors and a rear garden about forty feet in length and of sufficient size for Sowerby to build later his own museum and workshop[1] and to grow plants and apple trees (he was particularly proud of his espaliered French Codling apple[2]). To the front it overlooked the fashionable Apollo Gardens on the other side of Bridge Road. This was a very different environment from the smell and hubbub of Fleet Street, and Sowerby was to live there for the rest of his life.

Lambeth may have been a suburb but it was not a distant one as Hawkins disparagingly suggested in his letter to James Smith. A walk of only one and a half miles from Sowerby's house would be needed to reach the action at Covent Garden, or less than two miles to go to James Smith's residence in Great Marlborough Street, and access was easy by the new bridge. Nor was it a suburb in the sense of being predominantly residential with little civic activity. It had the Lambeth Palace (the London residence of the Archbishop of Canterbury), nearby pleasure gardens, Curtis's botanic garden, the residences of well-known people including, for a time, the

3/2 Mead Place in Lambeth. An old photograph. Sowerby's houses were the second and third from the left. The building was destroyed during the Second World War.

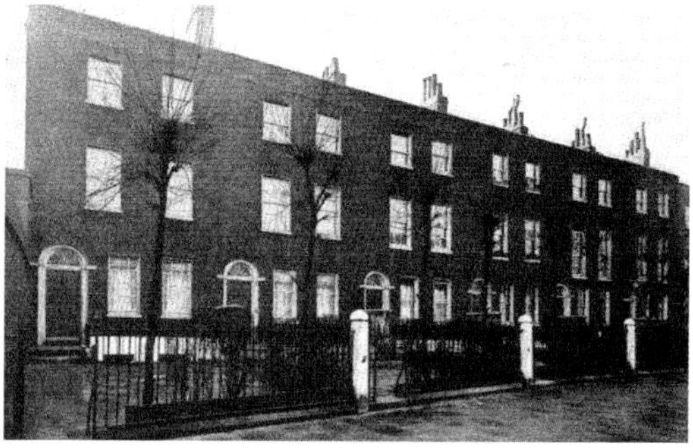

poet and artist William Blake, and John Lettsom with his grand house *Grove Hill* was not far away. The river frontage had timber yards, boat building and other industries including the Coade Stone Works.[3] In earlier years the two John Tradescants – father and son – had their renowned garden and museum collection ('The Ark') just to the south, which was to become the start of the Ashmolean Museum in Oxford. Their tomb is still intact in St Mary's Church garden adjacent to Lambeth Palace. Westminster Bridge had opened in 1750 and was a spur for significant development, of which Sowerby's house was a reflection. From the time that the Sowerbys moved into their house at the time of their marriage to his death in 1822 the number of houses in the Lambeth Church District had increased fivefold.[4]

Very close to Mead Place was the Asylum for Female Orphans established by Sir John Fielding in 1758 to help prevent prostitution and to ensure that the children 'are regularly and alternately employed in reading, knitting, sewing, and in the business of the kitchen'.[5] Sowerby was to become a supporter of the Asylum and even proposed a scheme of charitable giving to which the Queen, Dukes, Lords and other would donate: the Queen £20 down to shopkeepers £2.[6] The worthiness of the Asylum had even caught the eye of the naturalist and writer Thomas Pennant, who, in his *Account of London*, did not stint his lavish praise:

> On approaching St. George's Fields from Westminster-bridge are two charities of uncommon delicacy and utility. The first is the Westminster Lying-in Hospital... Farther on is another institution of a most heavenly nature, calculated to save from perdition of soul and body, the brighter part of the creation: such on whom Providence hath bestowed angelic faces and elegant forms, designed as blessings to mankind, but too often debased to the vilest uses. The hazards that these innocents constantly are liable to, from a thousand temptations, from poverty, from death of parents, from the diabolical procuress and often from the stupendous wickedness of parents themselves, who have been known to sell their beauteous girls for the purpose of prostitution, induced a worthy band to be found, in the year 1758, the Asylum or House of Refuge. Long may it flourish and eternal be the regard of those into whose minds so amiable a conception may have entered.[7]

3/3 (below) The male and female flowers of a cedar tree and the parts of mosses. In Sowerby's drawing book.

3/4 (overleaf) The first plate in Sowerby's drawing book, showing the parts of the stamen and different stamen structures.

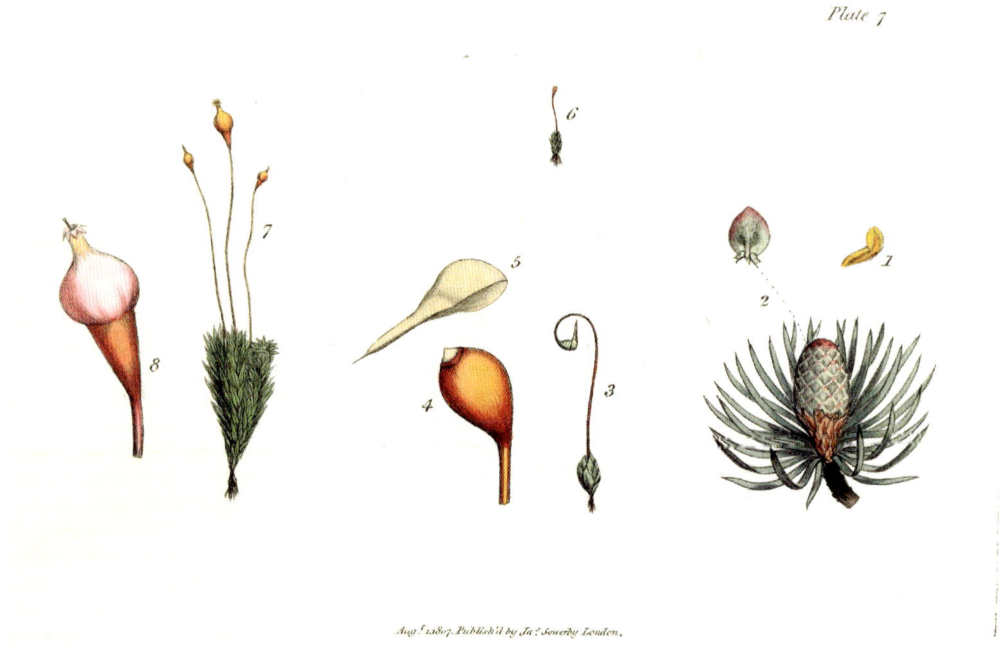

Sowerby's address was often written as 2 Mead Place, near the Asylum, Lambeth.

The years following his marriage were especially creative ones for Sowerby. His own small book *An easy introduction to drawing flowers according to nature* appeared in 1788. It is an attractive, informative work with six plates of his drawings demonstrating the art of illustrating different parts of flowers – calyx, petal, stamen and so forth – and dealing not just with flowering plants but with algae, mosses and fungi. It was for his students as well as a wider audience, for he felt there was a strong need. 'We have made use of Outlines, as more easy and distinct, for good Outlines are the first grand Principle of all Drawing', he wrote. It was published and sold by the Author at 2 Mead Place, and sold also by other booksellers in London, at a price of 4s. for the coloured version or half that for the uncoloured. He also embarked on a surprising publishing venture – his *Florist's Delight*. This was planned to appear in a series of annual numbers, each with six plates of the 'best Fancy Flowers' that florists could provide. Fancy they are with the folio-sized

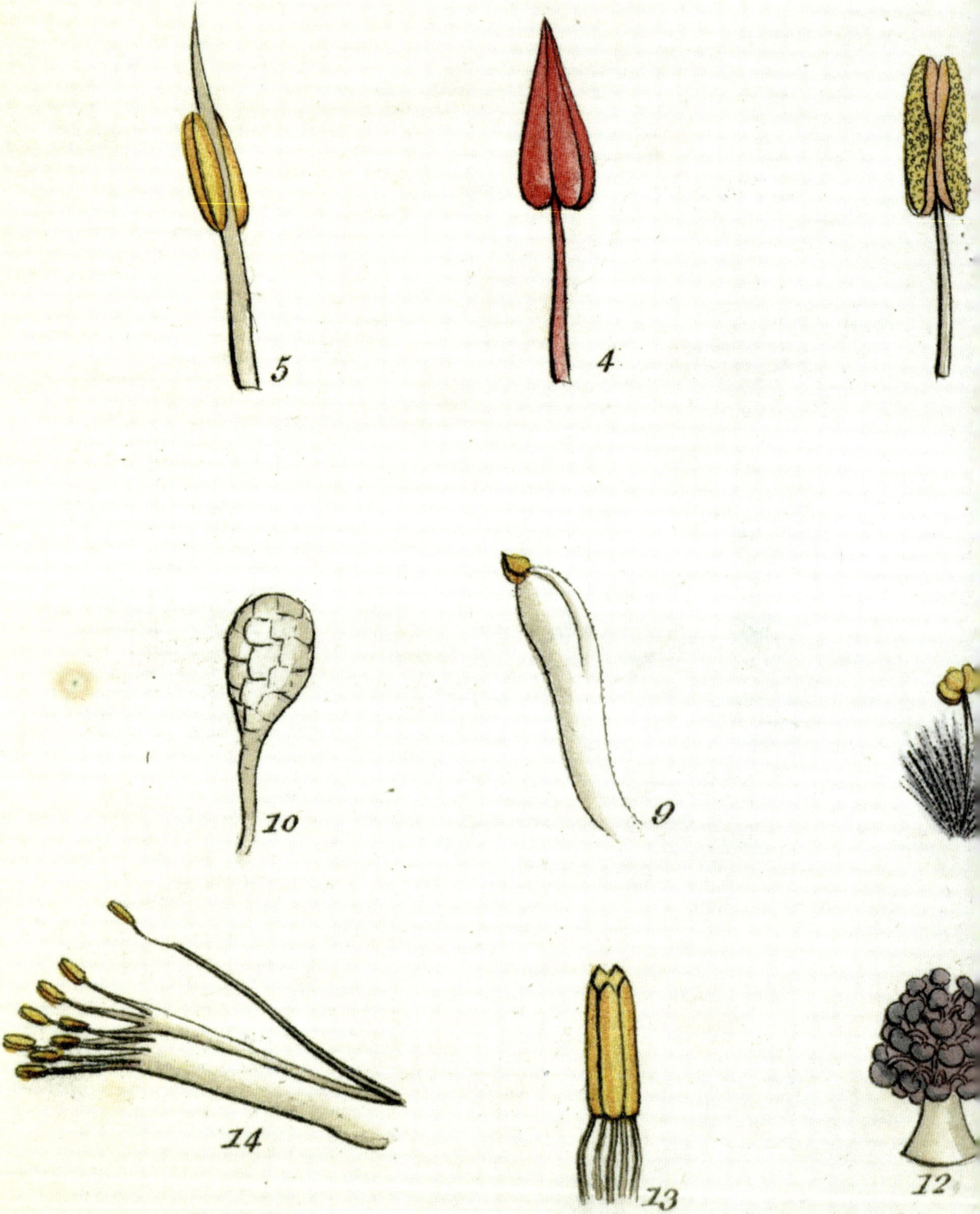

J Sowerby del: et sculp

Publish'd by J. Sow

Plate 1

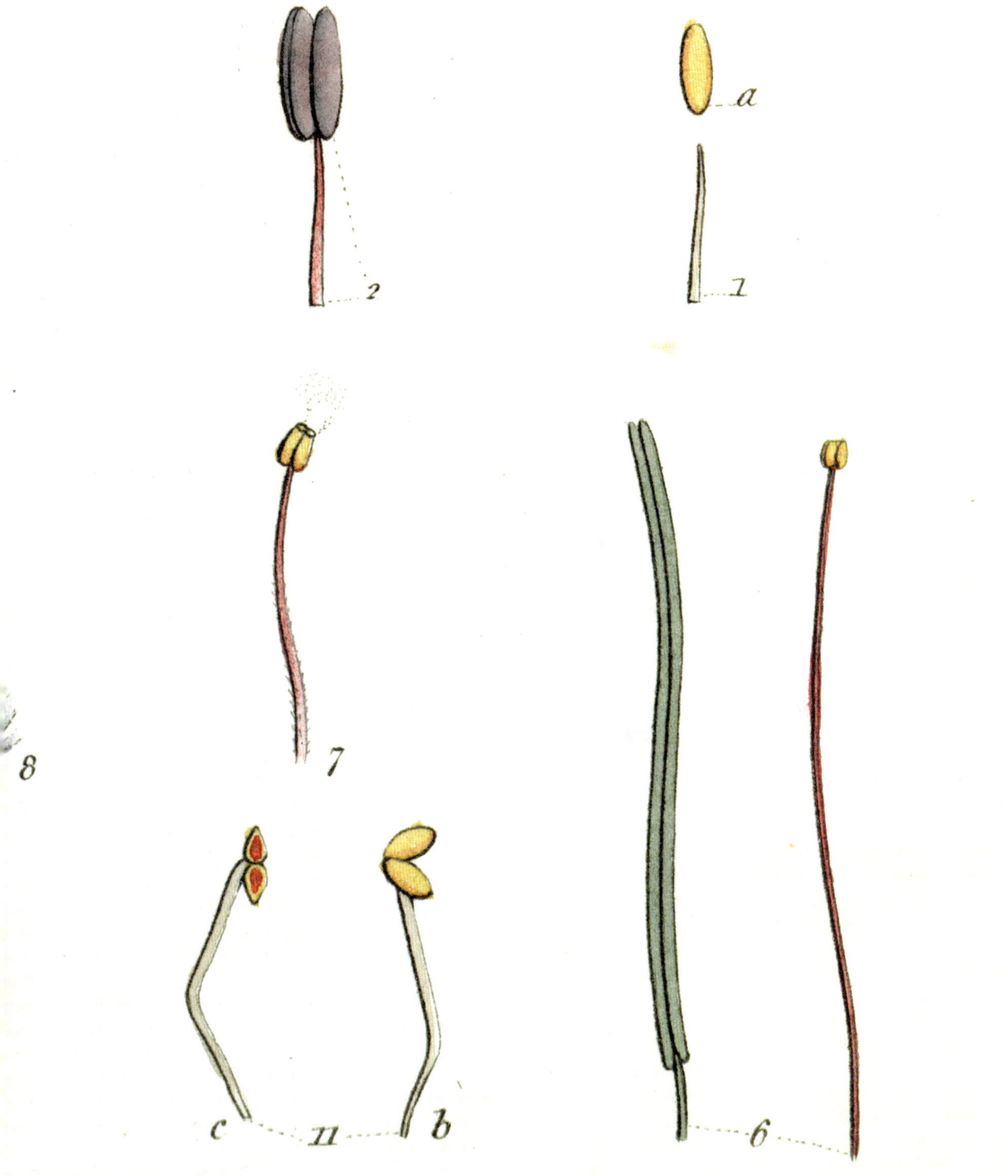

Nov.r 1, 1788.

plates in full and bright, sometimes almost garish colour. The choice of plants, apparently made by the florists, was to show the more flamboyant examples, such as Willett's Duke of Cumberland, a Polyanthus; Slater's Cheshire Hero, an Auricula in a Pot; and *Peregrinus apostolicus*, which was a 'bizarre' or broken-coloured tulip. The accompanying text gave details on the history of the plant and its desired appearance, stating, for example, that for a polyanthus 'the ground colour is most admired when shaded with a glowing and rich dark crimson resembling velvet, with one mark or stripe in the centre of each division of the petal'. It also provided information on planting, advising that the bizarre tulip *Castrum doloris*, for instance, should be cultivated 'by placing bulb on mix[t] of well rotted cow dung & fine mould'. These observations and instructions came from the florists; Sowerby was not a horticulturist and indeed he did not always care much for the results, as was revealed when he declared that 'speaking with strict propriety, the subjects of our first Number may rather be termed monstrous than natural productions, raised by fancy and minute attention to rival the simple flowers which grow wild'.[8] His contribution was to provide people with a lasting image of these specially cultivated and decorative plants, but he is likely to have also used drawings that he had made for earlier commissions. Some of the information is now particularly interesting because it shows the value of some novel offers made by the florists. One tulip, said to be very rare, was priced at one hundred guineas 'for the root', and the reader was told 'we will give the name in some future number; the present proprietor does not wish to have it made public at present'. This was a very expensive plant indeed. But others were much less costly – another tulip, *Gastrum doloris*, was 'much esteemed' and cost £3, while a *Primula auricula*, Redman's Metropolitan, was 'a cheap plant, yet it is much admired' and was 'capable of being well formed by the florist'. In this way the publication acted as an advertising medium for the horticulturist and florist.

The venture is, however, surprising because it repeated some features that had caused problems for William Curtis over his *Flora Londinensis* and of which Sowerby was aware. The numbers were in large folio and expensive, 12s. each, of which 6s. had to be paid at the time of subscribing and the rest on delivery of the work. It was unclear as to what might be included in future numbers and it was difficult to see that it would have wide appeal, especially with its rather strange opening sentence: 'As all the productions of the earth were designed for the amusement or convenience of its inhabitants, the most humble attempt to display the beauties of nature should not be despised by those, who, wrapped up in their own Pursuits, think every other insignificant.' The first number appeared in 1789, the third and last – with the title revised to *Flora Luxurians* – in the spring of 1791. Other numbers were intended, as some plates for them are in the collections of the Natural History Museum, London, but the sales are likely to have been low and unsustainable. Sowerby was also working on a more meaningful publication; one that was more accessible, cheaper and with better focus. This was *English Botany*, to which we will return. Sowerby was probably learning the same lessons experienced by Curtis.

His drawing book and *The Florist's Delight* proved that he could prepare, write, publish, sell, as well as draw and engrave the plates of botanical works. He could run his own business and at the same time practise his preferred occupation of drawing and engraving examples of natural history. He might also be able to satisfy his increasing desire to promulgate knowledge of natural history, not just to the wealthier botanists but to students and a wider, interested public. But these were early days and the economics of such a business were really unknown, or at least unsatisfactorily tested. Fashion, however, was on his side, driven in part by the example of the King and the influence of people such as Lord Bute and Sir Joseph Banks, and also by the development of Kew and other gardens which became home to some of the exotic and beautiful specimens being brought from distant lands. The life sciences were playing a more prominent role, emerging somewhat from the days of dominance by the physical and astronomical ones. This was so despite Joseph Banks, the naturalist, being initially seen as an unusual (and by some as an inappropriate) President of the Royal Society after the long line of physicists, physicians, mathematicians and astronomers since Isaac Newton (Sir Hans Sloane notwithstanding). The order being given to the natural world by the classifications of such people as Linnaeus and Antoine Laurent de Jussieu helped give the life sciences a better rationale and degree of respectability.

It was easier, of course, to participate in natural history than it was in these other disciplines or interests and participation could be by women, not just by gentlemen. People commissioned and purchased paintings and prints, drew illustrations, or made their own collections, sometimes employing others, if they were sufficiently wealthy, to do the arranging and cataloguing. These activities were enjoyable and useful. They were supported by the publication of numerous scholarly works. Nearly three times as many floras were published in Britain in the last forty years of the eighteenth century than in the previous sixty. More popular publications followed in their wake. Gilbert White's compilation of letters to Thomas Pennant and the Honourable Daines Barrington on *The Natural History and Antiquities of Selborne* (1789) is a reflection, albeit a famous and unusual one, of the times.

The fashion or interest in natural history also meant that the botanical world was making heavy demands of Sowerby. In just the three years

3/5a Watercolour by Sowerby of a moss species within the *Phascum* genus, used for Dickson's *Plantarum Cryptogamicarum*. see 3/5b.

3/5b A page from Dickson's *Plantarum Cryptogamicarum* with the engraving of Sowerby's moss drawing at the top.

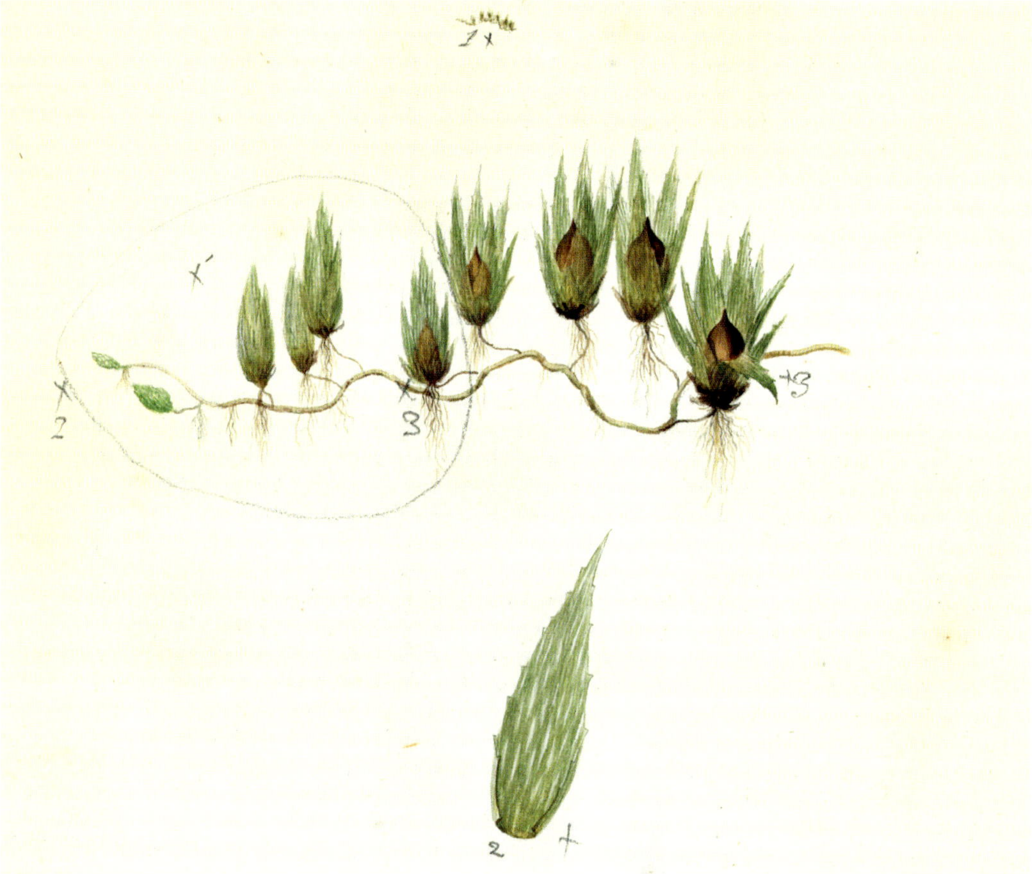

from 1789 to 1791 he was an artistic contributor to some eleven publications written by various authors, as well as to Curtis's *Botanical Magazine*, undertaking various other commissions and producing plates for books that were never to appear.[9] Most of those that did are major works, of significance to science as well as to people with aesthetic interests. James Dickson was one who employed Sowerby.[10] His passion for mosses, algae, liverworts and fungi led to his *Dickson fasciculus Plantarum Cryptogamicarum Britanniae* [Dickson's fascicle of British Cryptogamic Plants] issued over seventeen years (1785–1801). In Latin and dedicated to Sir Joseph Banks, it gave proper

Tab. VII

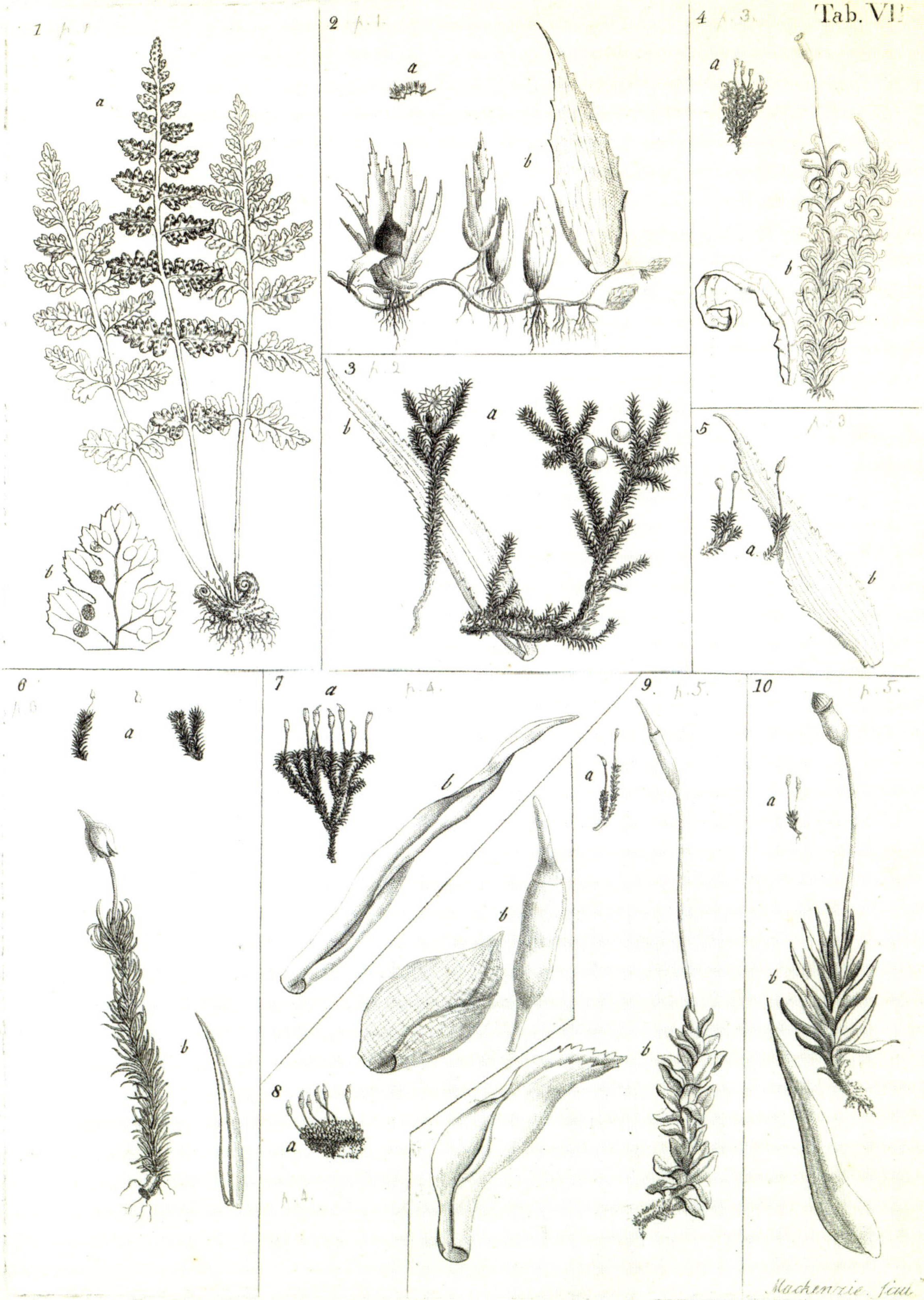

3/6 The lichenised fungus *Thamnolia vermicularis.* Watercolour by Sowerby for Dickson's *Plantarum Cryptogamicarum.* Fungi in this genus are sometimes called 'whiteworm lichens' from their appearance.

descriptions of numerous species of lichens and so forth, several of which were new, and contained twelve uncoloured plates of engravings from original Sowerby drawings. This is a scholarly and important work but it almost hides the essence of Sowerby's achievement. The engravings (by D. MacKenzie) while competent, were not finely done and used only parts of each original drawing. Inevitably, this, and the placing together of several small illustrations onto one plate, led to a loss of artistic and representational balance. Any observer might be impressed by them but not as much as seeing the originals (now held in the Natural History Museum), which masterfully depict over one hundred of those intriguing members of the plant world. Examination of just one drawing – that of a lichen (*Lichen vermicularis)* – reveals a consummate skill in the detailed drawing and the apparently effortless colouring to depict an almost feathery leaf-like association.[11] The plant seems real.

Sowerby's drawings are taxonomic; they contain the necessary detail, are confidently accurate and have smaller additional illustrations which reveal the intricacies of the plant's components. For example, his drawing of the moss *Bryum lineare* is accompanied by seven smaller illustrations, beautifully drawn and with fine shades of colouring.[12] Any subsequent loss of detail in the engraving is obviously to be regretted, more especially as taxonomists still consult the work because many species were described for the first time.[13] But it is, nonetheless, Sowerby's depictions of his keen observations that have helped give longevity to Dickson's scholarship.

If Dickson was of humble origins and poorly educated, Charles Louis L'Héritier de Brutelle was the opposite, although both shared a driving enthusiasm for botany. L'Héritier was an aristocratic and wealthy Frenchman who had been Superintendent of Parisian Waters and Forests and had acquired a love of botany. He was an ardent follower of the Linnaean system of plant classification and, like others of sufficient means and interests, established his own herbarium – one that was to contain around 8,000 species by the time of his untimely death, by murder, in 1800. He had been introduced to Sir Joseph Banks by letter in January 1783 and became a frequent correspondent.

On 19 September 1786 L'Héritier arrived suddenly and unexpectedly in London, to avoid having to return to the French authorities an important collection of dried plants which had been made by the French botanist Joseph Dombey while seconded to a Spanish expedition to Peru and Chile. Under an agreement between the French and Spanish governments, Dombey was not allowed to publish anything about the collection, which had been handed over to the Jardin du Roi in Paris. The French Government gave Dombey a small pension and cleared his

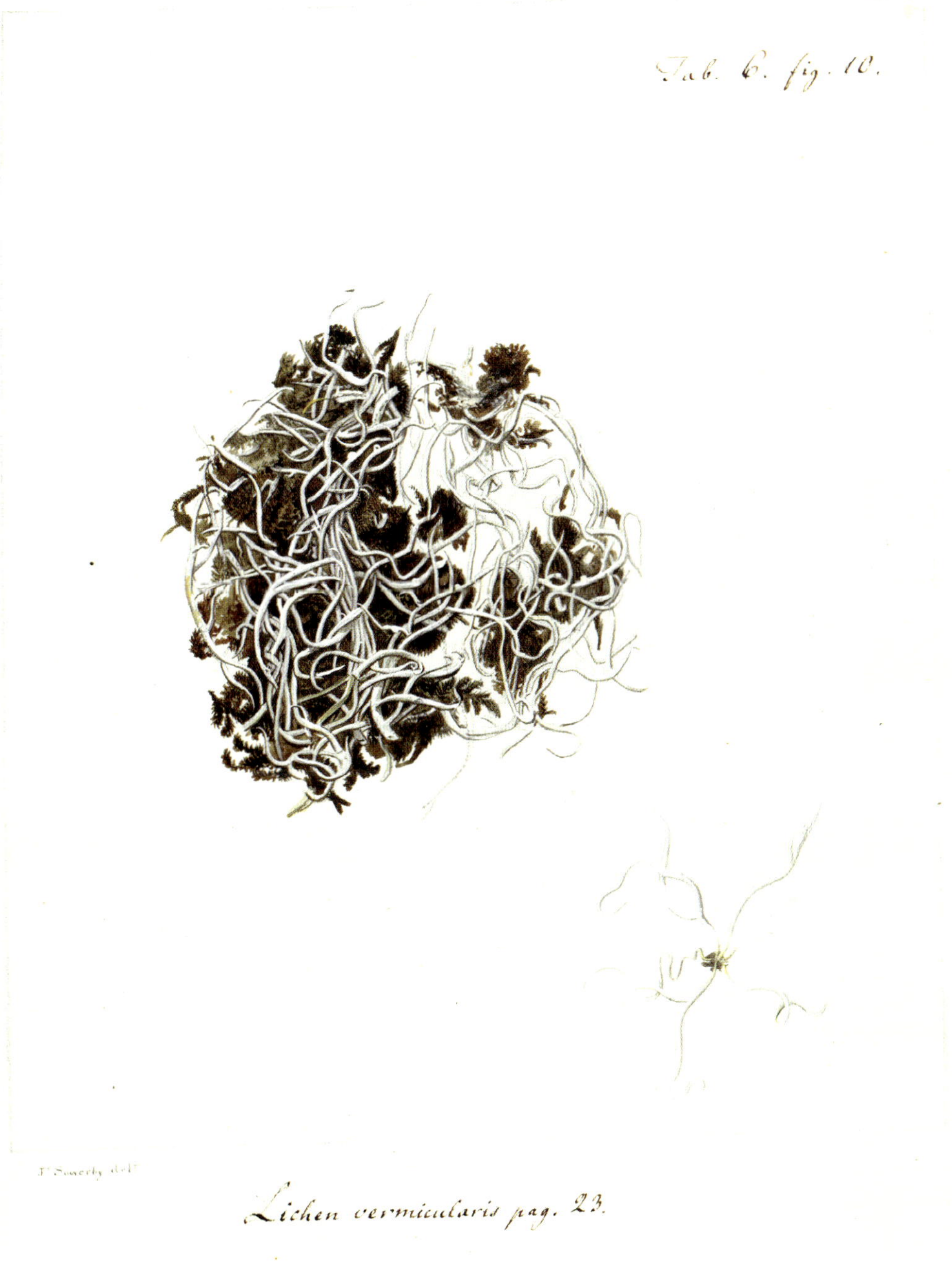
Tab. 6. fig. 10.
J. Sowerby del.
Lichen vermicularis pag. 23.

3/7 *Lightfootia oxycoccoides.* A member of the Campanulaceae family, named after the author of *Flora Scotia* and sometimes referred to as cranberry-leaved Lightfootia. From L'Héritier's *Sertum Anglicum*. It also featured in *Exotic Botany.*

debts for his pains, but L'Héritier, seeing an opportunity, persuaded the Director of the Jardin, Comte de Buffon, to pass the collection to him on the grounds that he would publish, at his own expense, descriptions of the new plant species. On hearing this, the Spanish Government protested and asked that L'Héritier hand it back immediately. Instead, the next day (9 September 1786), so as to escape the authorities, L'Héritier set off with the collection for London, via Boulogne, and on arrival made straight for Banks's residence in Soho Square.[14]

Banks was initially welcoming, and not much concerned that the French and Spanish authorities had been tricked, although he had been warned by James Smith that there might be trouble. He offered L'Héritier his hospitality but the initial plan for a stay of three months extended to an actual stay of fifteen months, with the Frenchman returning home in December 1787.[15] Joseph Banks had written earlier to James Smith showing his irritation that L'Héritier

> is still here he has received all sorts of civilities & actually has got his draughtsman over [the renowned Pierre-Joseph Redouté] who has drawn all the new plants as they come into flower a liberality which few countries would allowed him. I am a great Friend to such behaviour & tho I do not doubt that the moment he returns he will try to play tricks I can return his mischief upon his own head with interest he has not parted with a single specimen of Dombeys Herbarium... [16]

L'Héritier was enamoured with the botanic gardens, especially Kew, and with the opportunities that Banks's own herbarium provided. He used his time enthusiastically studying and preparing his illustrated works for publication. For these he employed not only Redouté but others including Sowerby, who drew one of the plates for his important *Stirpes novae aut minus cognitae* [New or lesser known plants] which was published over several years (1785–1791) but never finished. L'Héritier's *Sertum Anglicum seu Plantae Rariorum quae in Hortis juxta Londinum* [An English Wreath, or, Rare plants which are cultivated English flowers or rare plants in the gardens around London, especially in the Royal Gardens at Kew] published from 1789 to 1792, contains ten Sowerby plates. Finally, his *Geraniologia* (1791–1792) is illustrated with forty-four uncoloured plates and descriptions of geranium, pelargonium and so forth, six of which Sowerby drew. The folio plate 15 is a fine example; it shows a framed *Pelargonium quercifolium pinnatifidum* but with a leaf spilling out of the frame. (These rare plates can nowadays be seen on websites of early botanical and other works.[17]) The quality of Sowerby's work is very high and stands comparison with that of Redouté – it is this quality that is paramount in the continuing

4.

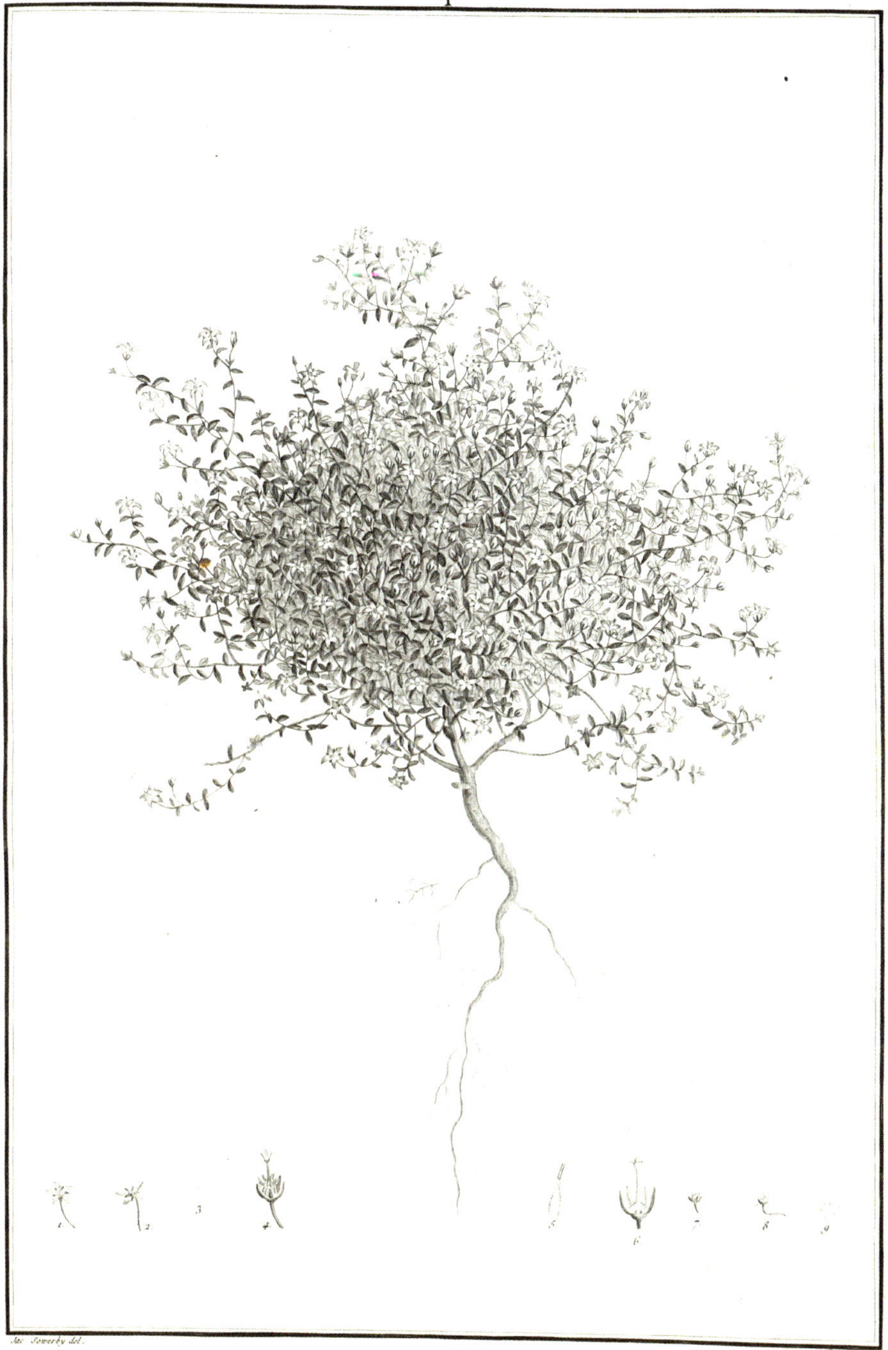

Jac. Sowerby del. Juillet sculp

LIGHTFOOTIA oxycoccoides.

15.

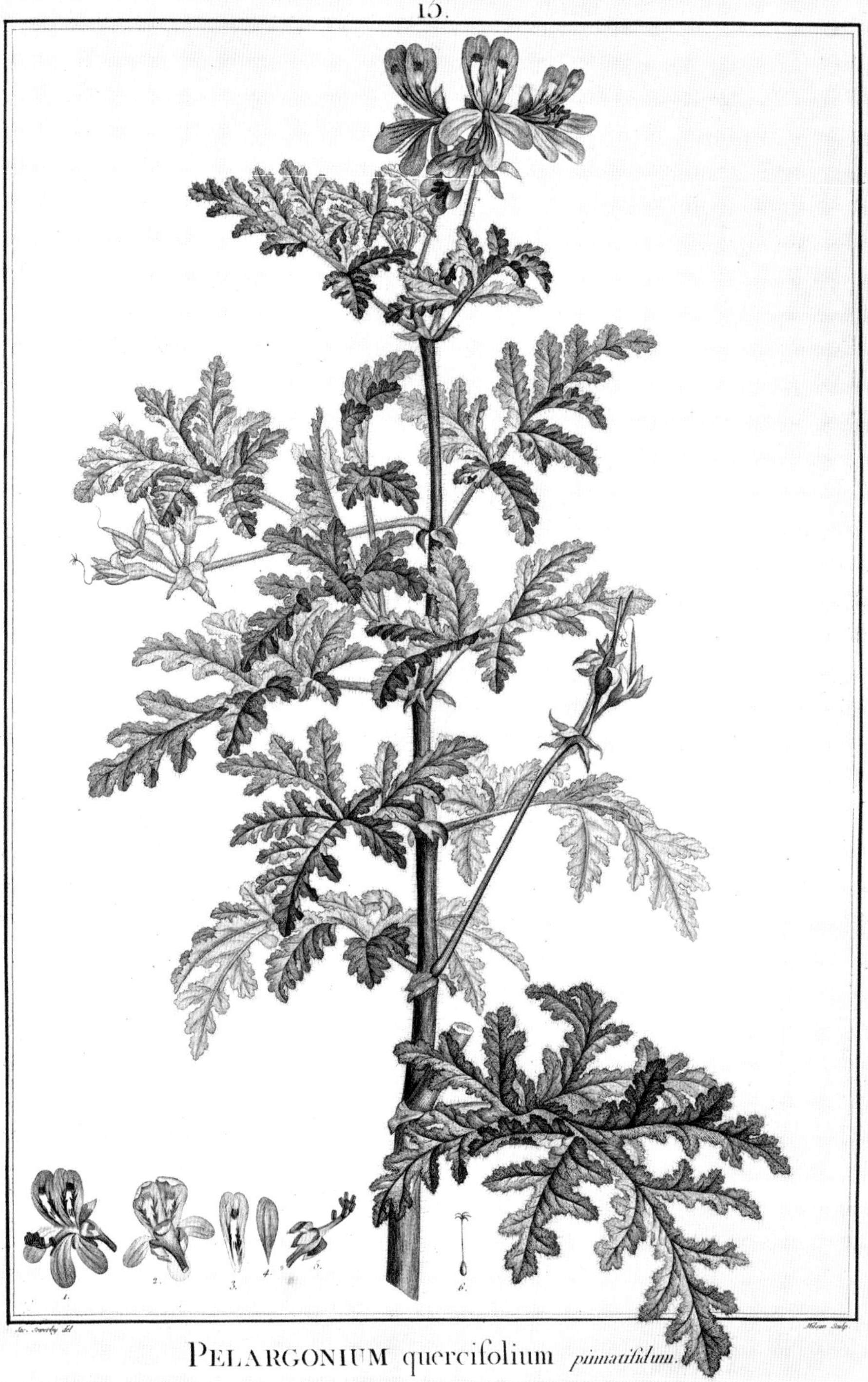

PELARGONIUM quercifolium *pinnatifidum.*

3/8 A commonly grown ornamental geranium (*Pelargonium quercifolium*) in L'Héritier's *Geraniologia*. The plate is an example of the elegance and accuracy of Sowerby's early work.

3/9 Daniel Solander. Stipple engraving by J. Newton, 1784, after James Sowerby. He was a Swedish botanist, colleague of Sir Joseph Banks and, among other contributions, Keeper of the Natural History Department at the British Museum for several years. Wellcome Library, London

3/10 William Aiton (1731-1793), oil painting. Artist unknown.

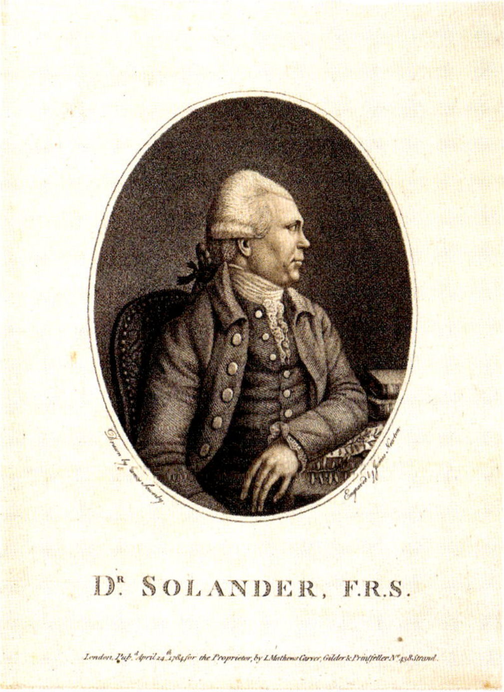

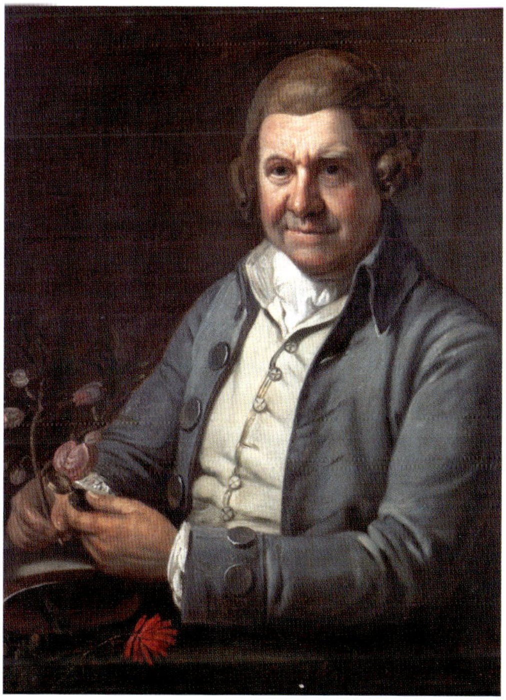

value of the drawings to taxonomy. Thirty-one of the thirty-five plates in *Sertum Anglicum* depicted species for the first time and several of them have never been illustrated subsequently.[18]

Although L'Héritier was an advocate of Linnaeus's classification it did not prevent him from making his improvements to it. In his *Geraniologia* he subdivided Linnaeus's *Geranium* genus into three genera based on the number of stamens; *Erodium* with five, *Pelargonium* seven and *Geranium* ten. Such adaptation was not always appreciated by other botanists, William Curtis, for example, suggesting that if the Linnaean approach had been strictly followed then the new grouping should be within classes, not genera.[19] Sowerby stayed out of this kind of debate but his direct knowledge of plants was developing quickly.

Because of his influence and breadth of knowledge, L'Héritier was also a contributor, alongside the Swedes Jonas Dryander and Daniel Solander, to the botanical descriptions in William Aiton's catalogue of the plants cultivated in the Royal Gardens at Kew, the *Hortus Kewensis*.[20]

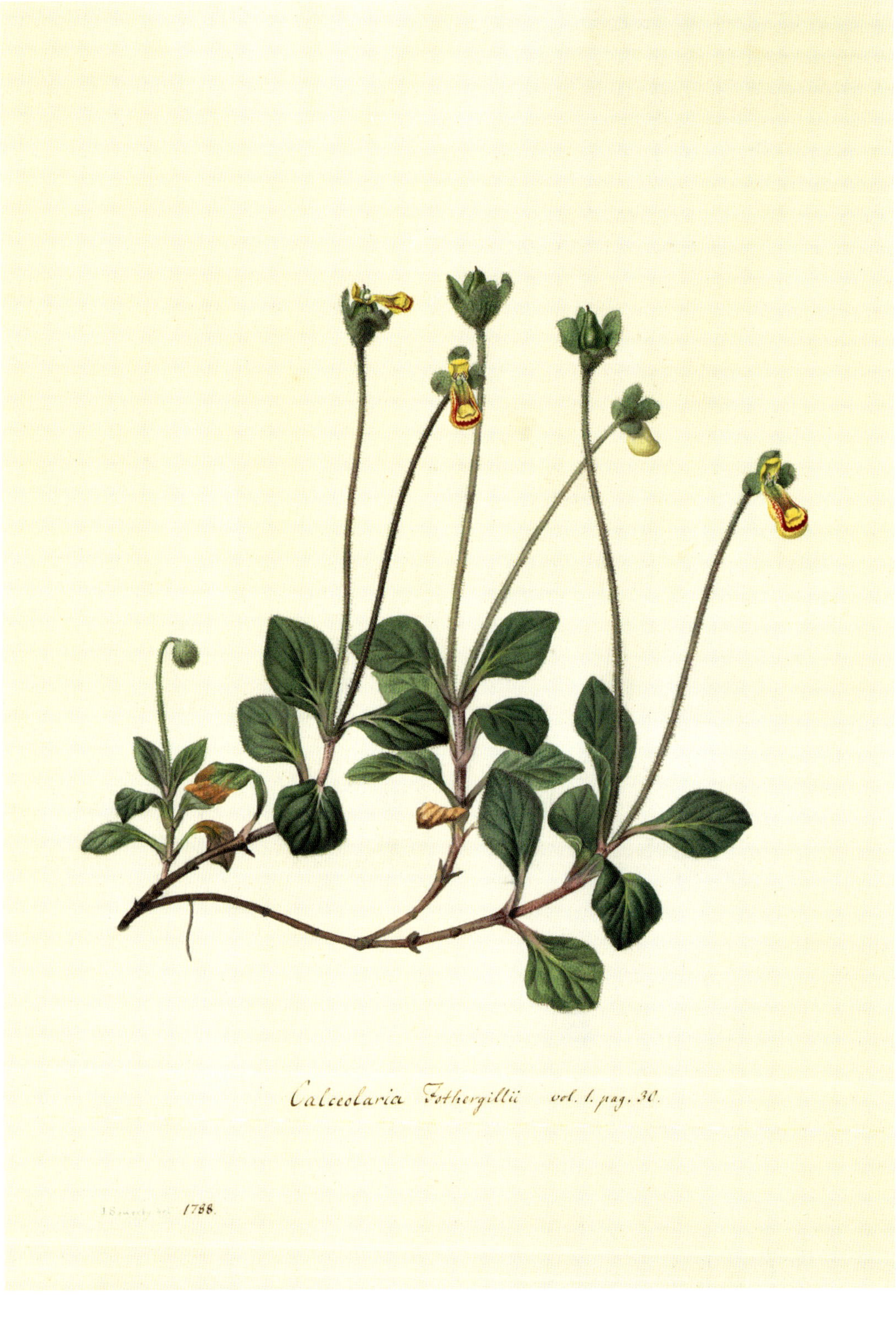
Calceolaria Fothergillii vol. 1. pag. 30.
1788.

3/11 Sowerby's original drawing of *Calceolaria fothergillii* for the first plate in *Hortus Kewensis*. This plant was introduced to Kew Gardens in 1777 by John Fothergill, M.D.

This three-volume work (1789) contains thirteen fine plates of engravings drawn by some of the renowned artists of the age – Francis Bauer, G. D. Ehret, J. F. Miller, F. P. Nodder – and three by Sowerby, who also engraved the final plate. The work was well-received; it showed the high status of the Gardens and helped set a standard to which gardens in other countries would aspire.[21] It also nourished Sowerby's reputation since undoubtedly it had a wide circulation in much of Europe.

Dickson, L'Héritier and contributors to *Hortus Kewensis* were active describers and cataloguers of plant species – tasks that were justified in their own right. It was new knowledge and it was sufficient for these men to describe and classify without embarking on any analysis of possible further relationships or to involve themselves with utility. These tasks were left to others, of whom William Woodville was one. A physician, botanist and Quaker, he was a leader in the prevention of smallpox, working as physician in the London Smallpox and Inoculation Hospital, where he had a botanic garden.[22] He published on the topic of the inoculation of smallpox in 1796 but earlier had produced his four-volume *Medical Botany*, a catalogue and description of plants based on the pharmacopoeia, or materia medica, of the royal colleges of physicians in London and Edinburgh. This was an impressive work of the time; it was very successful and considered the standard work on the subject,

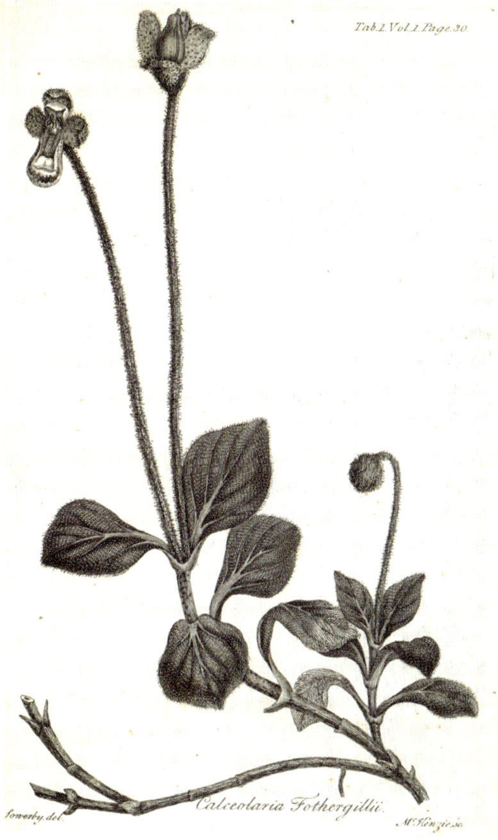

3/12 *Calceolaria fothergillii*. The first plate in *Hortus Kewensis*. The engraving used only parts of Sowerby's original drawing, shown in 3/11.

3/13 A tropical legume (*Dolichos pruriens* or *Mucuna pruriens*) which became infamous for the itchiness it produces on contact with the fine hairs of the seed pods. In *Medical Botany* Woodville mentioned several medical uses including as a drug to expel worms. Perhaps relevant to the times, he added that the colonist managers of plantations would give the medicine to their slaves who after the 'second dose, usually discharged an incredible number of worms, even to the amount of more than twenty at a time, so that the stools consisted of little else than these animals.'

continuing in publication for many years. Sowerby was still producing plates for reprinting and for the second edition towards the end of his life. Each plant has its plate (in coloured or uncoloured editions) with a description, synonyms, classification in the Linnean system and medical significance, sometimes with rather graphic detail. The text for deadly nightshade (*Atropa belladonna*) informs us that 'when a greater number of the berries are taken into the stomach, scarcely half an hour elapses before violent symptoms intervene; viz. vertigo, delirium, great thirst, painful deglutition, and retching, followed by furor, stridor dentium, and convulsions; the eye-lids are drawn down... '.

Woodville strongly recognised the need for accurate and helpful illustrations if his work was going to serve as an authoritative guide, as shown by his preface:

> A distinctive and characteristic knowledge of natural objects should certainly preced the consideration of their different properties and qualities; and with respect to plants, this knowledge is seldom to be adequately attained by a mere verbal description: accurate delineations therefore become necessary, and this department is committed to Mr Sowerby, an artist of established reputation, whose talents are not less conspicuous in the correctness than in the beauty of his designs.

And again, in the fourth volume of 1794, he wrote that the work 'owes much of the merit it may possess rather to the execution of the artist than to the compiler'.

This was Sowerby's strength – he combined accuracy with design. Specimens were not romanticised in their depiction, and people could rely on the drawings for scientific use, while they still display an intrinsic beauty. Care was always taken to provide Sowerby with good material, for example, for *Cucumis colocynthis*, or bitter cucumber, it recorded:

> Colocynth is imported for use to this part of Europe from Turkey, but it is yet unknown here of what place this plant is native. It seems to have been cultivated in Britain in the time of Turner, and the figure we have given here was drawn from a specimen of the plant produced by sowing the seed in a hot bed.[23]

The total number of plates over the four volumes reached 274, an important and significant undertaking. The success was reflected in the production of later editions over forty years.

The opinion that Sowerby was a draughtsman to be sought after was shared also by Richard Salisbury (formerly Markham); an unusual character within the botanical world. He and James Smith had met when students at Edinburgh, although Salisbury never took a degree. He became wealthy through a rather mysterious route by inheriting a significant

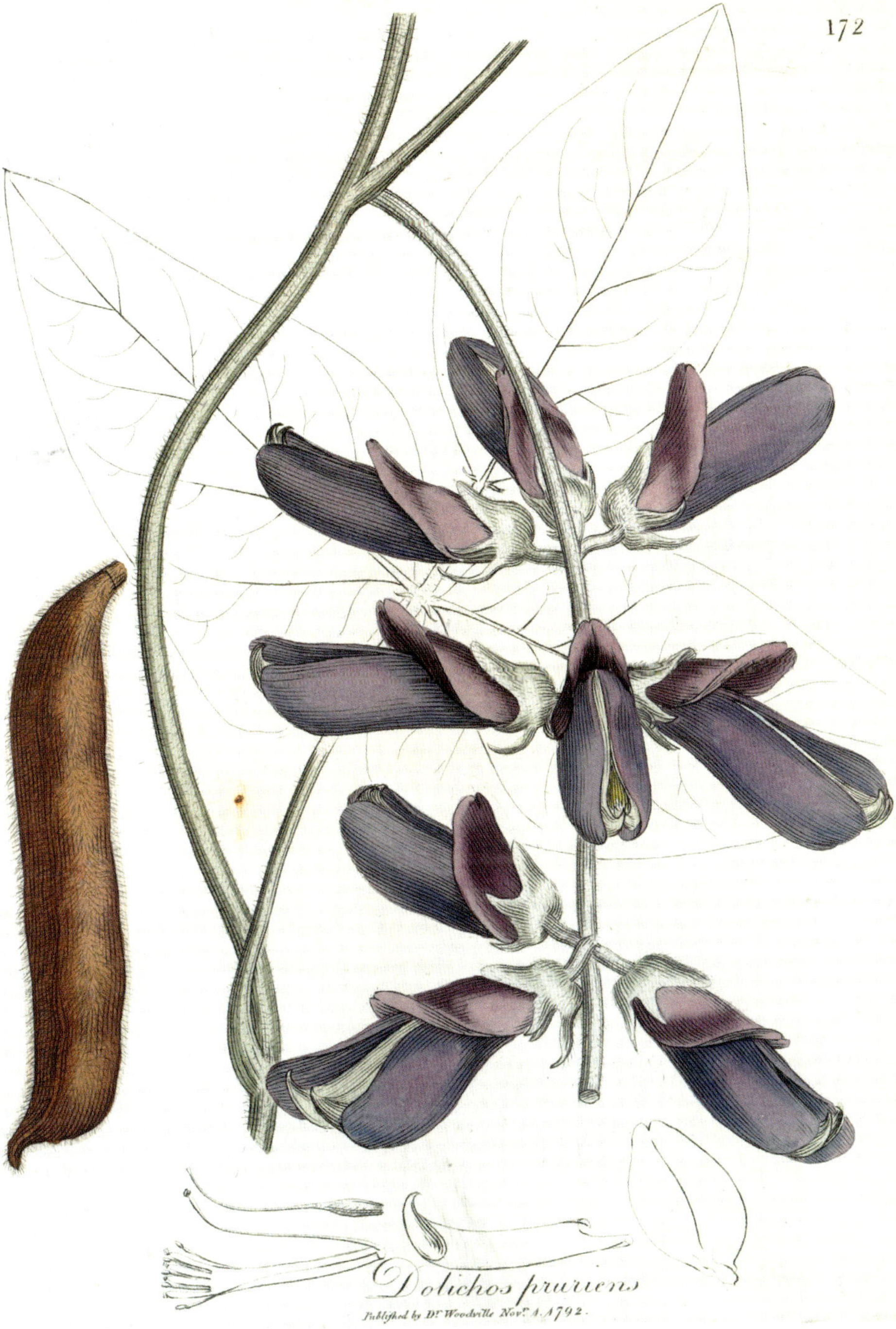
172
Dolichos pruriens
Published by Dr. Woodville Novr. 1. 1792.

3/14 The common but beautiful wood sorrel, *Oxalis acetosella.* Woodville said it was native to England and that in medicine its 'principal use is to allay inordinate heat, and to quench thirst; for this purpose, a pleasant whey may be formed by boiling the plant in milk.'

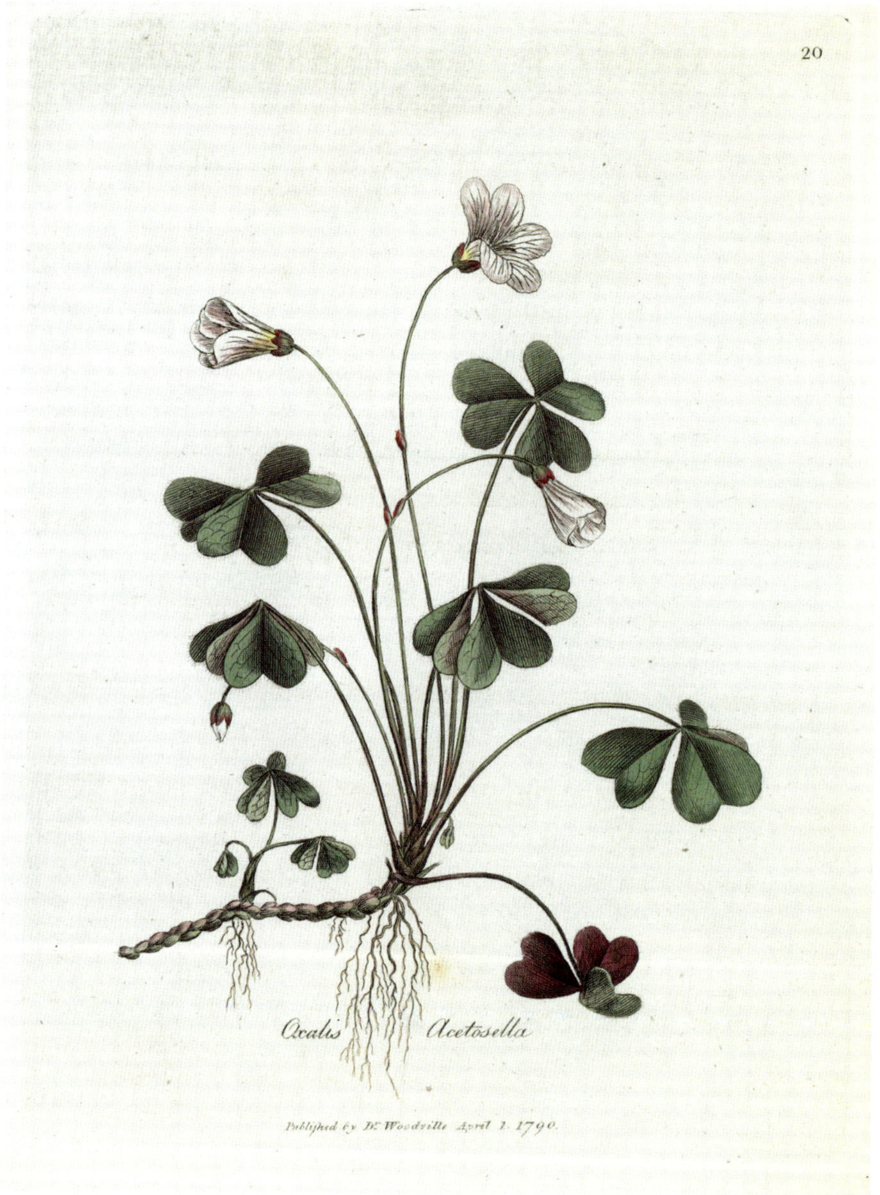

3/15 The nutmeg tree (*Myristica fragrans*), native to Indonesia and well known for its culinary uses. Woodville gave few medical uses in his *Medical Botany*.

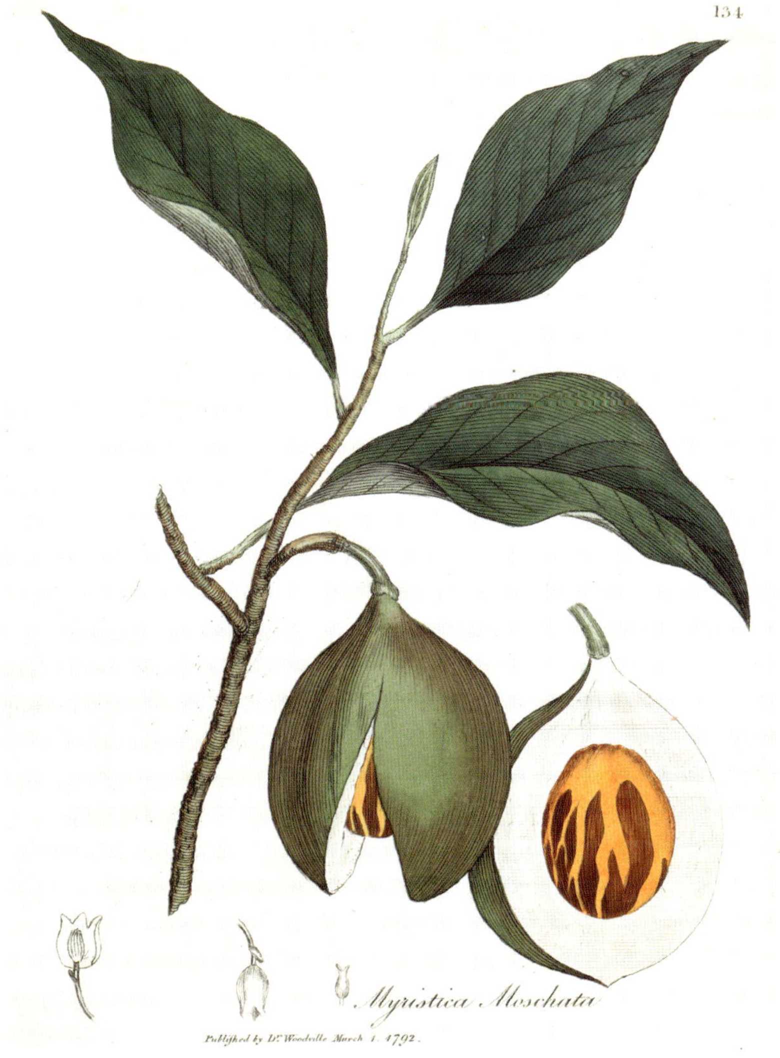

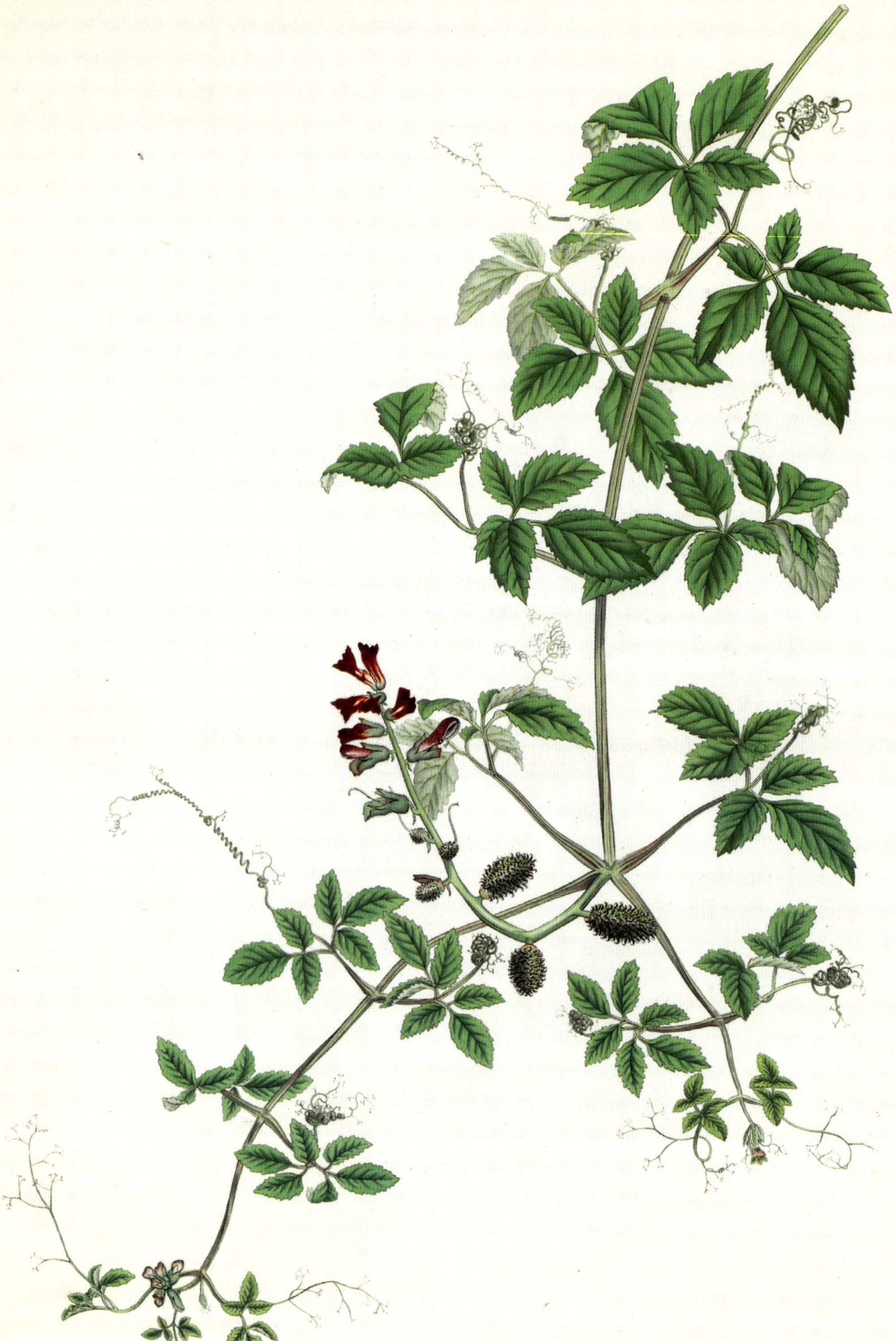

3/16 The decorative plant *Dombeya lappacea* (now *Tourrettia lappacea*). Collected initially by Joseph Dombey from Peru. The plate by Sowerby is in Salisbury's *Icones stirpium rariorum* of 1791 but the plant was in L'Héritier's *Stirpes Novae* of 1784. It is in the family Bignoniaceae with plants often cultivated for ornamental use.

fortune from a distant relative, a spinster, whom he had befriended through their shared love of flowers. The details of this woman are not known, neither are those of the fortune, but to receive it he had to change his surname from Markham so as to satisfy the spinster's desire that her ancient Welsh family name of Salisbury be continued. He then judged his wealth to be sufficient to follow the life of a landed countryman, living on his father's estate at Chapel Allerton, near Leeds, and pursuing his horticultural interests, in which he was rather competent. He was sufficiently well regarded in botanical circles to be elected to the Royal Society in 1787.[24]

Keen, no doubt, to be seen as active in publishing scholarly books on plants as were the other distinguished botanists of his day, he put Sowerby to work on the plates for his *Icones Stirpium Rariorum* [Images of Rare Plants]. This large folio work, issued in parts, contains ten beautiful, coloured illustrations of a variety of plants species – fuschia, lobelia, jasmine, veronica and so forth – accompanied by a Latin text.

'It is absolutely necessary for you to come down if I am to publish a Fasciculus – corrections & additions which can only be done by your own incomparable Pencil,' he wrote to Sowerby in January 1790. By November he 'is pleased' with Sowerby's drawings but 'mortified' that he had not yet engraved them. By January he decided that the first fascicle would sell well and at a cost of just over a pound, so he sent Sowerby £20 and asked him to attend to the colouring. By May 1791 the plates were finished and 'coloured perfectly to my satisfaction' but he was unable to pay Sowerby for, as he explained, he had lent £500 to a gentleman and was having difficulty getting it back.[25] Money was a frequent issue for Salisbury and, as a result, Sowerby suffered in turn.

The *Icones* fascicle appeared soon after and was clearly a work of some grandeur with its high-quality paper, large folio size, beautiful plates and a dedication to Sir Joseph Banks. Care had, as usual, gone into making accurate drawings and engravings. A pencilled note on one of the original drawings (plate 4, *Veronica heterophylla*) held in the Natural History Museum says 'there is an extreme fine Pubescence over the whole plant which must be shown in the engraving'. Salisbury was very pleased and it seems that a friendship of sorts between the two men was made, for Salisbury became godfather to Sowerby's third son, Markham, born a few months before publication. But Salisbury was to return later and then, as we will learn, in a belligerent frame of mind.

If the account of Sowerby's life during this period appears as a catalogue of books and drawings, that is because it almost was. It was a highly productive time, not only for the contributions to other authors' works but for the beginning of his own publications , and also for satisfying several smaller commissions

3/17 A 'new American Grass', *Agrostis cornucopiae*. Sowerby's fine plate (folio) accompanied John Fraser's privately published account, which includes a botanical description and experiments on its cultivation with a view to its economic use (1789). (*Agrostis cornucopiae* is a synonym for *A. perennans.*)

– a feature of his work that was probably always present. For the botanist and Scot John Fraser, he prepared one beautiful plate for his private publication *The New American Grass* (1789). Fraser had travelled frequently in North America from where he had introduced several plants to England, some of which feature in Curtis's *Botanical Magazine*.[26] By way of contrast in numbers, Sowerby made over one hundred fine watercolours of plants growing in the grounds of Dr John Lettsom's house 'Grove Hill' at Camberwell.[27] This was lucrative work and Lettsom was a reliable payer. So too, no doubt, were others, especially members of the aristocracy such as the Duchesses of Portland and Northumberland, the Marchioness Rockingham and Lord Lewisham, who all used Sowerby's skills. A short letter to him is probably typical and would act as a reminder about his status, if any reminding was needed:

> I am desired by Lady Rockingham to say that Her Ladyship have a Lilium in flower which she wishes you to draw, if therefore you can come on Monday it will be in a proper state, if however you cannot possibly come on Monday I imagine Tuesday will do.[28]

Mead Place must have been well set up as an artist's workshop with a good stock of candles for the winter evenings, the kerosene lamp still being some years off. Sowerby employed other artisans to do some engraving and colouring but with the start of a family he needed to maximise his income and so did much of the work himself. He was still beholden to others for work, although a greater degree of independence was now not too far away. Mead Place was indeed home to new arrivals. Three children were born to James and Anne within about five years of their marriage; James De Carle arrived on 5 June 1787 and George Brettingham less than a year later on 12 May 1788. Markham Valentine followed on 14 February 1791. The names of the first two obviously reflect Anne's family name, while Markham is because of the Richard Salisbury connection, it being his former name, and Valentine is for the date of birth. Salisbury was godfather to Markham, and it is almost certain that James's friend and brother-in- law, Robert Brettingham De Carle, was to George. The first two sons were to become highly integrated and important members of the Sowerby business, each in different but complementary ways; Markham was to suffer an untimely death.

Sowerby had not long settled into Lambeth when Curtis decided to leave it but not before Sowerby had made a watercolour picture of the garden. For ten years Curtis had striven to make his garden a success, as indeed it was, but the conditions were far from ideal, seemed to be getting worse and drove him to some despair. So in 1789 he moved away as a result, he said

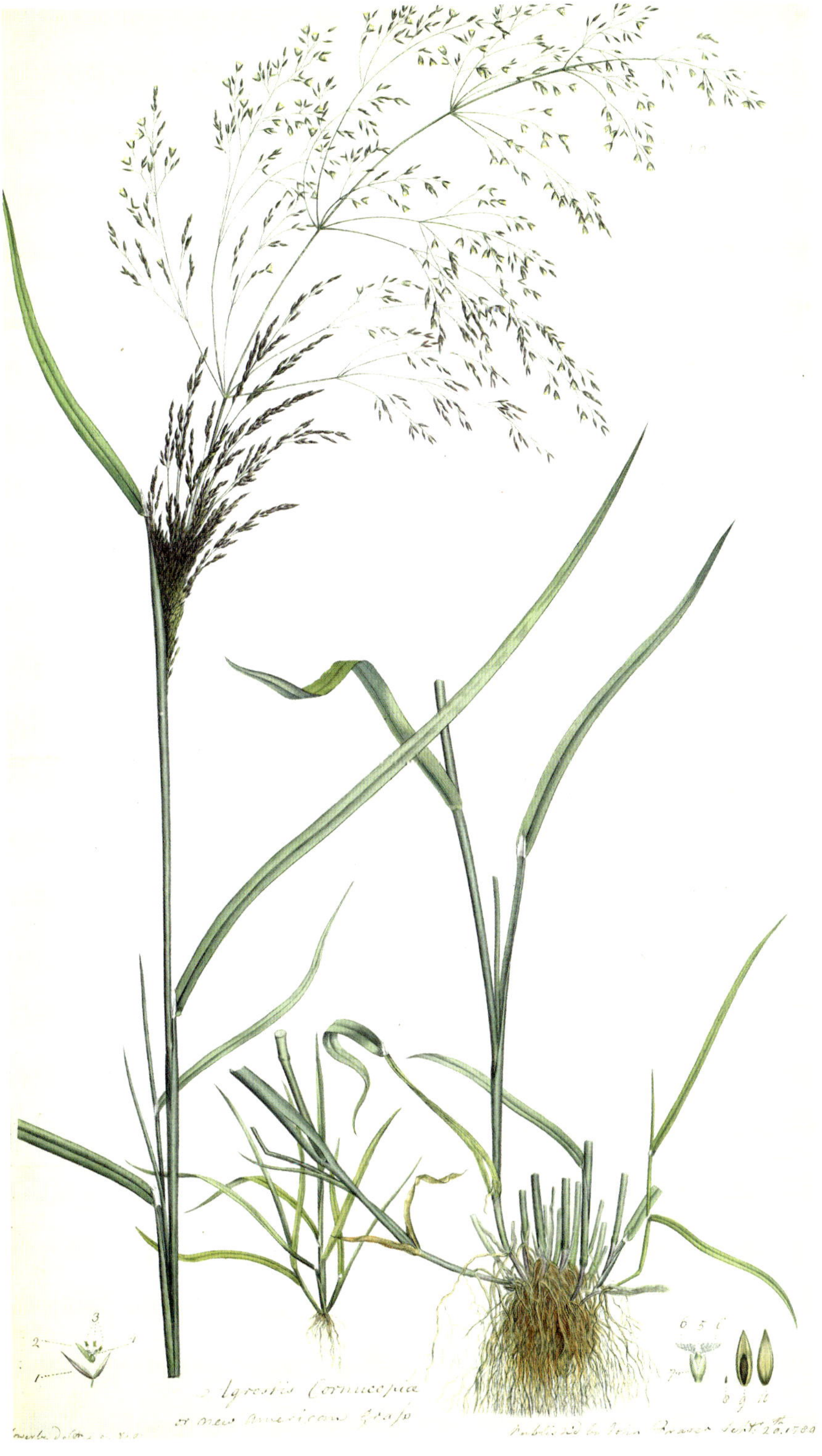
Agrostis Cornucopia
or new American Grass

of the smoke of London, which, except when the wind blew from the south, constantly enveloped my plants, and shedding its baneful influence over them, destroyed many; and in a greater or a less degree, proved injurious to most of them, especially the Alpine ones. In addition to this grand obstacle, I had to contend with many smaller ones, which became formidable when combined: such as the obscurity of the situation, the badness of the roads leading to it, with the effluvia of surrounding ditches, at time highly offensive.

He went to a new site at Queen's Elms Brompton, just one and half miles from the Hyde Park Turnpike and where over seven acres of land were available to him. Here he laid out different sectors for medicinal plants, British grasses, trees, shrubs, agricultural plants and so forth.[29] Work still continued on *Flora Londinensis* and the *Botanical Magazine*, with Sowerby providing over sixty drawings and some engraving, until 1790, when the start of Sowerby's own publication, *English Botany*, was seen by Curtis as a direct and unwanted competitor and led to an estrangement.

Curtis was not the only departure from Lambeth: Sowerby's brother-in-law, Robert Brettingham De Carle, died on 7 February 1791, at the early age of forty-two. He had been a successful member of the Coade Stone works, being 'an eminent modeller, whose skill was unrivalled' and had acquired some wealth and property, possibly helped by his late father, either through gift or inheritance.[30] Robert had married soon after coming to London from Norwich but there is no mention of his wife in his will. Instead, his will is more a reflection of his connections to the Sowerby side than to his family in Norwich. He wished 'to be put under the ground with as little expense as possible' and besides leaving just ten pounds to a godson and one shilling for his younger brother John, who was a successful marble mason, the rest of his estate went entirely to Sowerby's second son George Brettingham. He was only two years of age at the time but this early inheritance was very likely to have helped in his later entrepreneurial activities and is of particular interest because of the tensions that arose between him and his father later in his life around the time of his own marriage.

The will is also unusual in the added Codicil. Robert wished that only six people – no more, no less – attend his funeral: four Sowerbys including James and Anne, Samuel Davy the father of his godson, and his own brother John.[31] James was appointed the sole executor but the will was not proven until 29 October 1794, some three and a half years later. Did Sowerby simply take a long time to fulfil his task or were there were some claims from the De Carle family? For the latter there is just a little evidence. The probate document has a sentence in small writing, seemingly as a later addition, to the effect that Robert's father was the only person entitled to any of the estate if the deceased had died intestate but even he had predeceased him.

Anne would certainly have missed her brother. She drew strength and support from her Norwich relatives, especially in times of poor health and in later tribulations with the loss of at least one of her children. James would also have lost a close friend who had shared some of the trials of eighteenth-century life but there is virtually no preserved correspondence or other clues to give us a greater insight. One item, however, is a present Robert gave to James, before his marriage, of a rather fine stoneware jug made by him in 1781 and decorated with leaves, fruits and grasses – a portent perhaps of Sowerby's developing interest in natural history. The jug was made in 'the distant suburb' of Lambeth but has crossed the river and is now in the collections of the Victoria and Albert Museum, South Kensington.[32] On the north side of the river at the time was James Smith, who in the few years before Robert De Carle's death had been working closely with Sowerby for his various botanical publications. This was a burgeoning partnership rather than the close family friendship Sowerby had with Robert, but it was a prime driver of the direction that Sowerby's life was now to follow.

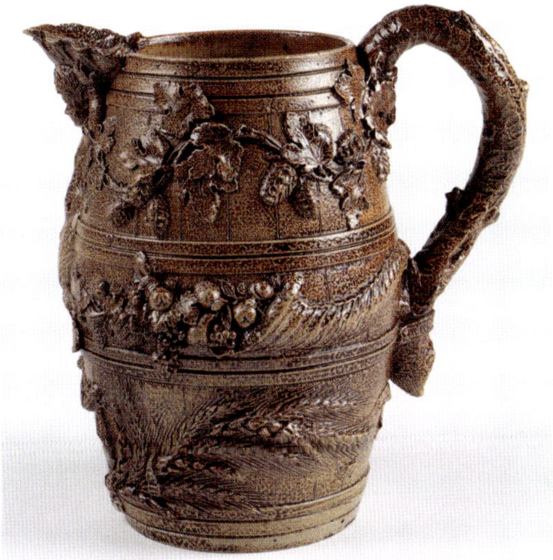

3/18 The stoneware jug made by George Brettingham De Carle for James Sowerby (1781). © Victoria and Albert Museum, London.

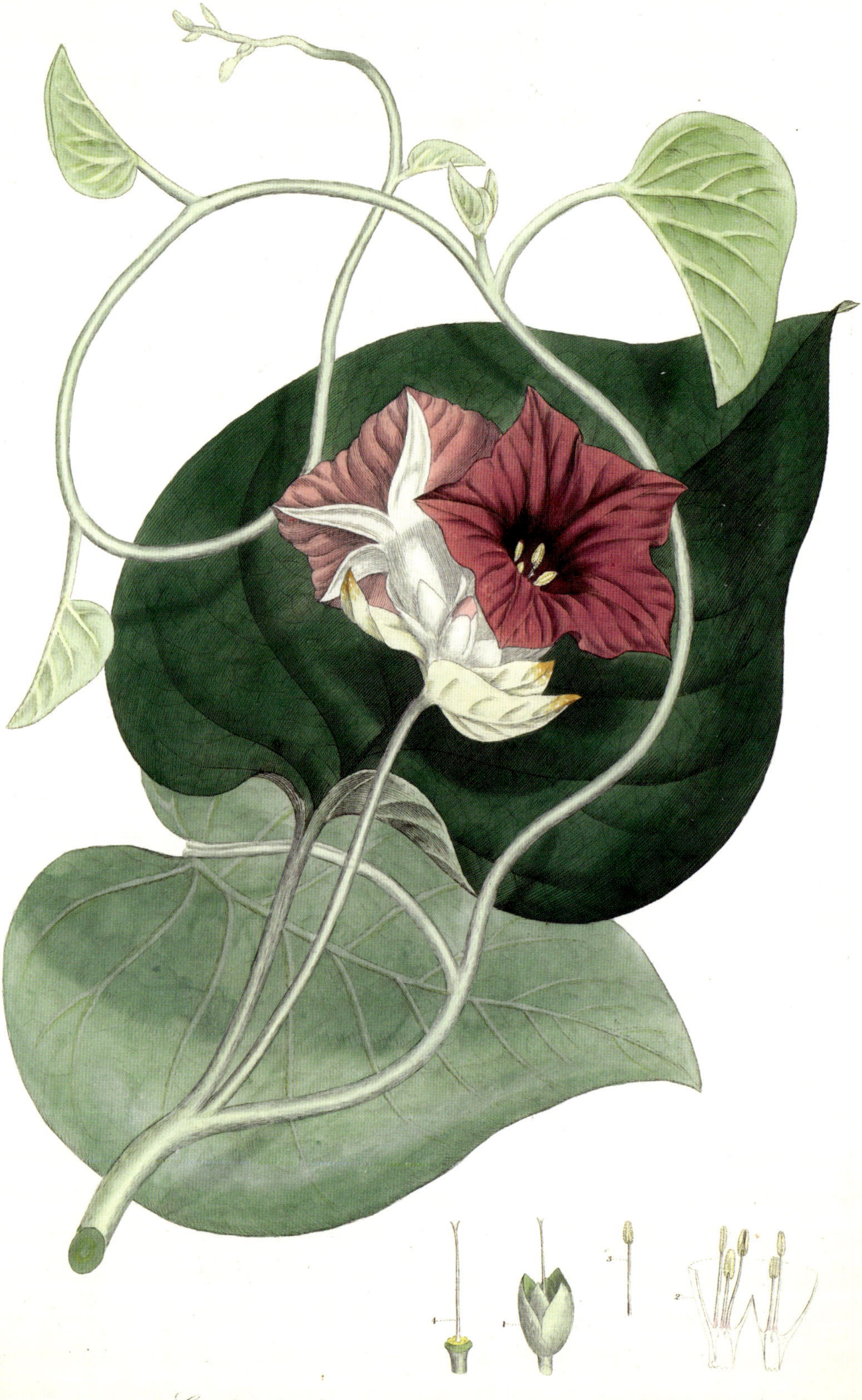

Convolvulus speciosus

CHAPTER FOUR

The two Jameses

The two Jameses – James Edward Smith and James Sowerby – were born within three years of each other but with contrasting backgrounds and opportunities. Circumstances had brought them into the same sphere of interest and activity, with each contributing sometimes similar, sometimes strongly different, but complementary skills. Both were hard working, disciplined and ultimately dedicated to natural history. Their personalities were to determine much about the style of their working partnership as well as its long duration; a partnership that was unusual in that both wished, at different times, to be in some control of the other. It was not to be a simple case of the patron and the artist, or writer and designer. It worked through a more complex set of interactions which for the most part were highly creative and successful but still had to deal with the prejudices and social perceptions of the time. Smith would never really be able to see Sowerby in any way as an equal.

Smith's wealth and his acquisition of Linnaeus's collection gave him an immediate entrée into the inner circle of botanists both in Britain and the rest of Europe. Within a year of the collection's arrival he had been elected a Fellow of the Royal Society (1785) and had been working with its President Sir Joseph Banks and his botanical librarian Jonas Dryander on its arrangement. Sowerby had no such entrée. Smith was also a natural communicator, later

4/1 *Convolvulus speciosus* drawn for James Smith's *Icones Pictae*. It demonstrates Sowerby's skill in depicting accurately the nature of the plant within a confined space. (*C. speciosus* is now *Argyreia nervosa* var. *speciosa*.)

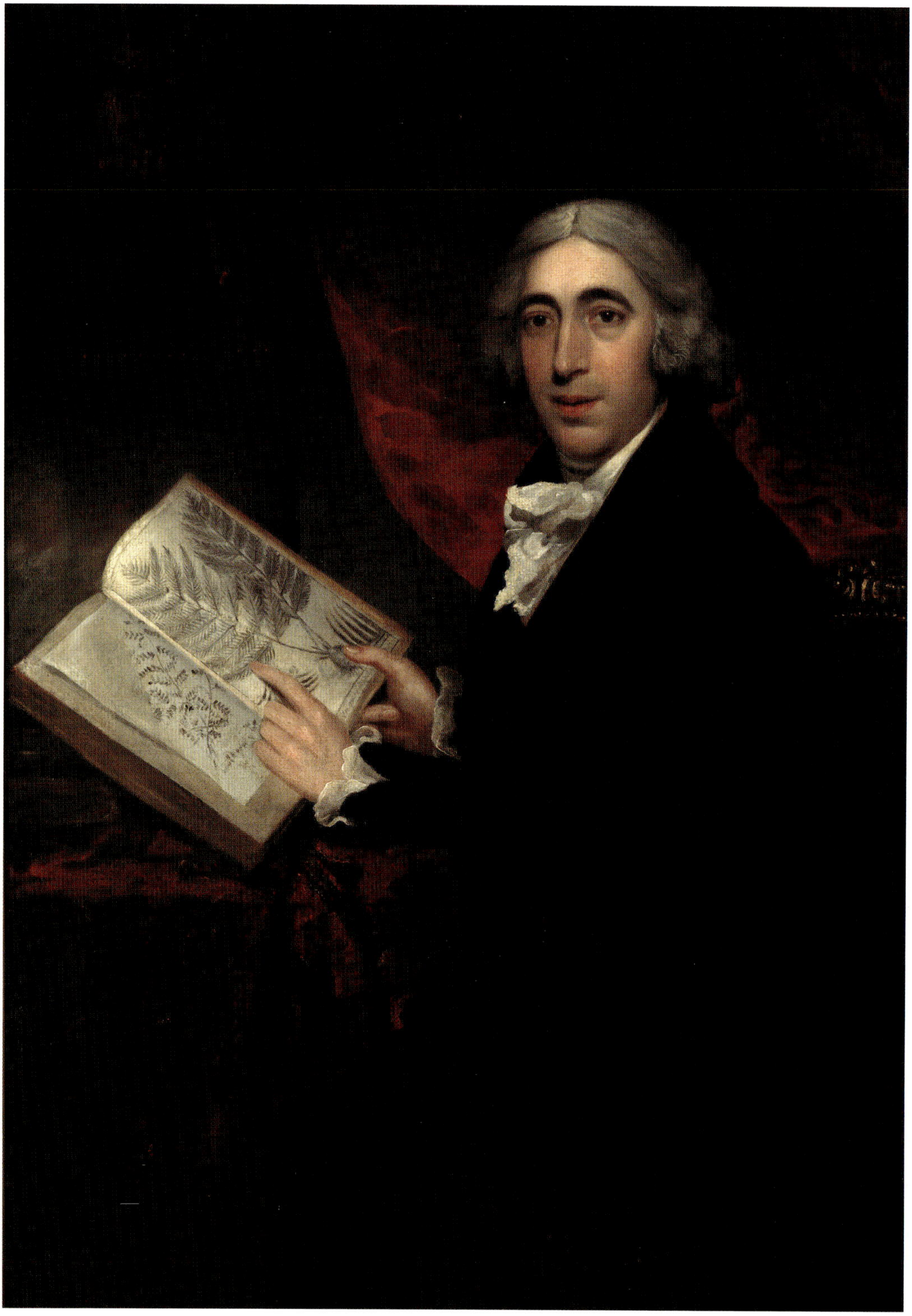

4/2 James Edward Smith. Oil painting by John Rising, unknown date. Credit: The Linnean Society.

giving frequent and good fluent lectures to an eager audience; Sowerby was more retiring. Smith wrote easily and quickly; Sowerby was never much of a writer preferring, as he said later, his drawing pencil to his pen, although each had to handle a mass of correspondence. Both men were to be married, Smith with no children, Sowerby with many.

While Sowerby was settling into married life and earning a living at Mead Place, Smith embarked on a European tour, financed by his father, to Italy, Holland, France and Switzerland. It lasted seventeen months, with his return in November 1787. The tour was very important for him; it consolidated his credentials as an influential and active scientist. He obtained a medical degree from Leiden University, met distinguished medical and botanical gentlemen, made new friends, collected numerous specimens, and hugely increased his knowledge of Europe's botany, institutions and culture, all of which was to aid his future dealings with foreigners as well as to enhance his own self-importance.

On his return he moved with the collections to Great Marlborough Street in the west of London. With the aid of others including Banks, the entomologist Thomas Marsham, and the botanist Samuel Goodenough, he immediately set about establishing a new learned Society to promote 'the cultivation of the Science of Natural History in all its branches'. This was the Linnean Society and its first general meeting, at his home, was on 8 April 1788. It started with thirty-six fellows, sixteen associates and about fifty foreign members.[1] The effective association with the Linnaean collections and the invitation to membership of many distinguished foreigners gave the Society an immediate status. This was serious business and it has remained one of the principal natural history societies in the world. Sowerby became an associate member in 1789 and soon after gave a paper on the flower structure of different species of *Passiflora* (passion flower). The printed version, published in the Society's *Transactions* in 1794, contains plates of three of Sowerby's fine drawings, clear taxonomic descriptions and his admiring comments on the beauty of the colours. Sowerby's vast experience in accurate illustration must have given him as good a knowledge and understanding of detailed plant structure as that of many botanists of his time. Indeed, he had earlier made drawings of different passion flowers, including one for the *Botanical Magazine*. He was made a full Fellow in 1793, contributing scholarly papers and discussions over many years, as well as aiding the smooth running of the Society and its publications.

Smith had published rather little by this time. His renown was gained principally through his Presidency of the Linnean Society and his ownership of the Linnaean Collection but he

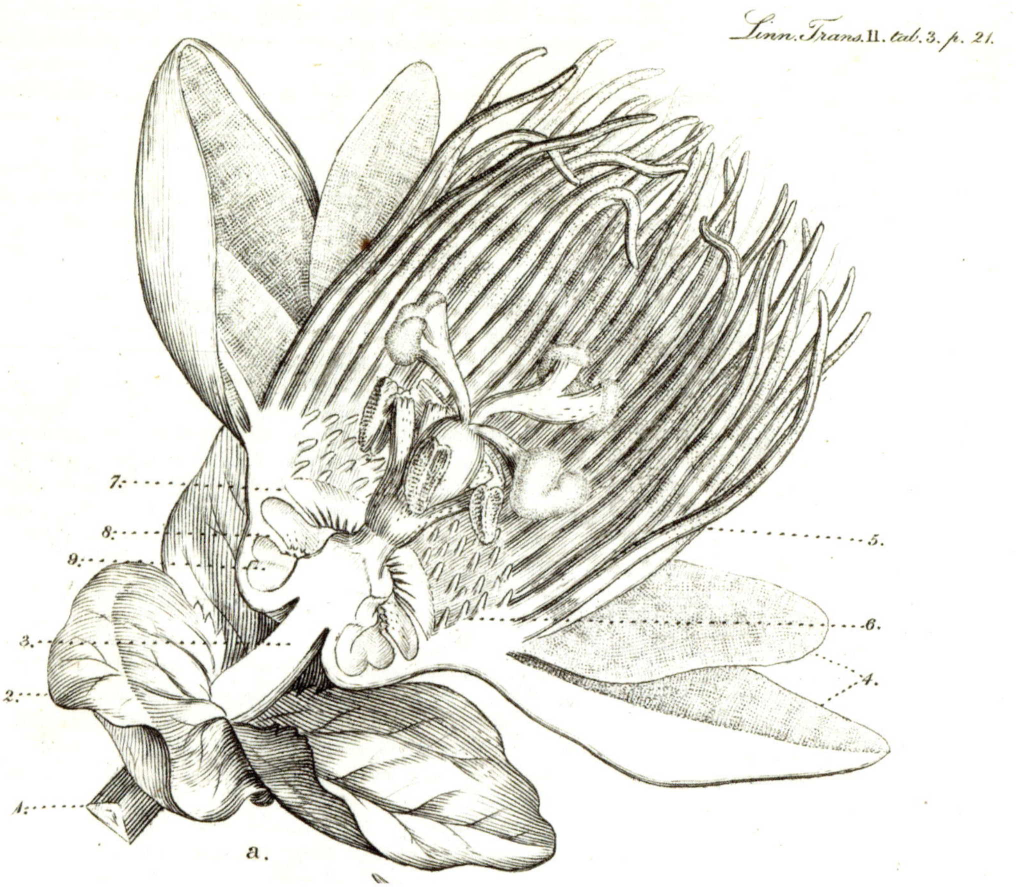

4/3 The passion flower, *Passiflora quadrangularis.* Sowerby's drawing of the structure of the flower for his paper presented to the Linnean Society (1790, published 1794).

4/4 (opposite) *Passiflora coerulea.* Engraving from Sowerby's drawing, published in the first volume (1787) of *Botanical Magazine.*

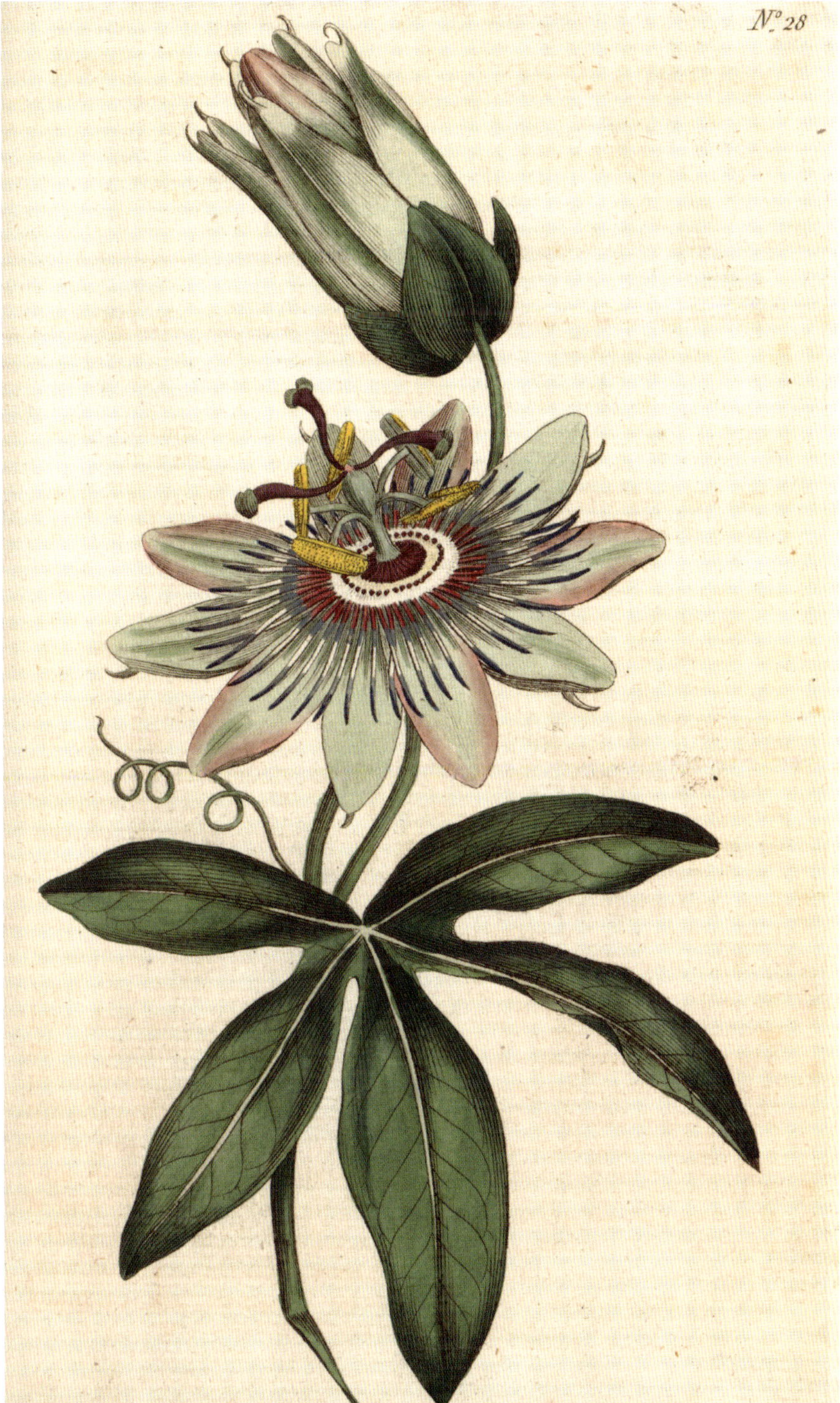
No. 28
Publish'd as the Act directs by W. Curtis Botanic Garden Lambeth Marsh

4/5 A fine depiction of *Begonia isoptera*, an Indonesian plant. There are over 1,500 species of begonias. James Smith validly published the description of this species in his *Plantarum Icones*, although Jonas Dryander had described it earlier.

moved quickly to redress the balance. In the space of three years from 1789 to 1792 he published three works, employing Sowerby as illustrator for all – it was a forging of their partnership. The works were informative since they dealt with some unusual or new species but broke no other new scientific ground. They were in a similar style and with the same intended audience as several works of the period. The first with the long title *Plantarum icones hactenus ineditae plerumque ad plantas in herbario Linnaeano conservatas delineatae* [loosely translated as: Figured plants hitherto not published or unknown mostly from Linnaeus's herbarium] was published in three fascicles over the years 1789 to 1791 with descriptions in Latin and uncoloured plates for seventy-five species of plants, some described for the first time. Sowerby is not mentioned as the artist or engraver but we know that he was; this may well be because some were drawn by Smith himself [2] and several of the others are based on drawings by another artist which were initially sent to Linnaeus.[3] The fact that only three fascicles appeared when so much could have been done suggests either that the work had limited appeal to other botanists or that Smith had decided to devote his energies to other projects.

The second publication *Icones pictae plantarum rariorum, descriptionibus et observationibus illustratae* [Coloured figures of rare plants] is in large folio and is more interesting. Smith wrote enthusiastically to his botanical friend Edmund Davall in Switzerland, 'I rejoice that you are satisfied with Sowerby. I like him much. We are about a work of superb cold. [coloured] plates, 6 in a fasciculus, to be dedicated to Lady Rockingham. Three plates are done, viz her Euphorbia.'[4]

In his preface he is also enthusiastic about Sowerby:

> The great number of new and rare plants however which abound in the English gardens have long demanded the attention of a scientific botanist to rescue them from oblivion; and I have been the more readily induced to take a share in the general labour by the persuasion of Mr Sowerby, one of the best botanical draughtsmen this country ever possessed, who was desirous of being engaged in some work worthy of his abilities. I should not have joined in the present undertaking with any person of whose skill and accuracy I have been less satisfied.

The purpose was to give accurate and elegant figures with scientific descriptions of the plants. Furthermore

> The most rare and beautiful plants only will be admitted into this work... It is desirable that the unlearned should be enticed into

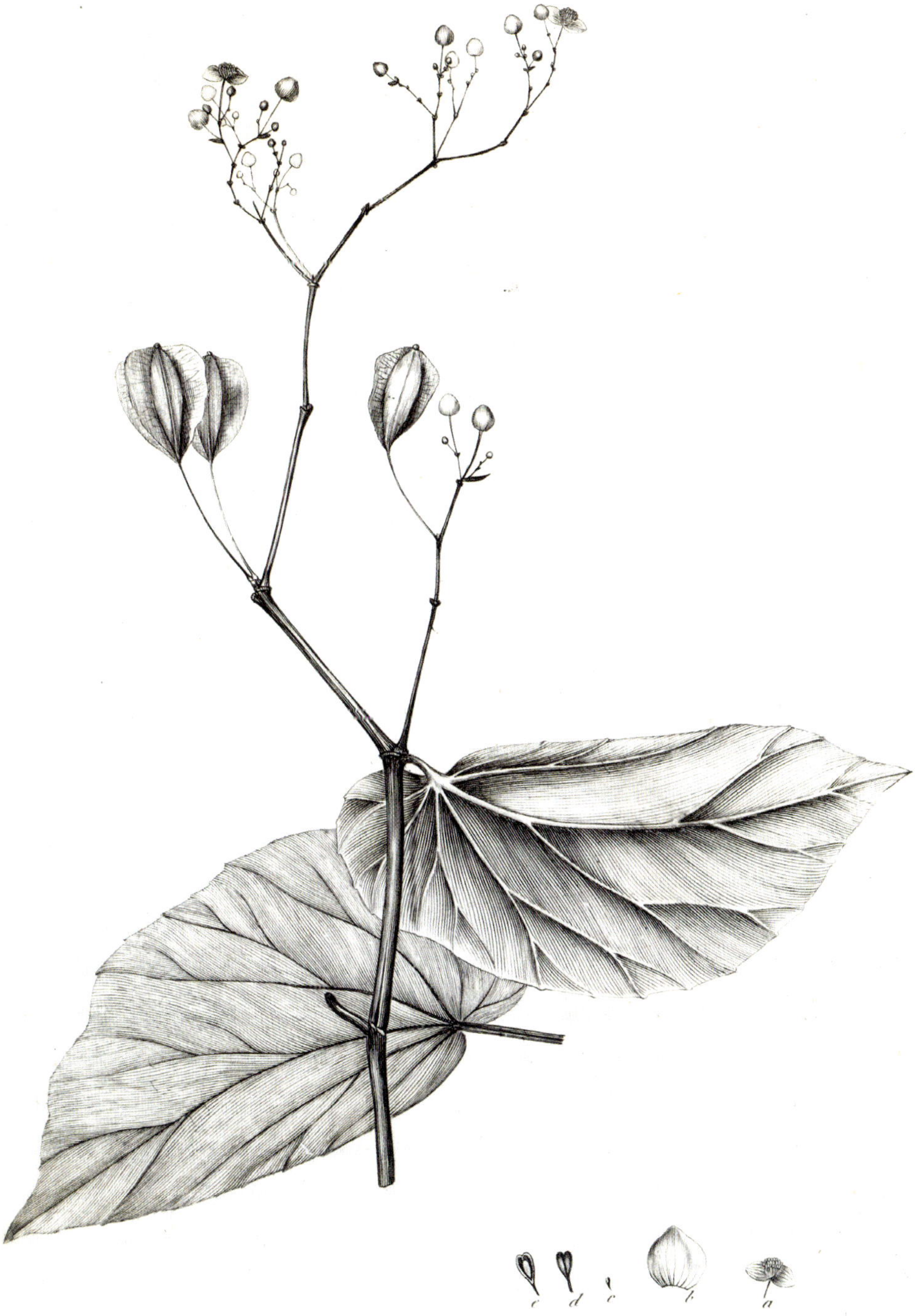

4/6 *Justicia coccinea.* A plant of the West Indies, which James Smith wrote could grow to the size of a small tree in a hot house. From his *Icones Pictae.*

> a familiarity with scientific and classical language, so that at length the learned and unlearned may join as it were by common consent, and, as they cultivate the same science, use the same terms.

This was not then intended solely for other botanists, and so as to enhance its audience it was available also in English for an extra payment. Indeed, it seems likely that the Marchioness was a patron or at least heavily involved. She was the wealthy widow of the Whig Prime Minister Charles Watson-Wentworth, the second Marquess of Rockingham, living at Hillingdon House, near Uxbridge, Middlesex, where she had a fine collection of plants. Her rapturous love of her flowers brimmed over in her (dictated?) letters to Smith but she was also anxious that the work would be sufficiently botanical:

> Lady R^{m} understood from Doc^{t} Smith that his intended Work was to be only of Plants never before colour'd in any book which of course excludes the Catesbea as it is already in Catesbys Work tho but ill done but if the specimen she sent is well executed by Mr Sowerby she mean't to have it to go along with her other Plants: Doc^{t} Smith does not mention the Euphorbia Wallinea but she imagines that will be acceptable to his book as she never heard of its being in any publication; and she is more interested about that than any...
>
> Ld. R^{m} cannot conclude without just repeating that Doct Smith's intended Dedication of the Work was a Compliment she could not wish to refuse, but from the fear that anything of sort address'd to her would be a real burlesque upon Botany to obviate that, she flatters herself it may apply more to her Plants, as a Collection which was once amongst the earliest and choicest of this Country.[5]

The drawing and colouring of 'her' Euphorbia (*E. punicea)* are magnificent, as they are for many of the other plants such as *Passiflora lunata* (or two-flower passion flower), *Hedera capitata, Portlandia grandiflora* (this one a double-page plate), *Oxalis versicolor, Convolvulus speciosus* and others.

All the fascicles, the third and last appearing in 1793, emphasised the coloured figures – this was the unstated (except perhaps in the letter to Davall) *raison d'être* of the publication. And it is the Sowerby plates that have remained as exemplars of botanical art and been given the recognition they deserve. They can be readily seen in all their fine coloured detail on a Missouri Botanical Garden website – a wonderful resource supported by the Andrew W. Mellon Foundation.[6] Here they state that the plates by Sowerby 'are some of his most accomplished work'.

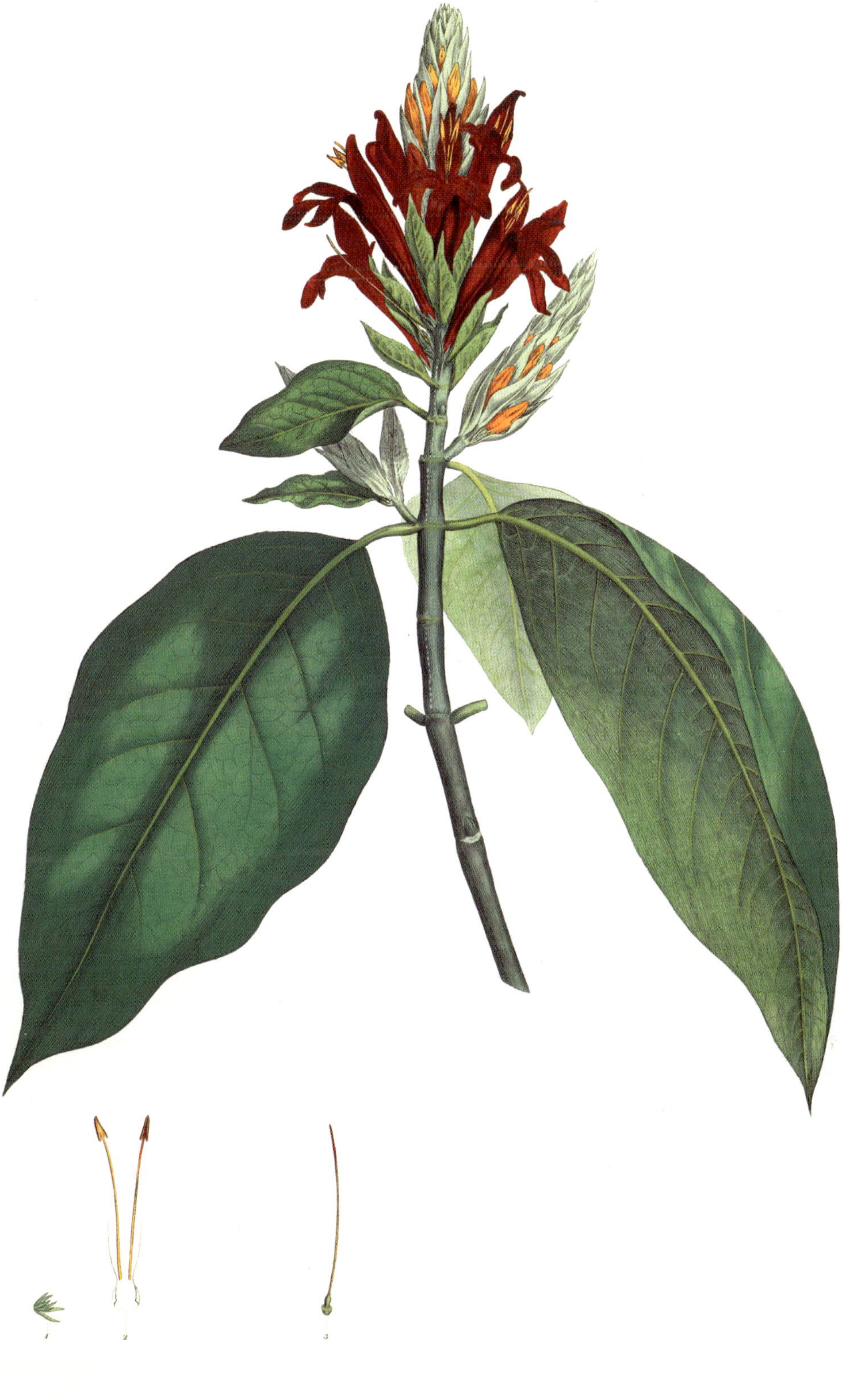

Justicia coccinea

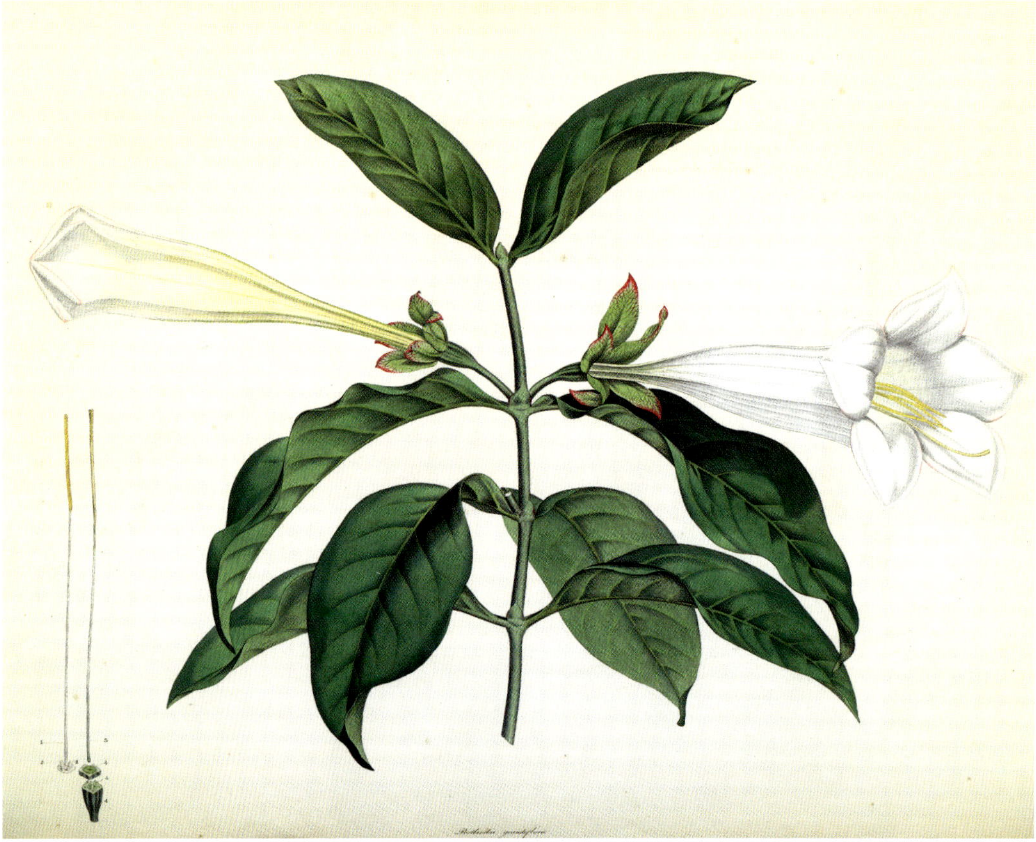

4/7 *Portlandia grandiflora*, which featured as a double-page spread in James Smith's *Icones Pictae*.

The recognition has not, however, always been the case. A later-generation Sowerby (William) had in January 1873 offered a copy of *Icones Pictae* for sale to the renowned bookseller Bernard Quaritch who replied to thank him for the offer but 'fear the book is absolutely without value'.[7] In 2006 a copy sold at an auction at Christie's London for nearly £4,000.[8]

Lady Rockingham was more than pleased to have individual paintings by Sowerby as well as those appearing in the Smith *Icones*, and had no compunction in treating Smith as a go-between in commissioning or vetting some illustrations of her own plants. She

> would be very pleased to see the two Paintings that Mr Sowerby has finished of the Portlandia & Catesbea & if he should have any painting of his own upon Vellum wishes he would send it at the same time for her inspection; for if he paints flowers equally well upon Vellum as upon paper, she possibly might prefer it.[9]

Sometimes she was bold enough to have Sowerby visit her, then feeling the need to confess to Smith:

> Without disfiguring the Azalia I could not send a specimen fine enough for a Drawing, therefore I have had Sowerby here; & I have also set him to make a drawing of a plant, a size he never did before which is both extravagant & perhaps ridiculous, but really the plant has flower'd in such magnitude, & is so very beautiful an object, that I could not resist having it drawn the size of life.[10]

The enthusiastic correspondence with Smith continues but within a year or two she shows clearly her irritation over Sowerby's lack of focus on her own requirements:

> This morning found the Passiflora suberosa out in flower & a beautiful little thing it is. It never flowerd here before, & I doubt whether it is apt so to do, therefore I shall like a painting of it, & shall send a good Specimen to Sowerby, if he really will do it well & quick; I think this & the Coripssis, done upon Vellum will make two elegant Drawings to go on each side of the Catesbea that he drew before; but I have not yet heard a word from him about the dimensions of the Datura & I cannot help fearing that he bewilders his head with understanding too many things at a time.

It was perhaps easier for a Lady to deal with a gentleman than go direct to an artisan' and no doubt poor Sowerby was filling his head with many things – too many things – as this was a period of remarkable output, some for other's works, some for other's whims and some for his own publications, but all for the necessity of earning money for a rapidly growing family. And not all botanists were prompt in paying Sowerby the monies he was due, including Davall, about whom Lady Rockingham was as enthusiastic as was Smith.[11]

Edmund Davall was an Englishman, living in Switzerland, with a keen interest in botany.[12] He had met Smith when visiting London and

4/8 The Barbary ape (*Simia inuus*) as depicted in Shaw's *Linnaean Zoology.* The much-stylised representation suggests that Sowerby was using a rather inaccurate drawing.

they soon became close friends, writing affectionate letters to each other. Davall collected and studied the Swiss flora and was preparing his *Illustrations of Swiss Plants*, assisted in plant identification by Smith who also acted as an intermediary to Sowerby for the preparation of the plates.[13] The work was never finished but some papers and artwork are held in the Linnean Society in London. Davall writes to Smith in February 1790, 'I have added 6 drawings for my work for Sowerby with the original specimens from which they were made... I shall write to Sowerby as soon as possible give him my compts. & recommend the good execution of the plates.' Their correspondence shows that Sowerby would be an expected visitor at Smith's home to collect specimens, drawings or payments.[14] It also reveals something of the tensions among the botanical community and of the demand for Sowerby's skills, as Smith explained to Davall in November 1789 about Sir Joseph Banks:

> You ask if Sr. JB is steady to me. I believe that he & Dryander mean to be as civil to me as they can, & why should I enquire further? Everybody is not blest with an heart. My intended coloured work is not relished by Banks because he wants to set Sowerby about one on the Kew plants, & mine will forestall it. But we are on the best terms possible. I use them generously as I ought, & they are very communicative to me. Every body thinks us the best possible friends. Pray keep this strictly to yourself.

Smith was also establishing a working relationship with George Shaw who, like him, had studied medicine at Edinburgh and had come to London to practise it. Instead he became a keen botanist, and was finally appointed keeper of the natural history section of the British Museum. He was a co-founder of the Linnean Society, later being elected one of its vice-presidents, and a Fellow of the Royal Society (1789). He was to work with Smith and Sowerby in describing some of the Australian fauna and flora but his first involvement with Sowerby was the publication of his own *Linnean Zoology* in 1790, comprising, he claimed, 'a complete illustration of the zoological part of Systema Naturae of Linnaeus'. In fact, it dealt only with Class 1, the mammals, and represented the lemurs, bats and primates. It is a relatively slim volume of twenty-six text pages and eight coloured illustrations but these show an artistic departure for Sowerby – for not only are they are of animals but the animals are placed in their natural settings. The lemur is in a leafy tree with a cloudy blue sky behind; the Barbary ape reclines on a grassy bank with a relaxed and pensive posture, underneath which is written 'we have fixed upon a species of ape... sufficient to mortify the pride of mankind by the humiliating general resemblance which it bears

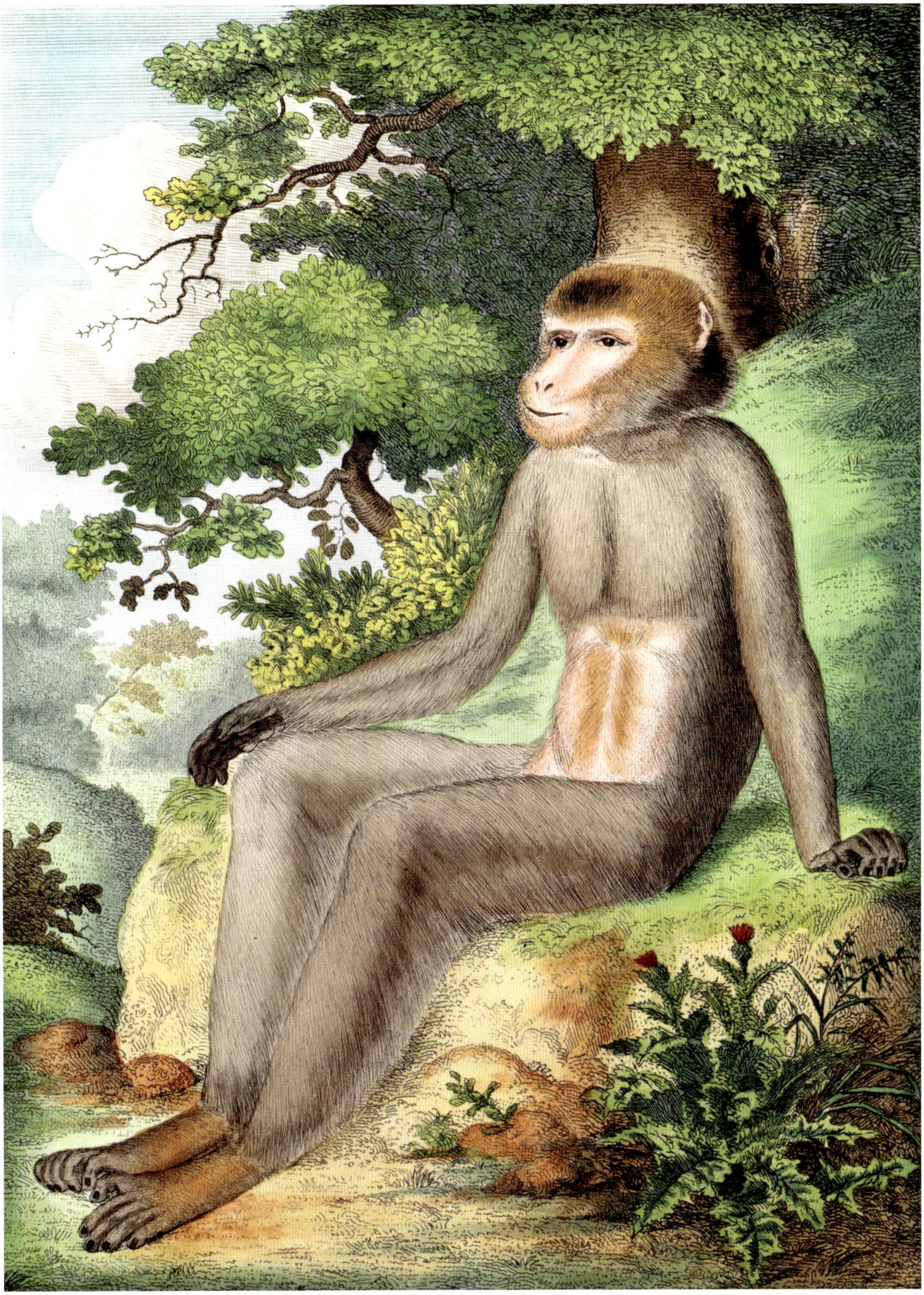

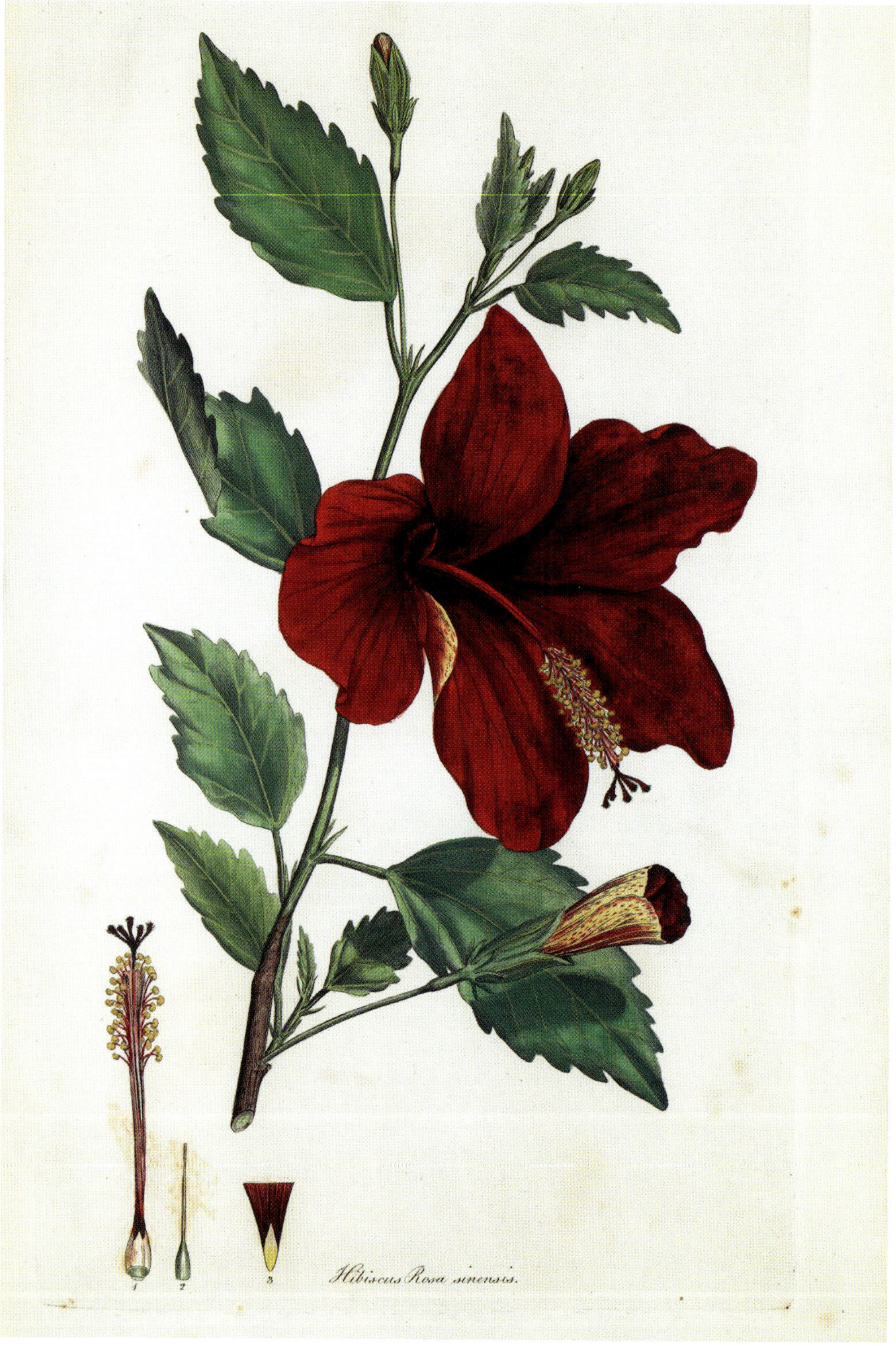
1
2
3
Hibiscus Rosa sinensis.

4/9 *Hibiscus rosa-sinensis*. James Smith wrote: 'This most beautiful flowering shrub is deservedly a favourite with the Chinese.' From his *Spicilegium Botanicum*.

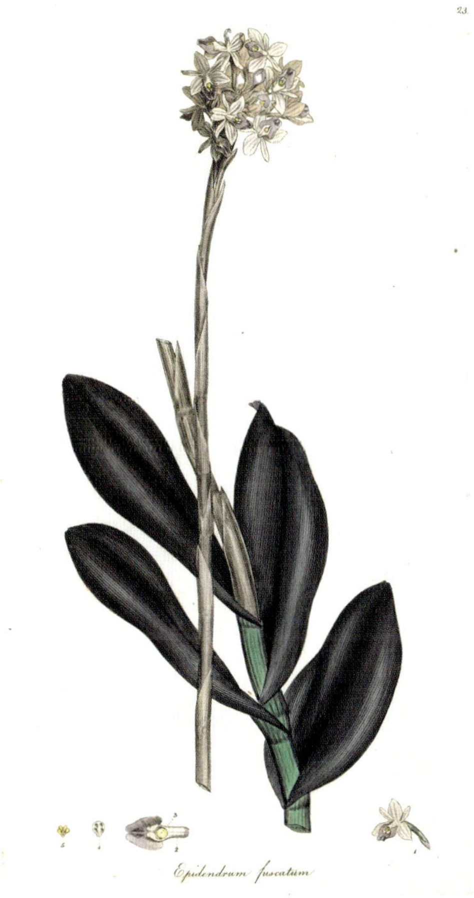

4/10 The epiphytic orchid *Epidendrum fuscatum* (now *E. anceps*), sometimes called 'brown epidendrum' on account of the colour of its flowers, and which grows in the American tropics. This drawing was based on a specimen from the 'hot stove of the Hon. Mrs Barrington'. From Smith's *Spicilegium Botanicum*.

to our own race'. It is difficult to judge Sowerby's accuracy in these plates. We neither know whether he worked only from other people's drawings nor if he had some specimens on hand for comparison. Incorporating the countryside into the figures would, of course, have posed little problem for him; he had had plenty of past practice in painting landscapes. A short review in the journal *Analytical Reviews* was pleased to see such a publication, especially as undertaken 'by a gentleman of Dr Shaw's classical taste, and an artist of Mr Sowerby's acknowledged talents'.[15]

We must return briefly to Smith's third publication with Sowerby around this time: *Spicilegium Botanicum* or *Gleanings of Botany* (1792), produced originally as two folio numbers, each of twelve species with their coloured illustrations. Its aim was to show 'coloured representation of uncommon vegetables as are not already sufficiently well figured by any author... generally though not entirely exotics'. It includes laurel-leaved *Cestrum*, drooping *Trillium*, yellow alpine soapwort, China-rose *Hibiscus* and many others including a fungus: the columnar puffball *Lycoperdon phalloides* (first described by James Dickson in 1785 and now named *Battarrea phalloides*). The work could be bought for 10s.6d. or with added English text for 12s. and was on sale at Sowerby's house, Mead Place. The plates are again of high quality with artistic flair and accuracy but it is difficult to imagine that the publication would have achieved much beyond appreciation of the fine illustrations – botanists were wanting something with a better focus than just gleanings. That something turned out to be the major collaborative and now famous work of the two Jameses – their *English Botany*.

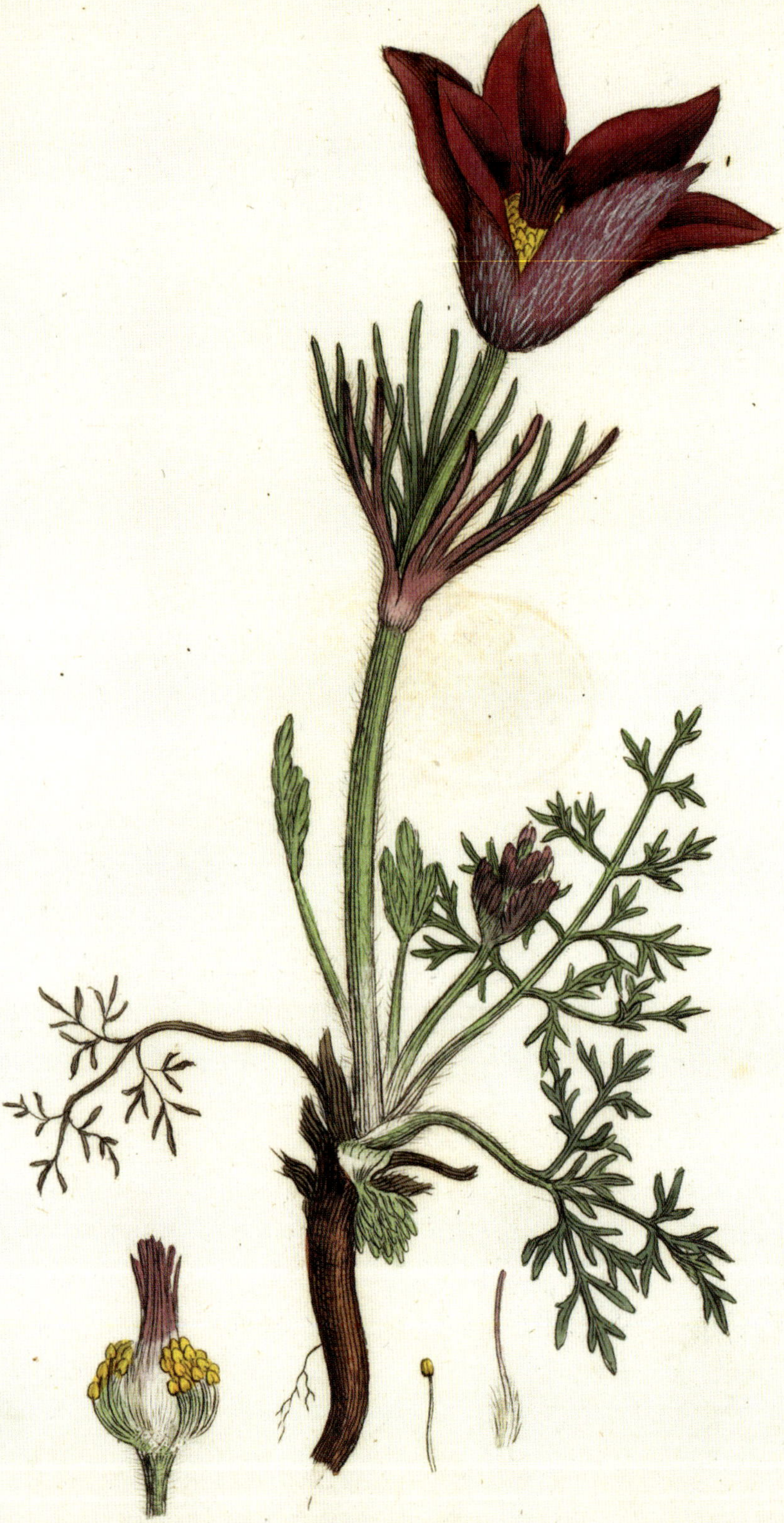

Js. Sowerby del. Apl 1 1792

Anemone pulsatilla

CHAPTER FIVE

Sowerby's botany

With his next major project, *English Botany*, Sowerby turned the usual accepted procedures and social etiquettes on their heads. Instead of the writer commissioning and paying the artist, Sowerby commissioned and paid the writer; instead of the proceeds of sale going to the writer they went to the artist, and instead of the writer finding a suitable publisher for the work, Sowerby published it himself. Sowerby, the artisan, was in control.

English Botany (or, to give it the full title: *English Botany, or, Coloured figures of British Plants, with their essential characters, synonyms, and places of growth.*) started publication in December 1790 and joined the ranks of floras and botanical books, serials and plates on sale being produced at the time.[1] But this was planned to be a serial publication with a difference and which, Sowerby must have hoped, would appeal to a wide audience, for it would be in English (not Latin), attractive to the eye while still botanically accurate, affordable, and its scope would be all British plants (despite its title of *English Botany*). It is surprising that 'no successful attempt has hitherto been made to illustrate them by original figures on a cheap and compendious plan', said the preface to the first volume. It would show that British plants were every bit as good as exotic ones:

> A knowledge of the plants of our own country is in many respects preferable

5/1 The pasqueflower, (named as *Anemone pulsatilla*). This plate appeared early (1792) in *English Botany* (volume one). It was described to be 'of great beauty'. Sowerby's original drawing is shown overleaf.

5/2 Sowerby's initial drawing of the pasqueflower with some pencilled comments by James Smith. It was used by Sowerby for his engraving and the partial colouring was a guide to the colourist. See previous page.

to that of exotics, as it can be much more completely attained, and is on several accounts more directly useful. Nor are the humble productions of our fields and woods deficient in real beauty, elegance or singularity of structure, in which respects some of them indeed vie with the most favourite exotics. Not to mention the indispensable necessity for those, who are occupied with the rural economy of any country to be well acquainted with its native vegetables; the study of them, as a mere amusement, has this eminent advantage over exotic botany, that is doubles the pleasure of every journey or walk, and calls forth to healthy exercise the bodily as well as the mental powers; while persons who know no such pursuits either sacrifice their health to sedentary employments, or are content to submit, without an object, and without enjoyment, to toil and labour for exercise, and for exercise alone.

But it would not be just an artistic work, it would have botanical credibility:

> He will spare no pains to have his botanical characters and synonyms accurate; and as he is fortunate as to have access to the first sources of information, he hopes this part of his plan will be executed so as to deserve approbation.

These 'first sources of information' had not been initially revealed; the covers of the early issues had only Sowerby's name. Botanists knew that Sowerby had plenty of experience in observing, identifying and depicting plants but they must also have suspected him of insufficient scholarship over matters of synonyms, classification and previously published descriptions. If a publication was to be respected it must have a respected author or co-author. Who was this and why were the readers not being told?

James Smith had already written to his botanical friend Thomas Woodward:

> Sowerby is abt publishing a work on English plants on the plan of Curtis's Magne, price 1/- title English Botany... between ourselves I write it for him gratis till he sees whether it will answer or not. 'tis in his name. I think I have hit on a good motto for it, & quite an original one, from Virgil –
>
> –'quos ipsa volentia rura Sponte tulere sua.' ['what produce the fields of their own free-will will bring forth.'[2]] The 1st No. will appear in a day or two – it contains Veronica spicata... I think I shall be found out as the author.[3]

So Smith was the author of the descriptions but a somewhat reluctant one, at least regarding any possible risk to his reputation. It is unlikely that Sowerby would have wanted to withhold

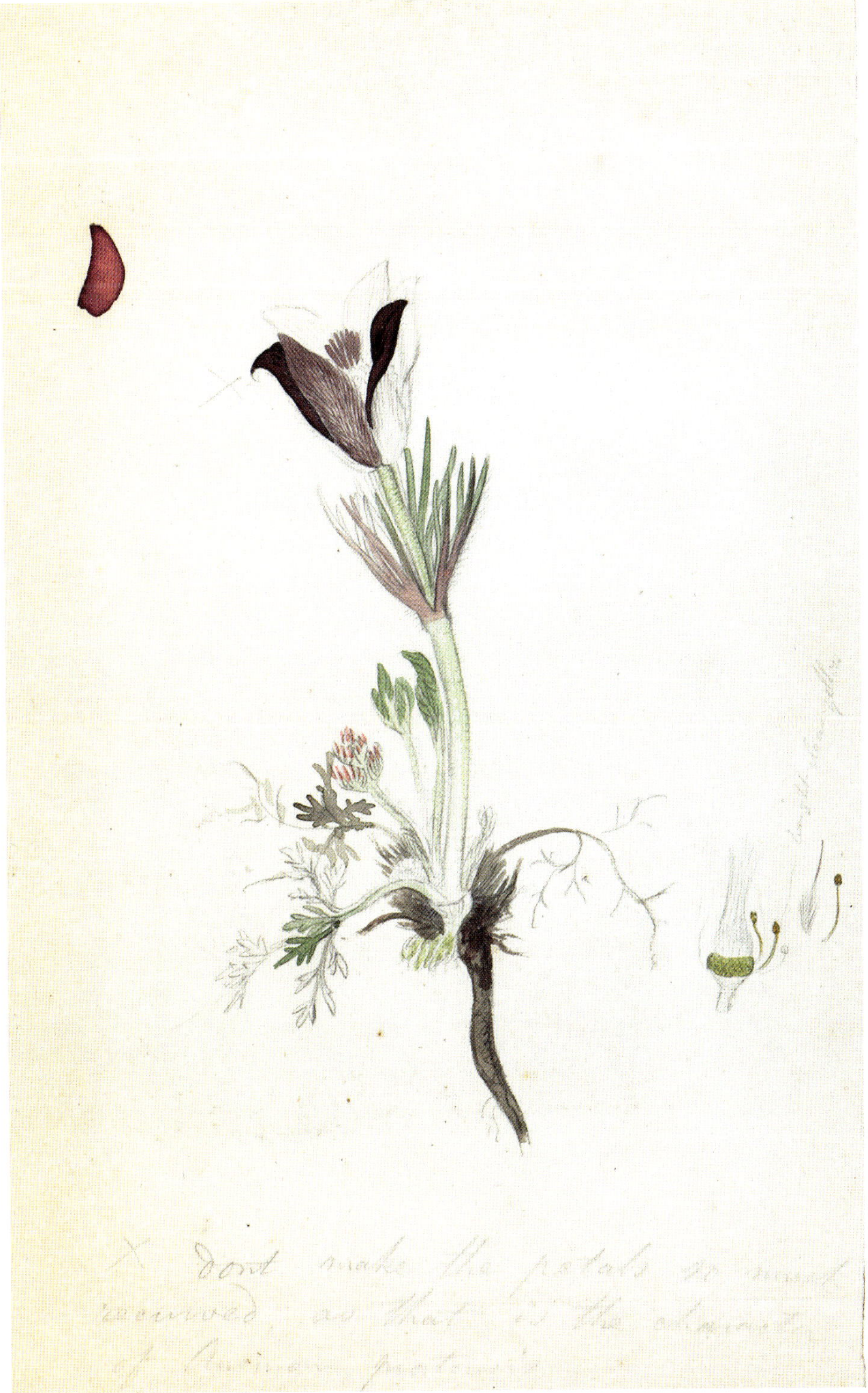
X dont make the petals so much
recurved as that is the character

2469. Menziesia coerulea

Aviemore

Dr. Smith

too sharp.
see Tr. of L. Soc
v. 10. t. 30. A f

This should have
rather more of
a blue tinge, as in
Fl. Dan.

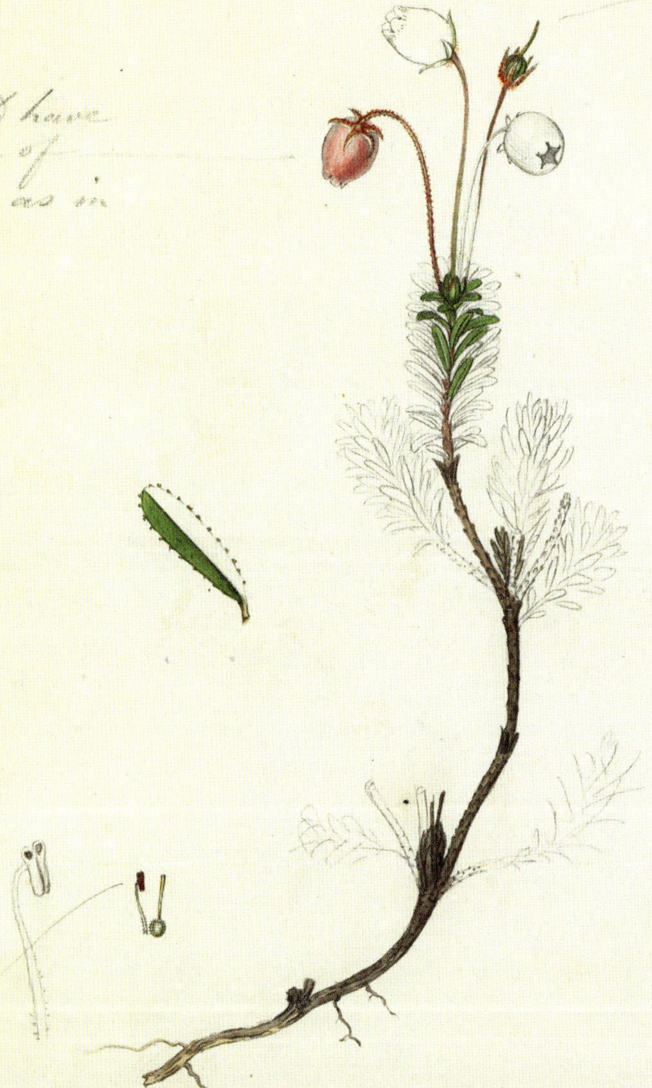

coloured by guess
is it right
yes.

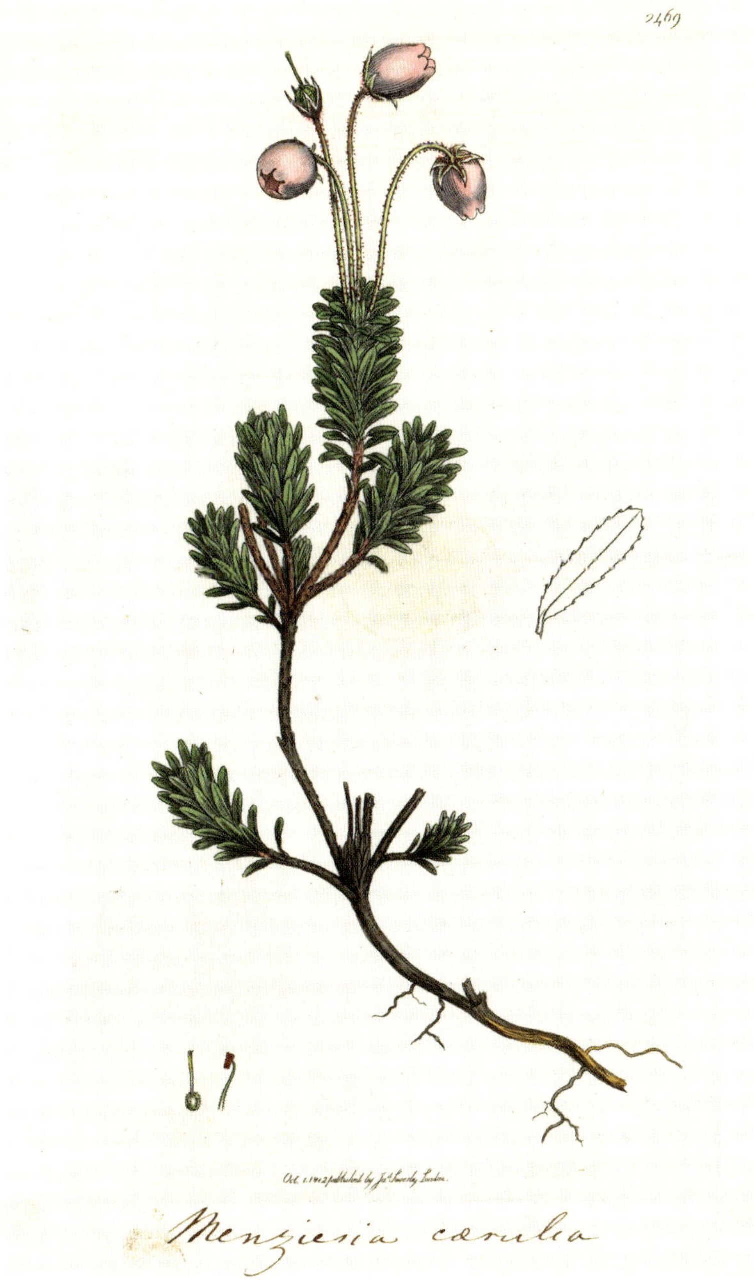

5/3 Sowerby's working drawing and the final printed, hand-coloured plate of the 'Scottish Menziesia' (*Menziesia caerulea*) in *English Botany*. This was plate number 2,469, of a total of 2,592 plates. James Smith and Sowerby had much satisfaction in 'announcing this as a British plant, on account of its rarity and beauty'. This bushy evergreen shrub was recorded as growing on dry heathy moors in parts of Scotland.

5/4 A detailed hand-coloured engraving of the 'dwarf thistle' (*Carduus acaulis*, synonym of *Cirsium acaulon*), which happened at that time to occur abundantly in Norfolk. *English Botany*, volume 3.

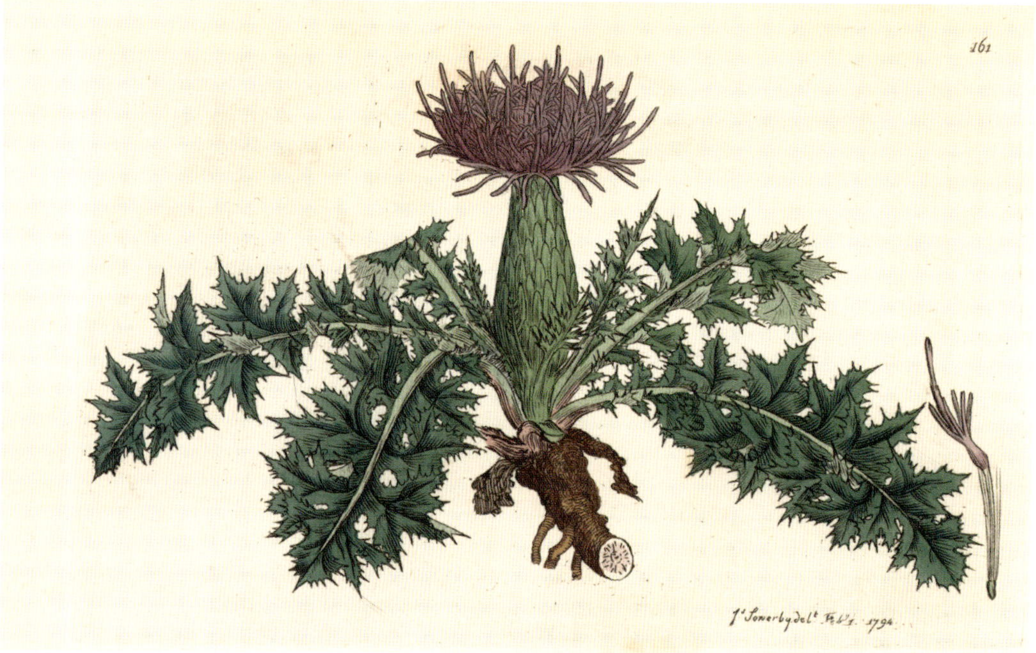

the writer's name even though Smith said so, for this particular name would have added status to the work. In any event and as we will see, Smith was later to admit that it was he who wanted his name to be initially suppressed. Was this mainly because he was not in full control; because this was Sowerby's project and he had not worked in this way before? Surely he had full confidence in Sowerby's artistic contributions and he knew what he himself would contribute? It was this latter aspect that concerned him, despite him thinking that he would soon be 'found out'.

He was worried to be associated with a work that might not be sufficiently scholarly. He later wrote to his botanical friend Edmund Davall in Switzerland: 'I long *very much* to hear yr. sentiments upon English Boty. wch. I now avow as mine. You know its aim is to make botany pleasing & popular, so you must pardon *trifling*.'[4] This early unwillingness by Smith to throw his wholehearted support into the project was to give him pause for thought on many a later occasion, more especially as he was always very conscious of his own status and recognition.

English Botany appeared as monthly numbers – in octavo sheets usually with three species to a number, covered and stitched in a blue paper wrapper. One page of text accompanied a life-size coloured engraving which recorded Sowerby as the artist, and the date. Each plant was given its species and common names, its position in the Linnaean classification, and then a few lines on the general and specific characters. Synonyms with references to other botanical works were cited. Finally, various observations such as on habitat or localities, comparisons with closely related species, and on some uses or even taste were included. The result was an easy-going and clear text accompanied by a fine illustration showing most, if not all, the plant including its root, together with smaller figures of enlarged parts of the flower, such as the stamens. They were often exquisitely drawn and coloured.

Plants were published in a sequence that initially must have appeared random, and it would remain this way until the project's end. Possession of a good specimen, preferably in flower, was the prime driver, but such availability rarely followed botanical classifications. Sowerby could not wait to amass a suitable number of coloured engravings for a group of related plants before publishing. But if the work had a kind of spontaneity then that was part of its charm, and it meant that a large community could be involved in the whole enterprise. Purchasers, in effect, became contributors; *English Botany* was to some extent their own and its popularity was enhanced.

The first reception was certainly positive. Thomas Woodward wrote on Christmas Day to Smith and after some words about his health and rheumatism, continued:

> I have got the first No of English Botany, & am charmed with it. I have recommended it to everybody I have seen who buys things of this sort, & I am sure when they have once got it they will go on with it. The paper, print & execution of the plates is so much superior to Curtis, that there can be no fear of its succeeding, to say nothing of your part, in which I am sure you will be found out.[5]

Woodward was a good and frank friend to Smith, sharing equally in the same letter his views on *English Botany* as on his own constipation and his taking of mercury, a not uncommon practice of the time:

> I am myself surprised at the effect of Quicksilver, knowing it can only act mechanically, & it seems impolite to account, how so small a quantity taken daily can remove any obstruction, if such a thing exist; yet facts are facts, & most undoubtedly not only my complaint has not returned, but my digestion is better & my bodily habit much more regular.

There was no doubt that Smith could trust this friend to speak his mind! He thought highly of his botanical skills and named the fern genus *Woodwardia* after him.[6]

Woodward advised that Sowerby should always use good quality paper, unlike Curtis in his *Botanical Magazine*, which had engendered criticism.[7] He was concerned that *English Botany* might not continue but Smith was able to assure him just six months after its commencement that it 'sells amazingly, & is by no means likely to stop'.[8] The success was undoubtedly due to the style and accuracy of Sowerby's drawings – plants could be identified using those alone – the accessibility of the text, the relatively low cost at 1s. per number (the same as Curtis's *Botanical Magazine*) and the focus on British plants. On opening the first issue, subscribers saw a fine illustration of a lady's slipper orchid (*Cypripedium calceolus*), rare at the time and which was to become even rarer, and the opening sentence: 'If the beauty or scarcity of a plant, or the singularity of its structure, entitle it to our notice, the Ladies Slipper certainly merits the first place in a work on British plants.' So started a work that was to continue for twenty-four years and contain 2,592 hand-coloured illustrations.

By the end of 1792, two years after the start, twenty-four numbers had been issued with a total of seventy-two descriptions and plates. They comprised the first volume and with it came a new confidence and desire for development. It was decided to double the publication rate and to additionally include ferns, algae and mosses – the non-flowering plants or Linnaeus's class of Cryptogamia but excluding the fungi. Woodward had suggested at the start that doubling the rate would improve sales and now highly approved of the alteration, still thinking that it would increase purchases by botanists while conceding that in his own town of Bungay in Suffolk the demand had dropped from five copies to two on account of the increased cost; but then the previous buyers 'were such as only bought them as pretty pictures'. What Sowerby thought of this last remark is not recorded. Woodward also approved of the inclusion of the cryptogams but somewhat inconsistently said that not more than one or two should be in a number 'for fear of disgusting the florists or the Ladies'.[9]

The longevity of the publication, its popularity and its subsequent fame are all things that Smith failed to anticipate, or even slightly dream of, at the outset. When a letter, signed simply 'RG', appeared in the widely circulated *Gentleman's Magazine* of February 1793 complaining about the representation of a geranium species in the twenty-fifth number, Smith felt compelled to not only respond with a good defence but also to declare authorship and reveal his earlier somewhat churlish approach:

> However averse to literary altercation, I think it incumbent on me to reply to your

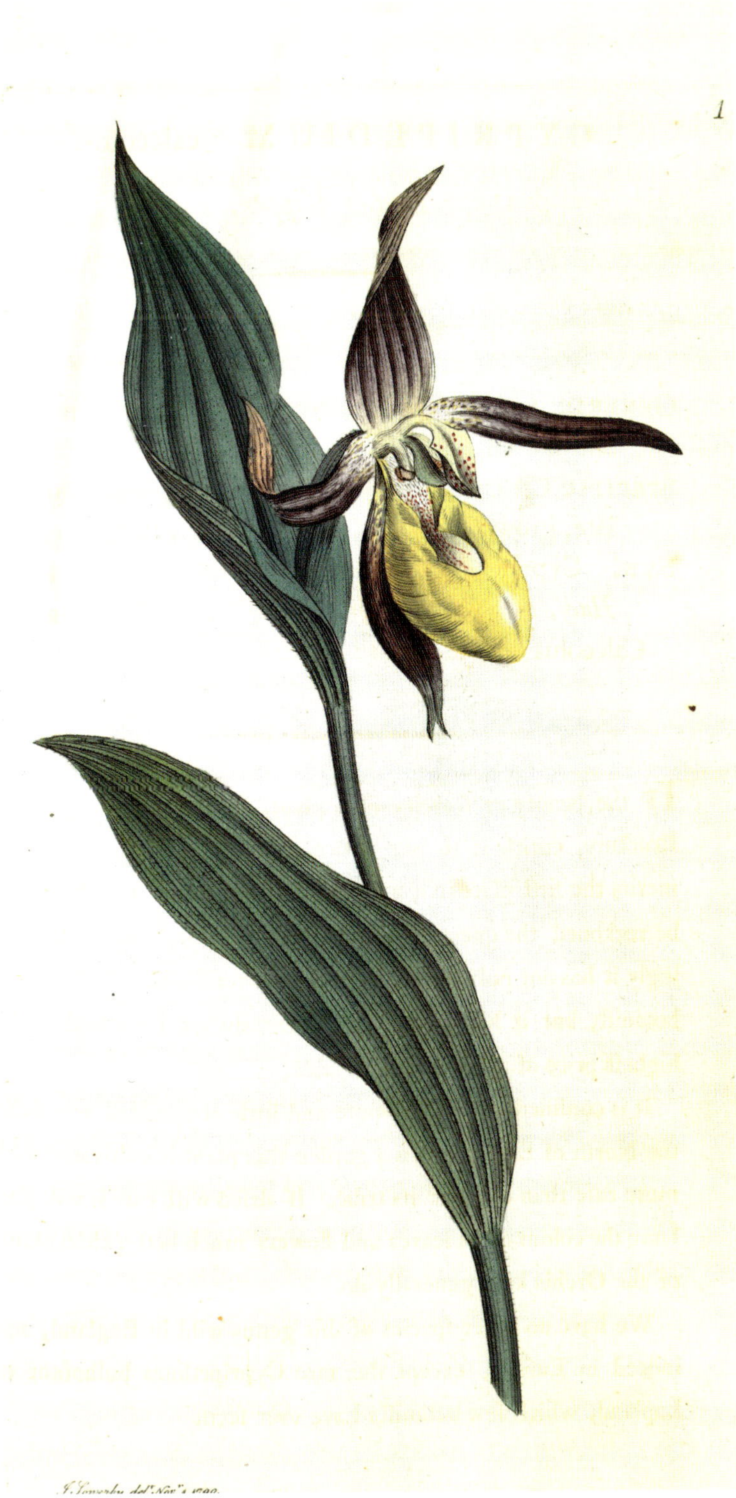

5/5 Lady's slipper orchid (*Cypripedium calceolus*). The opening plate to Sowerby's *English Botany*.

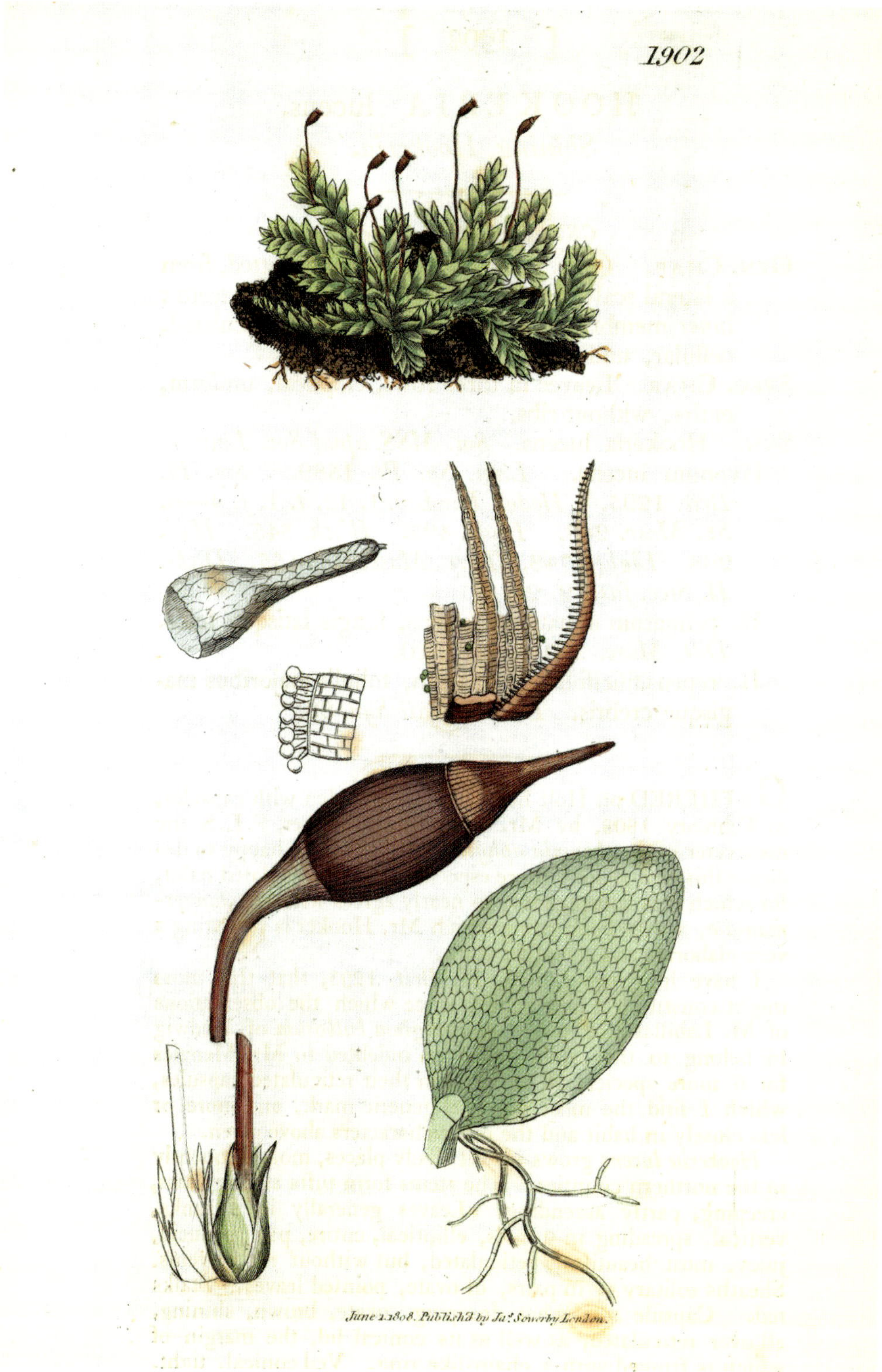
1902
June 1.1808. Publish'd by Jas. Sowerby London.

5/6 The moss 'shining Hookeria' (*Hookeria lucens*). Gathered for *English Botany* by William J. Hooker in Norfolk and named after him by James Smith. Hooker was a renowned botanist who later became director of the Royal Botanic Gardens at Kew.

> correspondent R. G. in your Magazine for February, respecting number XXV of English Botany. Mr Sowerby having all along been pleased to submit that work to my correction, I wish to be considered as answerable for any faults there may be in the letter-press, which I hope will not be very many. His excellent pencil deserves to be supported with accuracy and utility on my part; and having found the work generally approved, and capable of being eminently serviceable to the science I have taken more pains than I at first intended, wishing to render it as worthy of the public acceptance as I could.[10]

It is perhaps worth noting that the same issue of the *Gentleman's Magazine* contains a virulent criticism of the French National Assembly in permitting the murder of the French king, Louis XVI and of the state of French Republicanism. Meanwhile British botany was continuing in its most gentle way, with just a little questioning about detail.

Smith did little else to declare his authorship of the text until he produced a preface for the fourth volume (1795), saying rather disingenuously that he had been surprised by the success of the work and that he 'chose rather that the work should make its way by any merit or utility it might be found to possess, than be indebted to other performances for a name'. But less than two years later he was still unhappy about the risk of not getting due recognition, which his earlier stance of hiding his name had, of course, made very likely. Now he wanted his name on the covers of each issue, for he had learned that a few, possibly important, people knew nothing of his involvement, and so he wrote to Sowerby proposing this. Sowerby must have replied somewhat strongly about the nature of the publication and his own role, although this letter no longer exists, for Smith then wrote a defensive letter attempting to clarify his proposal and to calm matters between them:

> As to the alteration of the covers, I only meant that when any more are printed my name might be put on with yours, that people in general might know at first sight who writes it – I meant it as much for the good of the work as any thing else, because I have met with 2 people (Lady Hume & a gentm at Cambridge) who not happening to read the preface to the 4th. vol. did not know I had anything to do with it. – You cannot suppose I meant to claim the property of the work in any sense. However, to prevent this, when any new covers are printed let them be 'Published & sold by the editor Jas. Sowerby etc. etc.' The editor is the person who has the property of a work. Or take any other means you please to secure it. I wish to do every thing to satisfy you. Only don't call yourself the author, that is the writer, because that contradicts the title

page of every volume, & it is not true. You will now I trust understand me – if not I'll explain it when we meet – I know you mean rightly.[11]

It is not difficult to understand any frustration that Sowerby must have felt over Smith's approach. Here was a work with an emphasis on the illustrations, as made clear in its title and in Sowerby's initial preface. Smith did not want to be named at the outset; now he seemed to be declaring sole authorship and in a way that appeared to minimise any credit for the botanical value of the illustrations, let alone the botanical contributions that Sowerby often made to the text. This was a disagreement of its age. Artists weren't given the same status as people who wrote the words. Niceties over the use of the term 'author' were important but the term 'editor' also failed to give a true indication of Sowerby's role. Smith was unable to see beyond these boundaries but he was conciliatory enough to allow Sowerby a degree of freedom in the final decision, so as to ease Sowerby's distress. It would not, however, be the end of Smith's sensitivity, and regret, over the authorship issue. The saying 'a picture is worth a thousand words' is as true as ever in the world of plant identification and a Sowerby plate is a fine rendition. The coloured flower has the detail for linking to the Linnaean classes; the correct portrayals of the leaf, stem, and often the root are further significant aids for correct assignments. The true worth of *English Botany* is in its plates. It is right that it should continue to be known as 'Sowerby's English Botany' to this day (librarians and cataloguers notwithstanding).

Initially Smith had written without payment but Sowerby wanted to pay him half a guinea for the description of each species and this soon happened. Later, in 1799, Sowerby proposed an increase to one guinea, which was also accepted.[12] Neither of the two Jameses needed to spend much time or money on specimen collecting, for they received plants from numerous botanical friends as well as new acquaintances who wished to get in on the act. If a British plant grew in a garden or neighbourhood then it was an important fact for recording in *English Botany*, even more especially when the donor was named and praised. Smith had a beguiling style. 'We are obliged for wild specimens to Mr John Mackay of Edinburgh, a most diligent and skilful investigator of the vegetable kingdom, by whose communications we have often been enriched,' he wrote, and wrote similarly for many others.[13] Specimens were contributed by the famous, the rich, the gardener, the clergy, the enthusiast, and those who would like to exchange specimens for their own desiderata. Lord Viscount Lewisham, to whom the work had been dedicated by Sowerby, was a frequent contributor, as were their old friends Thomas Woodward, James Crowe and James Dickson, as well as the Sussex botanist William Borrer,

5/7 An elegant depiction of the amphibious *Persicaria*, sometimes called water knotweed. The seeds, according to William Curtis, were a favourite food of water fowl. (Named in *English Botany* as *Polygonum amphibium*.)

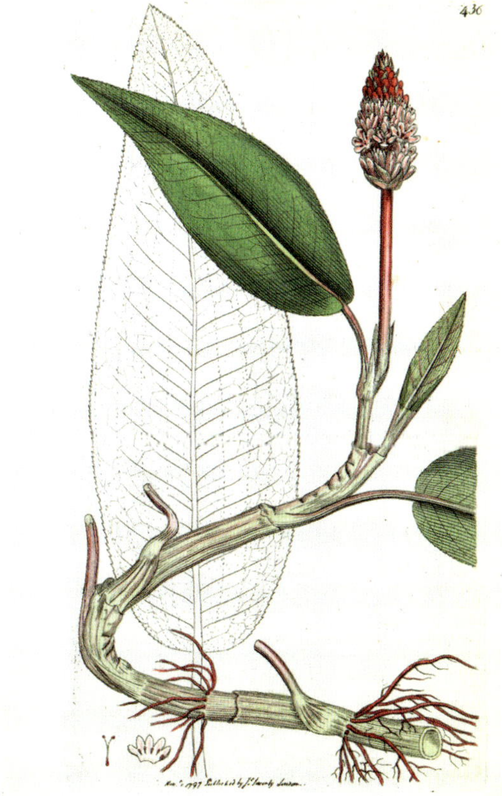

the surgeon Sir Thomas Cullum and Richard Salisbury. Charles Lyell (father of the renowned geologist of the same name) was a regular donor.

One particularly zealous correspondent was Charles Abbot, a teacher, botanist, entomologist and vicar in Bedfordshire. Numerous letters and specimens were sent over many years to Sowerby who, in turn, encouraged him by sending a list of desiderata for him to collect and despatch. Abbot really desired his name in print – initially pretending it to be for his wife's benefit when he wrote 'pray remember me for some plant in E. Botany as soon as you can as Mrs Abbot likes to see my name in that book' but owning up just a month later: 'I candidly own it is my wish to appear as often as possible in E Botany. Indeed you missed a fair opportunity of mentioning me in your last number as it was from me that you received the ampthia specimen of the Centunculus.'[14] Abbot and his wife were keen collectors, they had a herbarium and in 1798 he published his *Flora Bedfordiensis* with coloured plates provided by Sowerby. Money was a problem and he was forced to sell his three volumes of *English Botany* to a clergyman, and to ask Sowerby to tell him with candour how long he could wait for the payment of his account. He was to remain active in natural history to the end of his life, newly identifying and distinguishing some plant species, developing a fine collection of insects, and often sharing his views and discoveries with Sowerby. Abbot was just one of many contributors, each with their own set of interests and problems.

In the small space given over to the text Smith did not always confine himself to taxonomic matters; over the years he wrote about beauty, of medicinal aspects or about occasional uses. He informs us and admonishes others over the aesthetics of alpine cotton-grass (*Eriophorum alpinum)* with respect to the hairs of the seeds:

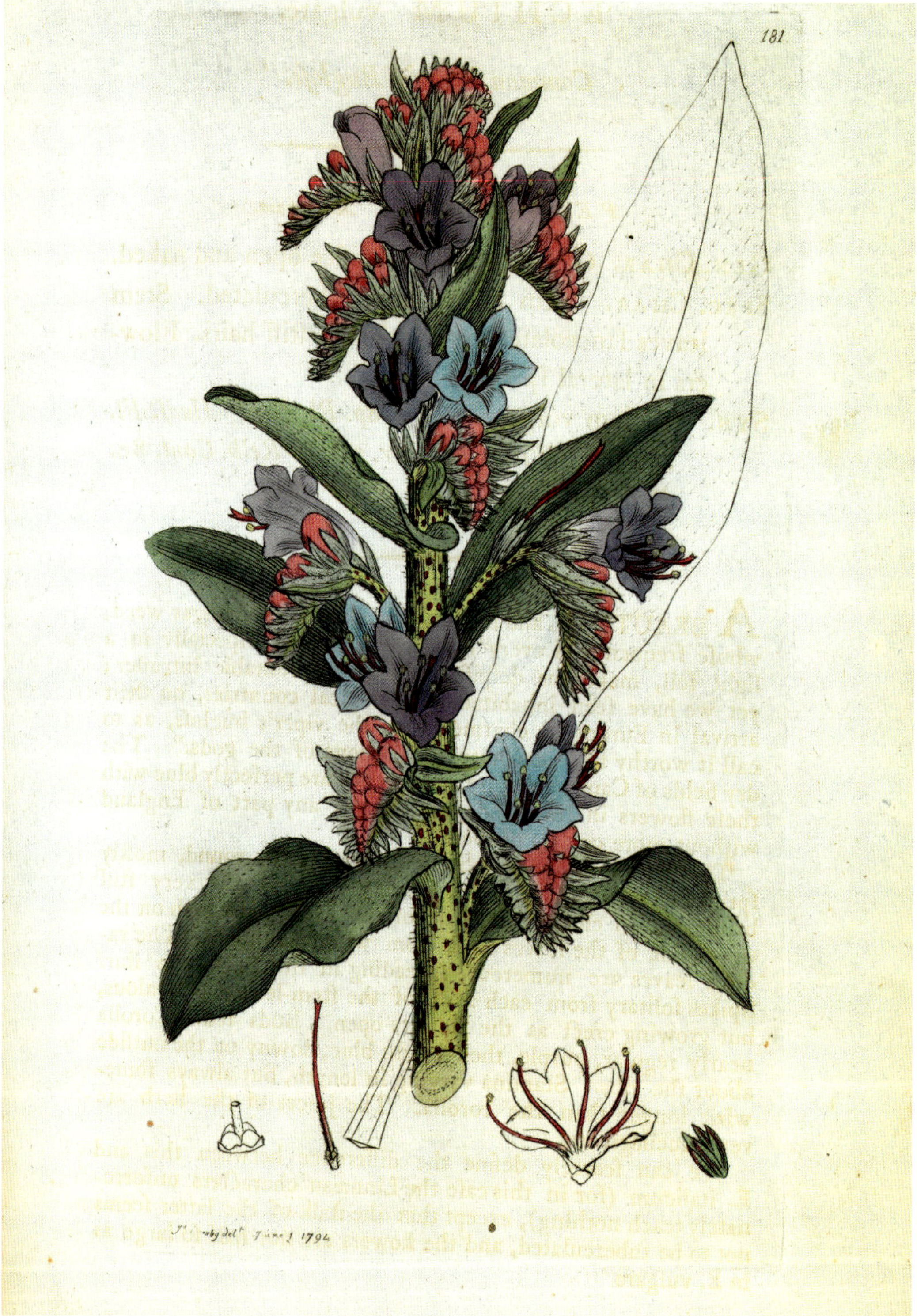
181
June 1 1794

5/8 Viper's bugloss (*Echium vulgare*) described by James Smith as 'a beautiful and magnificent, though very vulgar weed'. Also that some 'inhabitants of tropical countries... call it worthy to "decorate the gardens of the gods"'. Such a plate would have meant a lot of work for Sowerby's colourists.

5/9 Dawson Turner. Engraving by Miss D. Turner after a drawing by J. S. Cotman, 1810.

> Their extreme whiteness and their delicate texture give an elegant appearance to the plant, though it boasts none of those gaudy colours requisite to excite the attention of many, who really fancy themselves persons of taste and admirers of Nature, but who are as inadequate judges of their own talents, as of the universal elegance and wisdom of the creation around them.[15]

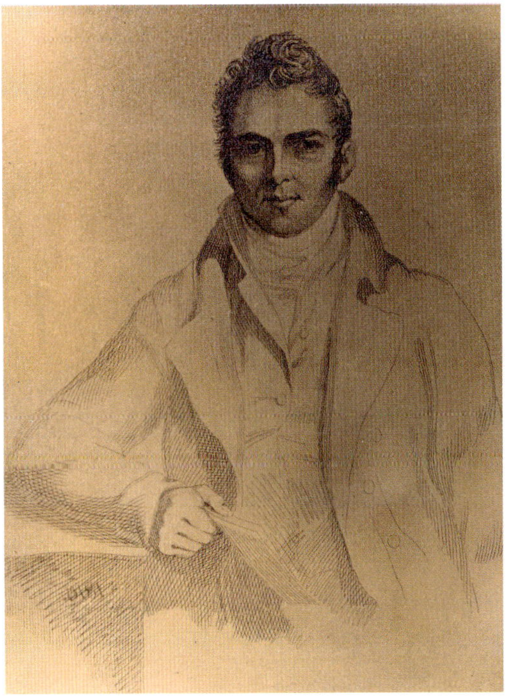

Biting stone-crop can be used in 'scorbutic and scrophulous disorders',[16] nipplewort in curing sore breasts, while for common henbane 'the seeds and capsules together, smoked as tobacco, are in some places a remedy for the toothache, but convulsions and temporary insanity sometimes follow its use'.[17] The unusual moss balls of the *Aegagropila* alga can be used 'to wipe pens upon',[18] while several lichens may be used in dyeing.[19] These are all typical examples.

Occasionally new species were described, new genera announced,[20] and Linnaeus was sometimes criticised or praised,[21] as were others, including Curtis, particularly about the identification of species. Smith took the opportunity to honour some of his botanical friends by naming a species after them, such as in *Lichen turneri* for Dawson Turner.[22] He included the botanical observations and collections made by Sowerby[23] and confessed his own few errors.[24]

Dawson Turner would no doubt have also been very pleased to see the introduction of cryptogams into issues of *English Botany* from 1792 onwards, for these were his current passionate interest. Turner was a banker in Yarmouth, wealthy from inheritance on the death of his father in 1794, and following in his father's footsteps by working in the Gurney and Turner's Bank. He had an early interest in mosses, lichens and algae and was later to publish numerous papers on them, as well as his later significant works on mosses (1804) and seaweeds (1808–1819). He had a full correspondence with Sowerby, sending him specimens and advice, sometimes

5/10 The original watercolour and the published plate (opposite) of the seaweed *Fucus spiralis*. This was published in *English Botany* on the strong encouragement of Dawson Turner. The location of the specimen and the name of the donor are written on the original, in James Smith's hand.

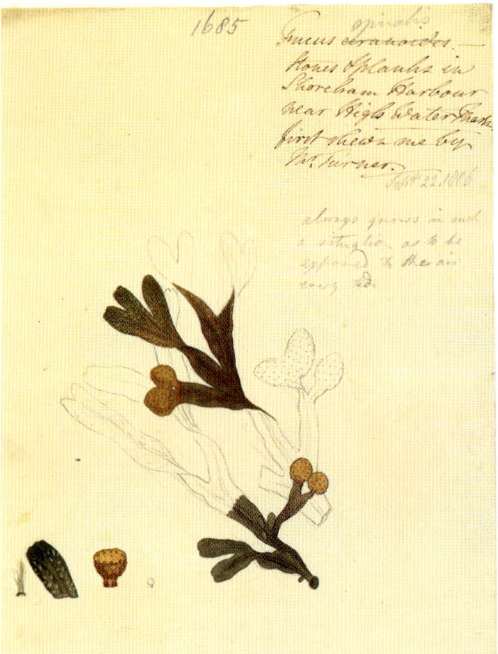

critical. Two figures in the latest number of *English Botany* were, he opined, 'a complete disgrace to the work – the woolly root being omitted' and suggesting that subscribers should have new drawings for which he would send proper specimens.[25] But more often he praised Sowerby on the drawings.

The Reverend Hugh Davies was another who was pleased about cryptogams, especially as the plates contained enlarged details of the plant's components, and he contributed much, often directly to Sowerby. He was rector of Aber, in Caernarvonshire, Wales, and had travelled with Thomas Pennant to the Isle of Man as well as helping several authors with their natural history works. Correspondence with Sowerby started in 1794 and would continue over many years; Davies was soon to become an active adviser on Sowerby's desires to amass a significant collection of birds, minerals, shells and other animals. But for now he was important because he was sending plants from Wales and with evidence of its somewhat distinct flora.

Turner and Davies were just two of the many who now corresponded with Sowerby. He had become a kind of clearing house, helping people with their identifications, sending and receiving specimens, even doing favours for people such as paying the London bills of his country correspondents.[26] He would do these things because he was asking favours in return and not just about plants but also of fossils and minerals. People were anxious to help even though they didn't purchase *English Botany* for themselves. Mr Fly, gardener to Lord Rolle of Bicton, Devon, wrote that he was sorry he could not always go out to search for specimens 'but you must consider that as a servant I am not at liberty to do as I please'.[27] Mr Salt, a cutler from Sheffield, sent Sowerby material including rocks and iron ore and thanked him for 'all his favours when he was in London' and a little later for offering to propose him for membership of the Linnean Society.[28] The Welsh naturalist Thomas Pennant wanted help with his identifications of fungi, having collected

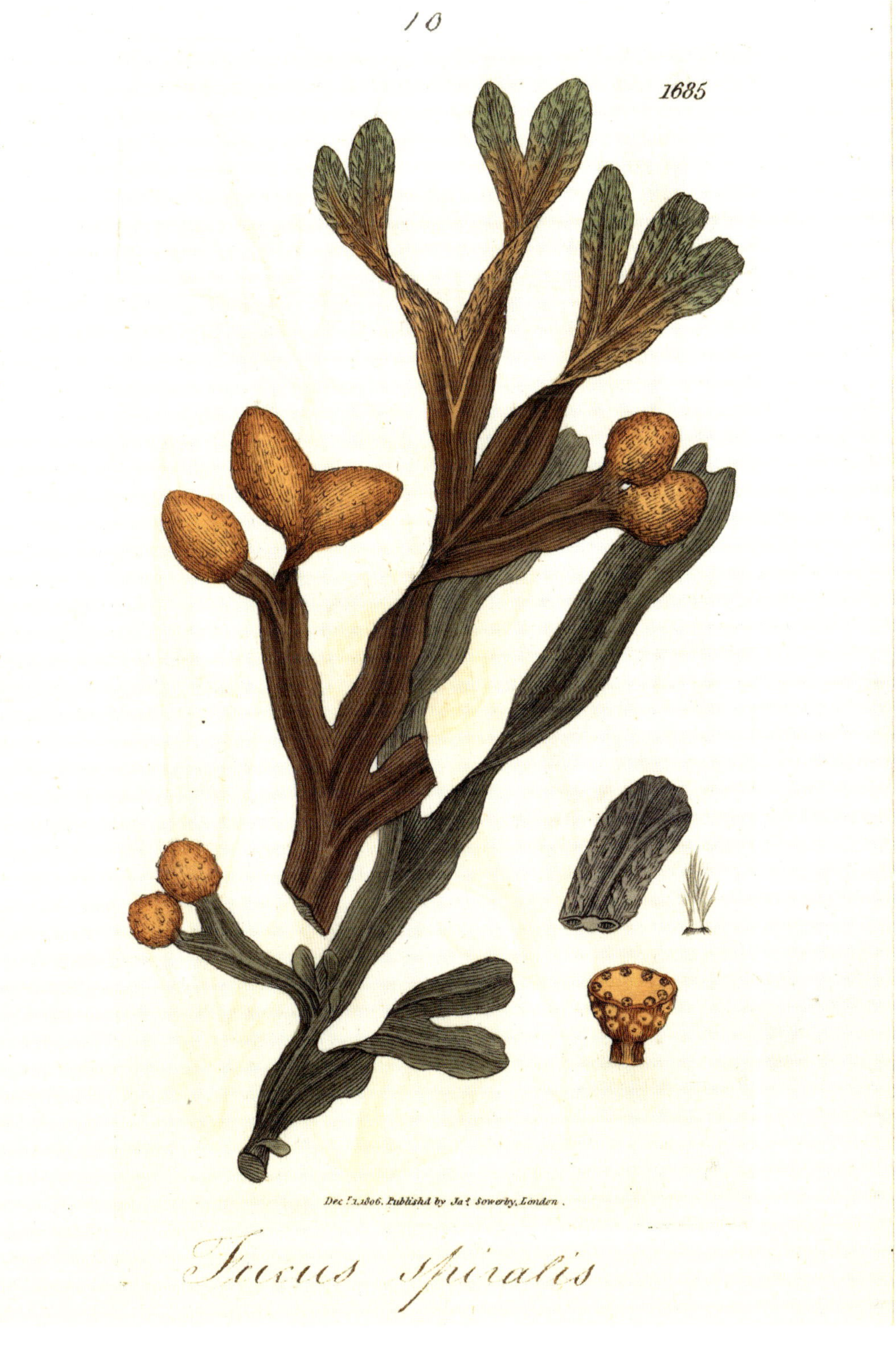
10
1685
Dec 1.1806. Publish'd by Jas Sowerby, London.
Fucus spiralis

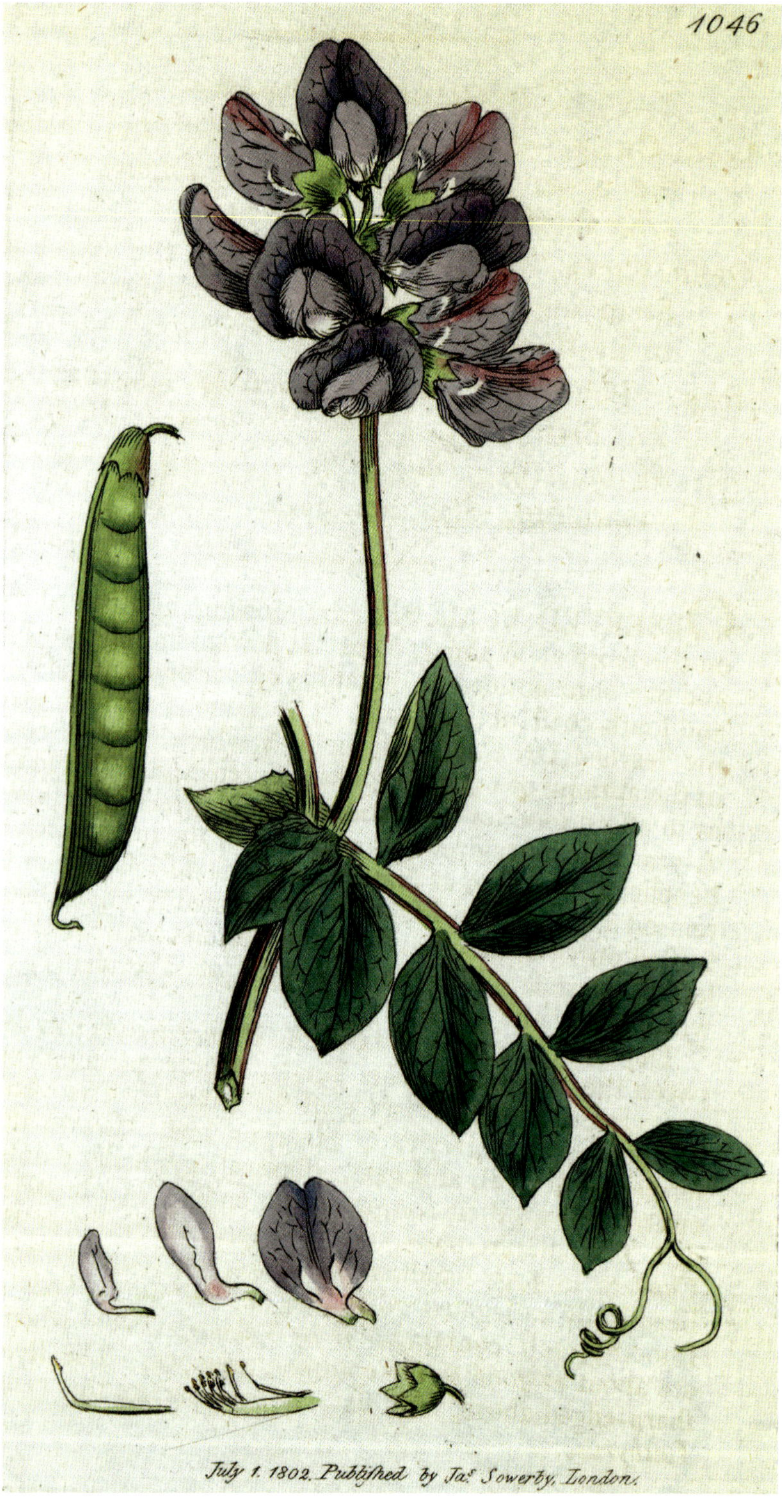
1046
July 1. 1802. Published by Jas. Sowerby, London.

5/11 The sea pea (*Pisum maritimum)* collected by Sowerby on the east Kent coast. Smith reminds the reader of how this plant is said to have saved the lives of people on the Suffolk coast during a famine in 1555.

many in the prolific season of 1797.[29] Lists of desiderata were circulated with Sowerby as a focus.

Mead Place was the epicentre of activity for the production of *English Botany.* Here Sowerby drew the plants using, where possible, freshly collected specimens, preferably in flower and showing the different parts including the root. He applied watercolours to indicate the appearance of the living plant with its varying tones and textures before sending the results to Smith for critical appraisal, suggestions and the writing of the accompanying text. On their return, Sowerby, probably with the original specimen in sight, carefully engraved (sometimes etched) the image onto copper plates, which were then sent to the printer. Several hundred copies of each would be printed most of which were then given, with an accompanying reference image or drawing, to colourists, some of whom would occasionally work at Mead Place, while a stock of uncoloured plates would be retained for possible later orders. Mr John Adams was, as a regular colourist possibly with his own assistants, charging one pence and a farthing (1.25 old pence) per illustration. Each month he coloured several hundred; a bill for September 1798 shows he executed 701 pages for each of the plates 499 to 510 of *English Botany*, for which he charged £44.3s.3d. At the same time he was colouring Woodville's *Medical Botany*, published by Sowerby and still selling well. Allowing for the costs of paper, printing, wrappers and original copper plates, as well as James Smith's fee, the income to Sowerby would be reasonable; if a large number were sold then his profit would be good, and this seems to have been the case.

Sowerby worked fast and usually achieved the desired result after just one attempt, although he would make several sketches to depict the detail of the plant's features. In the winter he worked by candlelight or possibly by the flames of the open fire. The colourists, whether family members or employees, would also have needed light and space. They would often apply just one colour to several copies which were then allowed to dry before the colourist moved on to the next tint. They needed to work quickly and effectively since they were paid by results. But in the early years Sowerby did not have the other family members to help him – Anne was not an artist and she was busy with their young offspring. But as the children grew they were trained in the several parts of the enterprise and acquired their own skills, making their contributions in conveniently different ways. James De Carle especially was, at an early age, to become a fine illustrator and an important contributor to the publications.

During these early years of *English Botany* Sowerby was a very busy and productive man. His workload was made no easier by Smith's departure from London to his new home in Norwich. Smith had married Pleasance Reeve, daughter of an attorney and merchant of

Lowestoft, Norfolk, in March 1796 but living in London was not agreeable to them and they moved to a house owned by Pleasance's father later the same year.[30] Preparing *English Botany* now had to be done by correspondence and the sending of the 'monthly' parcels and money.

The payment was often a point of concern. Smith would ask for some cash, claiming that he was in need of money that month. The one, five or ten pound note would be divided, each half being sent at different times, but if half was stolen they were in the same position as the thief unless they had special patronage at the banks. Occasionally Smith would be in more urgent need. In one letter he thanks Sowerby for a good supply of drawings before continuing, 'if you have a superfluous 10£ note I should be glad of it in the next parcel – it is hardly necessary to divide it – I will run the risqué – Only mark the N° and sum in y^r letter & let the parcel be safely booked'.[31] (This Sowerby did, sending Bank of England note number 512 on Saturday 18 April – it was safely received.) Visits by Smith to London became few, even though his beloved Linnean Society wished his presence to be much greater. This made Sowerby's skills and presence in London valuable to Smith in new ways, especially in his aiding the running of the Linnean Society.

English Botany was not, of course, without its critics. They would raise complaints either directly to Smith or Sowerby, or by letters to the *Gentleman's Magazine*. Some wrote about the rendition of a plant, others about the coverage of species or the omission of their favourite flower. One anonymous writer complained about the work being 'spun out to such a length and at so great an expense by the insertion chiefly of cryptogamous plants'.[32] Was this one of the disgusted Ladies that Woodward warned about? But we must anticipate a more serious complaint, which came later in the spring of 1806 from an awkward source – a friend of James Smith.

Richard Salisbury, a Vice-President of the Linnean Society, had been a student with Smith in Edinburgh and had worked with Sowerby on his own *Images of Rare Plants*,[33] but was now having a public spat with Smith about the text in *English Botany*. This, he claimed in front of other members of the Society, was plagiarised, frequently verbatim, from Linnaeus's *Genera Plantarum*, and he was further enraged by Smith's rebuttal. Unable to get his way verbally at the Society and finding that Smith would not make any concessions,[34] he went to the trouble and expense of publishing a pamphlet entitled *The Generic Characters in The English Botany, collated with those of Linné*.[35] In this he gave vent to his views, yet at the same time proclaiming that he 'did not wish to obtain a victory over him [Smith] on any subject about which they differed, much less so paltry a one as this'. The pamphlet attempted to give evidence of the plagiarism and took the opportunity to add further criticism of

5/12 The liverwort *Marchantia androgyna* featured towards the end of *English Botany*. This was first introduced to British botanists by James Dickson – see plate 2/5a for the specimen in his *Hortus Siccus Britannicus*.

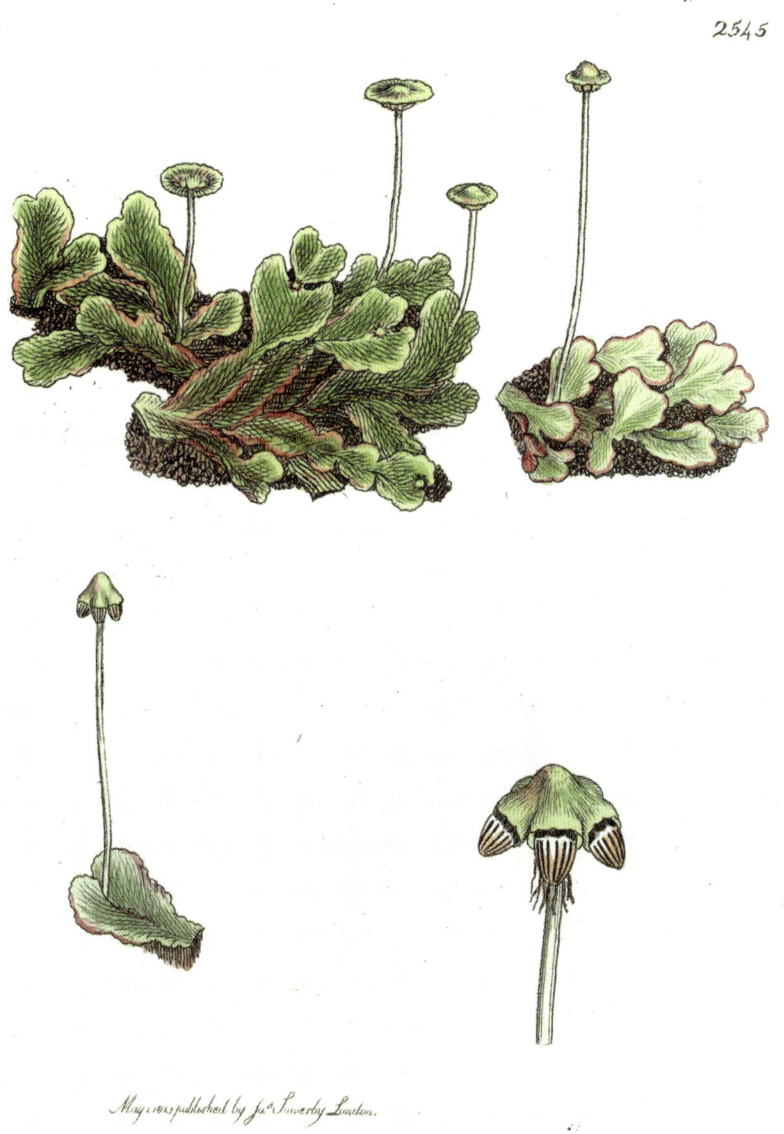

5/13 'Lords and ladies', *Arum maculatum*. Sowerby's clear but detailed plate in *English Botany*.

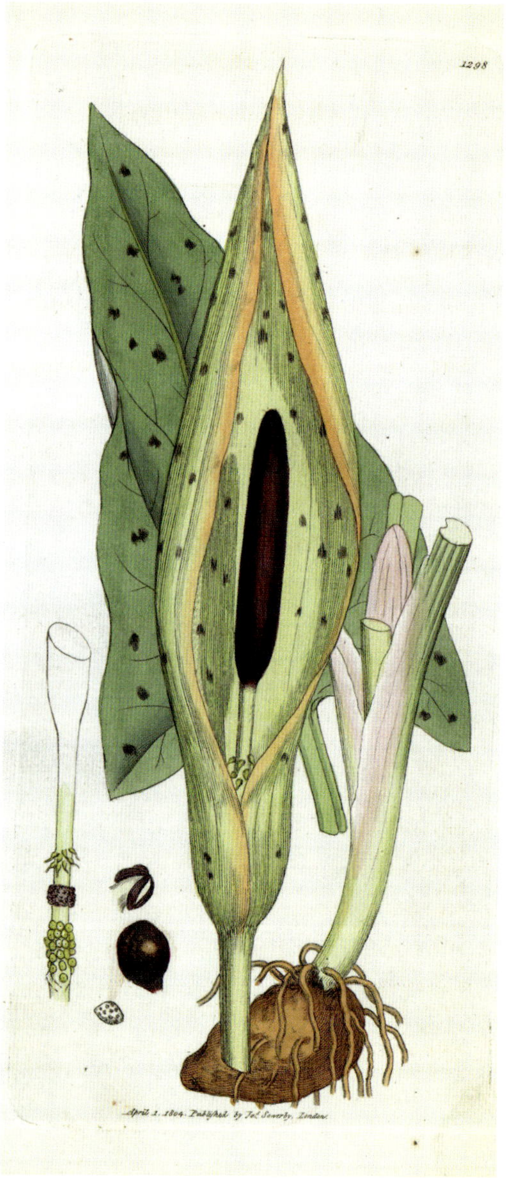

Smith such as 'the author seldom in any of his publications makes a remark on natural affinities without being in an error', while at the same time declaring his friendship.[36] Underlying the criticism was also the fact that Salisbury believed more in the efficacy of the 'Natural System' of classification by Antoine de Jussieu.[37] This was being adopted over Linnaeus's sexual system in continental Europe but Smith paid rather too little attention to it, being wedded to everything Linnaean.

Smith was badly stung by the attack even though he felt completely free of the accusation. Inevitably there would be many similarities in descriptive texts about the generic characters of a plant but Smith wrote much that was also fresh. None of this directly affected Sowerby. There was no criticism of the illustrations but he was upset by the episode, dealing, as he had to, with Smith's reaction, and from that point on he became more sensitive to any possible criticism from that quarter. Sir Joseph Banks attempted to reconcile the two men but to no avail. Sowerby seems to have been silent on the whole matter but we know from some of Smith's letters that he was despondent. Indeed, this had not been the first attack by Salisbury: one a few months earlier had been spitefully critical of descriptions and illustrations in another of Sowerby's and Smith's publications, *Exotic Botany* – to which we will come later.

One who was not critical of the content but offended by the mere existence of the publication was William Curtis. He saw the work as being in direct competition with *Flora Londinensis* and his declared aim that once his efforts on London were finished, he would turn to the whole of Britain. He spurned Smith and avoided the Linnean Society, but cast little blame towards Sowerby. He also failed to see the benefits that *English Botany* quite often brought to his own works, and he reacted by publishing *Flora Londinensis* in an additional smaller and cheaper format so as to be more competitive, but this did not last long. Smith was surprised that Curtis should have seen their undertaking as being hostile to his, but he surely realised it was truly a competitor in a market where many potential purchasers had limited funds. Relationships were never repaired. Smith retained his rather hard attitude even to the extent of, upon hearing of Curtis's failing health, suggesting to Sowerby they should produce a work on garden plants to replace the *Botanical Magazine* on its expected cessation, an idea that Sowerby did not like at all. And then when Curtis died in 1799 Smith wrote, 'poor Curtis is a loss to the science, though not to us. Tis pity he was so crabbed.'[38] Both *Flora Londinensis* and the *Botanical Magazine* continued after his death, each with its own history.

Sowerby's own family was growing and had to be encouraged in the realms of natural history. James De Carle was only about ten or eleven when Sowerby started taking him to Sir Joseph Banks's breakfasts at Soho Square, where friends would gather to experience the talk of natural history and the examination of specimens. Three more children had been born by this time: two boys, John Edward in 1793 and Charles Edward in 1795; and their first girl, Anne in 1797. With six young children, a busy perhaps somewhat obsessive husband, many visitors and a pile of correspondence to assist with, his wife Anne would have had a rather hard life. The little evidence available suggests she was a dutiful and happy spouse but her health was quite frequently poor. Soon she was to be much more than a housewife and mother; she became a key member of the Sowerby team and its outputs.

Sowerby had proved to himself and others that he could be independent in many ways. *English Botany* had been published for more than eight years and was going stronger than ever, with around 900 copies being sold monthly. Its future life seemed reasonably assured, despite the difficult times of the age with its increasing inflation and a weak economy. It had succeeded in ways where Curtis's *Flora Londinensis* had not. The two Jameses worked well together, even with their disagreements, but it was also clear that Sowerby desired greater degrees of independence. This matched his broadening interests in natural history, not just to include the fungi, but also animals, fossils, minerals and rocks. These interests were to take him beyond the limits of artistic illustration into travel and the building of significant and accessible collections.

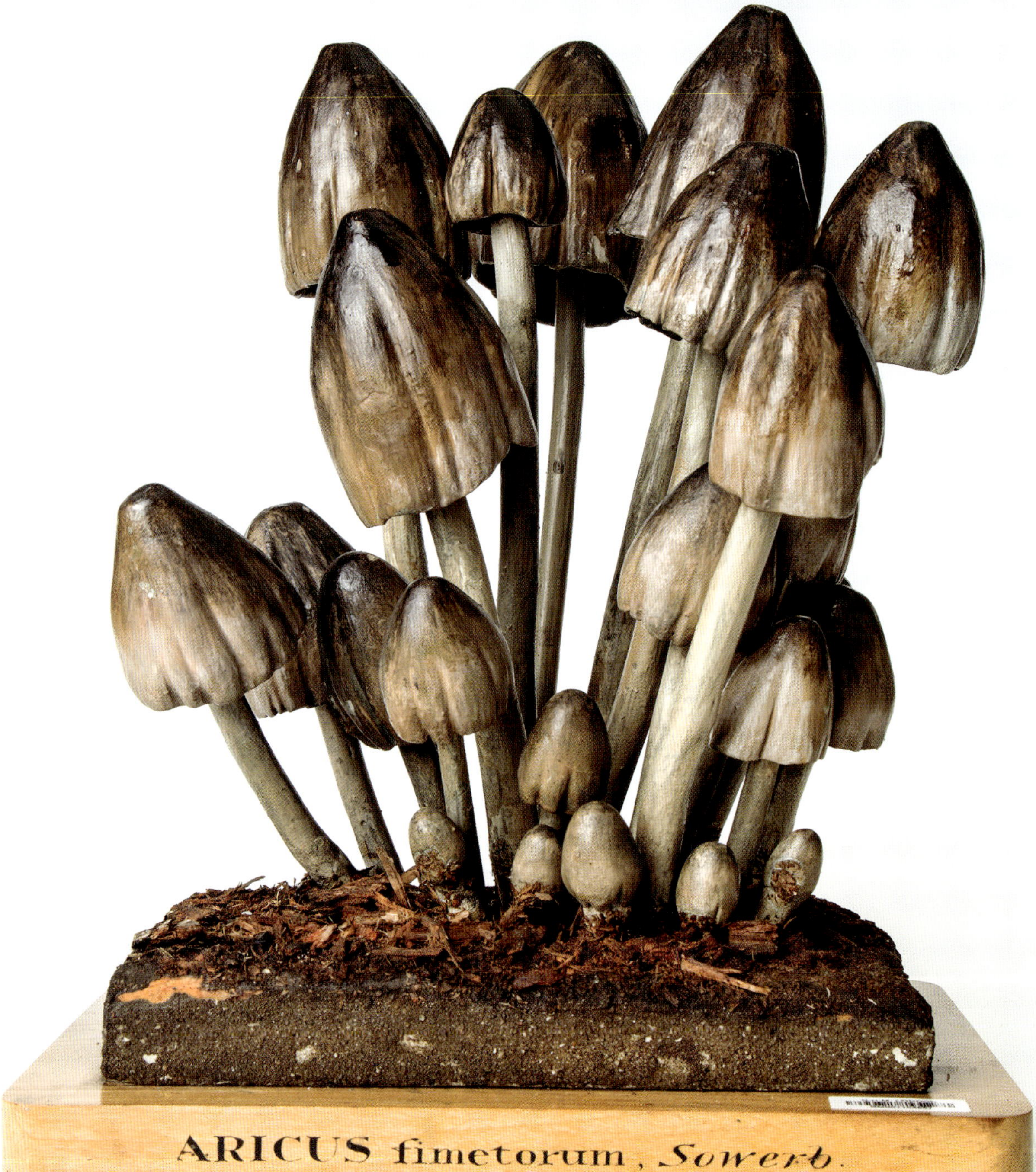
ARICUS fimetorum, Sowerb.

CHAPTER SIX

Models and museum

6/1 Model of the ink cap fungus (given as *Agaricus fimetorum* now *Coprinopsis atramentaria*).

Fungi do not feature in *English Botany*. James Smith had neither the same interest nor knowledge about them as he did of other parts of the botanical world, but then many botanists did not understand fungi. Spores had been identified only around 1730 and some of the first steps to a reasonable classification were made in 1801 by Christian Persoon.[1] The circle of fellow botanists of which Smith and Sowerby were part contained only a few who would proclaim themselves to be really competent mycologists. Some, such as Sowerby's friend Thomas Woodward, published papers on fungi but usually within a wider botanical interest, while the intriguing Mr Benjamin Forster, who was a man of the Enlightenment by being an inventor and scientist with wide interests including mycology, did not publish about them until much later.[2] There were also the problems of transport and of preservation. Many fungi are difficult to preserve, at least in a condition that is truly satisfactory for taxonomic study, and some modern methods including freezing were, of course, not available. William Withering had proposed methods for storing them in solutions with 'rectified spirit of wine', one involving lead acetate, another copper sulphate, but they were not generally adopted or satisfactory.[3] Drying was the best method for several species.

Sowerby, however, was interested – rather passionately. He not only collected, exchanged and studied specimens; he now undertook a

6/2 Sowerby's model of the dog stinkhorn fungus (*Phallus caninus* now *Mutinus caninus*). One of the few remaining models of fungi Sowerby made for his museum. The height of the model is about twelve centimetres.

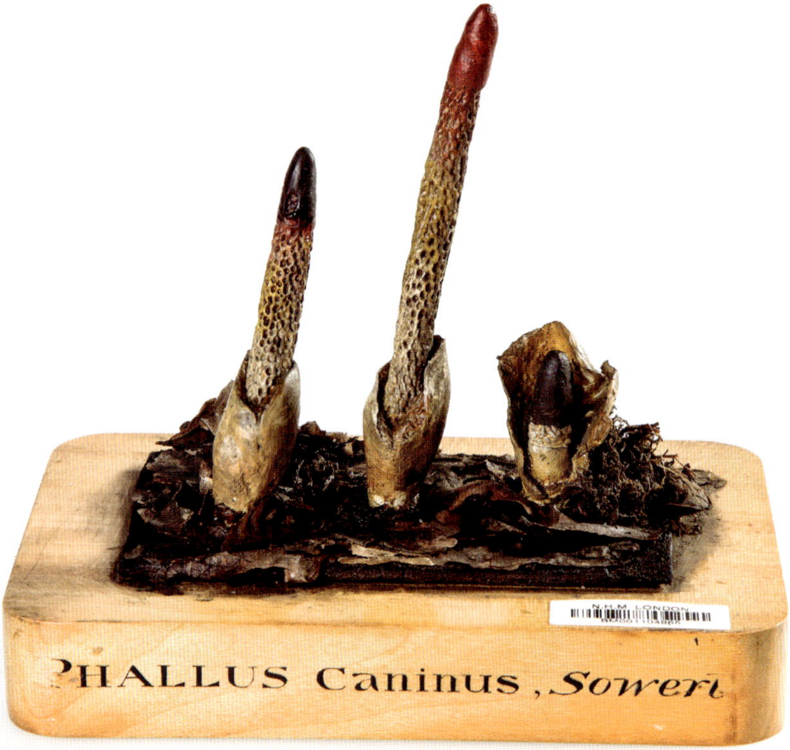

major project to make life-size models of many types, a laborious and lengthy task that had not been done to the same extent before. The models are not just simple ones of well-known and relatively robust species; they are of a variety of forms, sometimes in groups, placed within or on their growing medium whether soil, dried leaf litter or tree bark, and showing the detail of their structure such as the gills, pores and ring. Some even show the mushroom in the process of bursting from its universal veil. He laboured for almost twenty years (1796–1815) to produce nearly 200 models. It was an extraordinary undertaking. For the most part he applied his same aesthetic skills to their design as he did for his drawings, choosing to make them from several materials including unbaked pipeclay, judging this would allow for the necessary detail, including characteristics such as scales on a cap or a fragile ring of tissue on a stem. Pains were taken over the colouring to ensure a lifelike appearance, with appropriate nuances such as a glistening cap or shades of tone. Fly agarics, delicate mycena on a tree trunk, morels, boletus, shaggy ink cap, giant puffball, truffle, dog stinkhorn and cup fungi were some of the many which emerged from his modeller's hands. Mead Place was indeed a diverse workshop.

His aim was to show the ranges of form and colour and even something of the habitats of different fungi, presented in a way independent of the relatively short season during which many fungi flourish.[4] This was a well-intentioned project since identifications were a problem for many people, who sought Sowerby's help and instruction. Even the naturalist Thomas Pennant remarked on the abundance of fungi during the autumn of 1797 and the difficulty he was having with naming them,[5] and the physician and botanist John Hull asked Sowerby to check the descriptions of the agarics for his forthcoming *Pocket Flora*.[6] People were well aware of Sowerby's expertise as he had been collecting and studying specimens for at least ten years,[7] and was always active in seeking and sharing information.

But the somewhat limited usefulness of the models must have been known to Sowerby as he worked on the various types. Although they show the size, shape and some of the detail, they naturally cannot provide important information such as the colour or form of the spores, the smell and taste, or even the more detailed nature of the gills, and they reveal little about the wider aspects of the associated habitats. So what exactly was in Sowerby's mind when he started making these models and then devoting so much energy and time to them? How did he envisage them being displayed and used? They were to require a lot of space, they were too fragile for frequent or rough travel and there was no obvious home for them besides Mead Place. He clearly wanted them to be seen and to be instructional. His vision was the establishment his own museum, which would contain not just models of fungi but plants, birds, mammals, minerals, meteorites, fossils and paintings. Another major personal project was born.

A large room was built at the back of 2 Mead Place. His plan was to have proper display cabinets, cupboards, sets of specimen drawers and a fireplace. By 1796 he was starting to fill it with specimens and objects. His infectious enthusiasm, his network of potential donors, his willingness to ask, or even cajole, his naivety and his intention to further the cause of natural history all helped him to gain material of good quality. Lord Seaforth donated bird specimens,[8] James Brodie (MP for Forres in north-east Scotland) sent preserved animals including an

6/3 Sowerby's model of a small fungus which typically grows on bark (given as *Agaricus corticalis* now *Mycena corticola*).

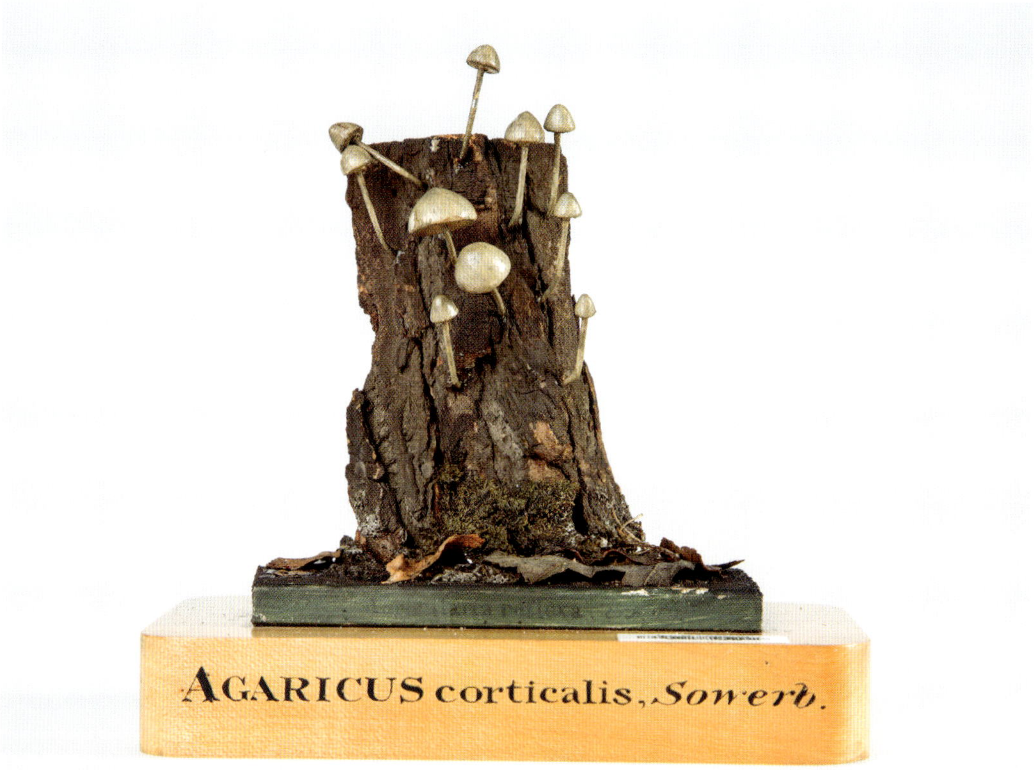

Aberdeen snake, with some, appropriately from a Scotsman, in a whisky-rich preservative,[9] and the naturalist George Montagu (often called the 'father of British Ornithology') sent fine shells from Devon.[10] Sowerby bought specimens, especially minerals. A certain collector, Mr Day, sold his minerals to him in 1800 for £90, and his mineralogical books a few months later.[11] Sowerby acquired fishes, insects, fossils, rocks and meteorites. Anything that was in the world of 'natural history' could feature in his museum.

A carpenter was employed to make show cases, cabinets, drawers and cupboards. This was a big project that occupied Sowerby until his death. Some twenty years after the start he was still making improvements by placing the British specimens in existing cases and non-British ones in new cases constructed by his carpenter

6/4 Model of the common morel (*Helvella esculenta* now *Morchella esculenta*) which is edible and much sought after.

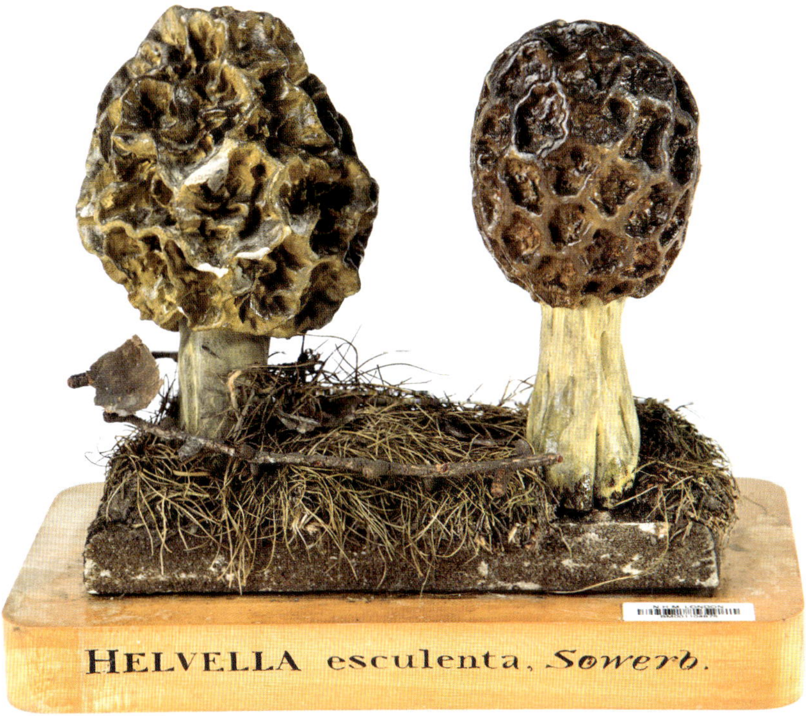

over six months.[12] People were invited to visit, others asked to come or introduce their friends, and some came to examine specimens and discuss their findings. In the typical spirit of free communication that was a hallmark of Sowerby's work and interests, the museum was open without charge to the public every first and third Tuesday of the month from 11am to 3pm. In the cold winter months a fire warmed the space.

People were ready with advice, sometimes suggesting that the whole idea of a museum was a thoroughly bad one. Once Sowerby had decided on something he was rarely deterred and the process of acquisition could turn, from our vantage point, into an amusing struggle to find specimens which might not even exist, or if they did, could only be obtained with difficulty on the part of his friends and acquaintances. If Sowerby wanted a dead small goat with fine horns then he would get one, or so he might think. And perhaps the generous botanist and clergyman Hugh Davies, in Anglesey, Wales, would help. Davies had already helped authors over their botanical problems, he had travelled with Pennant on his tour of the Isle of Man and he had provided specimens for *English Botany*.

The saga of the goat started in the summer of 1799 and is best told through the correspondence, albeit one-sided since we have only the letters from Davies to Sowerby. Davies had warned Sowerby about developing a museum soon after its start:

> I admire your zeal for Natural History; it is a delightful study but I must as a friend be excused if I offer you some advice concerning your design of making a collection of the natural production of Great Britain. I have myself collected, & assisted others for years. Quadrapeds & birds are not easily obtained; the preservation of them is attended with very great & constant trouble and expence; and when obtained of what use, equivalent to the expence and trouble they be to you, who reside so near the first museum in the world; Fish are likewise very troublesome of conveyance, and very expensive on account of the spirits and glass jars, I would advise you to decline them, unless a new subject should perchance to offer itself; a general collection of minerals is almost endless, difficult to be procured and troublesome to be conveyed.[13]

He was soon surprised to discover that Sowerby had quickly acquired a 'very considerable collection of Birds',[14] so perhaps the request for a goat skin and the rejection of his advice was no great shock. Nevertheless, he tried to dissuade Sowerby:

> In my opinion there is an insuperable objection to your admitting the Goat into your collection, its intolerably rank smell, which it is impossible by any means to get rid of; if your room was ten times the extent it is, it would be rendered absolutely insupportable by it.[15]

He was, however, happy to send some of the birds Sowerby has asked for.

Just over two months later, despite his advice, he was not only willing to help Sowerby but became fully committed to obtaining the best:

> As you seem determined to admit the Goat into your collection, you will not be disappointed; you say you would have the smallest; – do you mean a young one, or a female with moderate horns, or a veteran (which must be large) with such horns as would do a London Alderman credit? Determine this point, and let me know as soon as convenient, and I will endeavour to

> suit you: will it be necessary to use any salt to the skin in packing it? – will it heat in being packed up in a box so as to cast the hair, shall it be sent by waggon or by coach, if coach will take it; be pleased to answer these queries.[16]

Another month then passed and he must have been pestered by Sowerby for he wrote that 'as soon as I can procure a Goat at all to my mind it shall follow; a small veteran with fine horns is not to be gotten I fear'.[17] But another three months yielded no success and he had to explain again to Sowerby that his ideas were somewhat naïve:

> I trust you will believe me when I say that I have taken some pains to procure a goat skin for your collection, I have examined several herds myself: – I have spoken to the most extensively dealing Butcher in this neighborhood, describing the subject we wish for; his account agrees with my own observations that they answer the ends of the few persons who rear them at about two years old, and they very rarely keep them longer, and then their horns are not much longer than a man's hand, they are no object at all; as to their idea of their running wild among our mountains it is quite without foundation.[18]

It is unlikely that Sowerby ever got his goat or goat skin, or certainly not the one he had in his mind's eye. Davies's patience must have been tested but he remained a loyal and generous correspondent. He was not the only one. A Mr James Murray, a gardener at Richmond Park House to the west of London, offered a deer skin and asked whether a 'Nied Fawn' or a 'Mannil coloured Fawn' would be preferred, and of what age, and whether or not it should be alive, as though the museum might almost be a zoo.[19]

Private museums were not uncommon in London. Some became famous and long-lasting such as that belonging to the physician William Hunter (1718–1783). This was in Great Windmill Street and contained fine collections of coins, minerals, shells and insects, as well as anatomical preparations. It later, early in the nineteenth century, formed the basis of the Hunterian Museum in Glasgow. Others satisfied a more temporary need such as Rackstrow's Museum in Fleet Street with its grisly and titillating displays of human body parts, stillborn foetuses, animals with two heads and waxwork models.[20] But none were of a similar style and purpose as Sowerby's. His was to develop into one of considerable interest with a large range of specimens, some of much scientific importance, over many branches of natural history. Here was to be found his herbarium, numerous drawers of insects, several hundred minerals including native gold, diamonds and many meteorites, animals in spirit as well as dried fish, a chameleon, an otter, a seal, snakes and amphibians. The bird collection

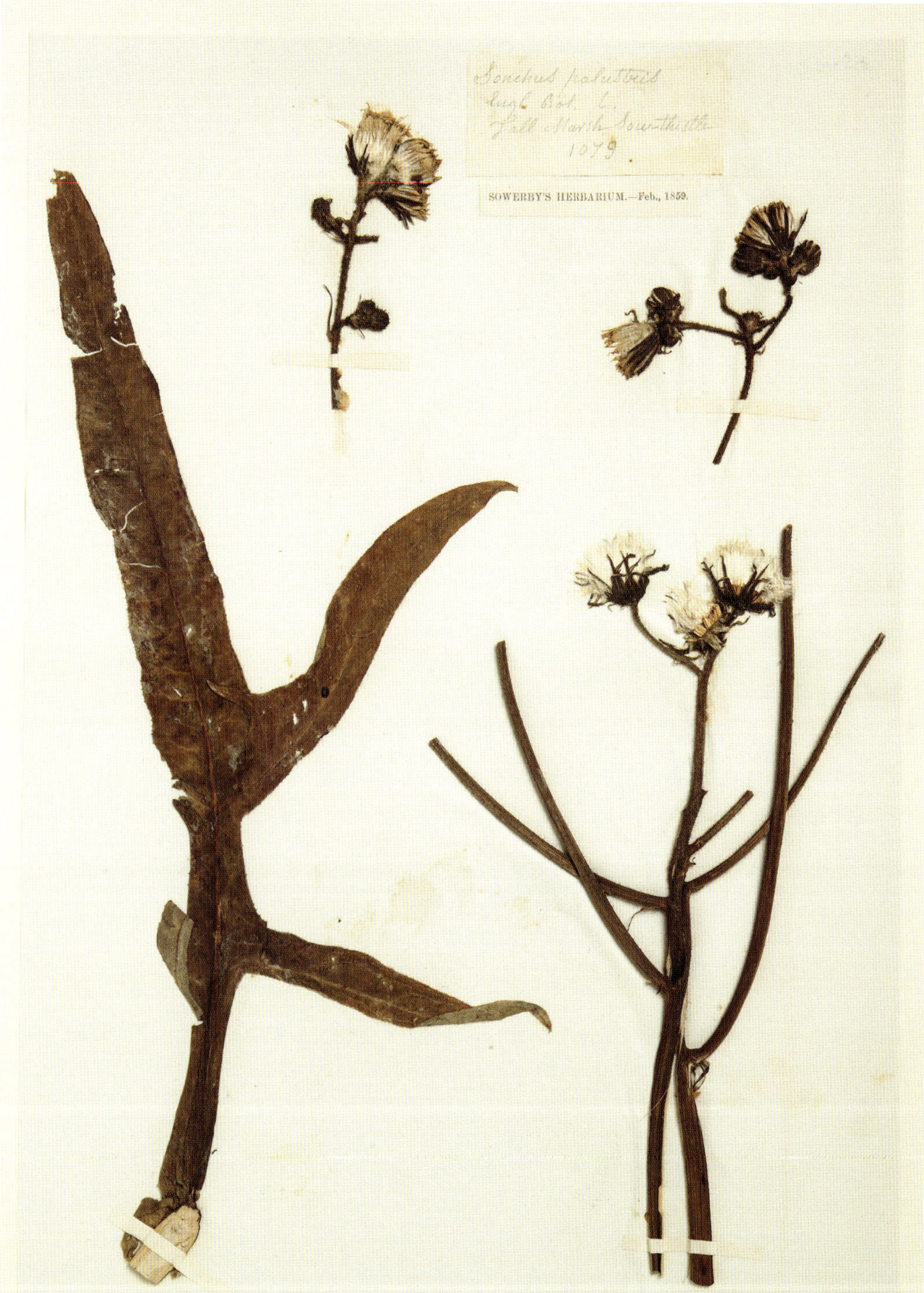
Sonchus palustris
Engl. Bot.
Tall Marsh Sow-thistle
1079
SOWERBY'S HERBARIUM.—Feb., 1859.

6/5 Herbarium sheet of the marsh sow-thistle *Sonchus palustris* from Sowerby's herbarium.

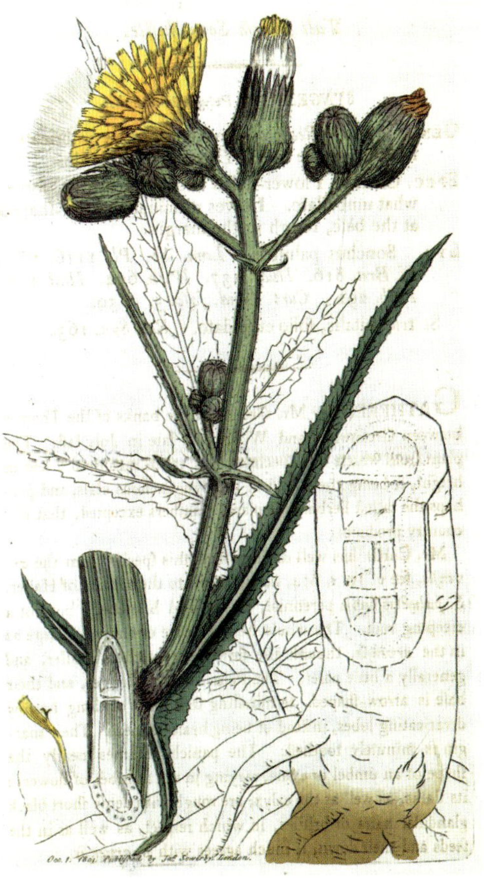

6/6 *Sonchus palustris*; the plate in Sowerby's *English Botany* (number 935).

was also extensive with some fine specimens including a golden plover, a red phalarope, an Arctic puffin and an osprey, as well as eagles and owls. It was visited and used by people of all walks of life.

A significant reduction in illustration work for other authors' publications released time for new activities for the museum and the models, but *English Botany* was keeping him as busy as ever it had done, and with some extra work occasioned by James Smith's move from London after his marriage back to Norwich. He was undoubtedly also teaching drawing and fulfilling commissions for botanical plates. Some publications were only just behind him. As early as 1790 he had illustrated a work that dealt with zoology rather than botany, *Speculum Linnaeanum* or *Linnean Zoology* by George Shaw, assistant keeper of the natural history section of the British Museum. Shaw and James Smith also commissioned him to prepare the plates for their separate but related works portraying some of the natural history of Australia (or New Holland as it was named then), which were published in 1793 and 1794. Each was a one-volume work with several illustrations: Shaw's dealt with such creatures as opossums, birds, a fish, a tortoise, a snake and a lobster; and Smith's with unusual trees, shrubs and plants, making it, in effect, one of the first floras of Australia.

One purpose of these works was to give the British public, or rather those who cared to purchase the books, a little insight into the natural history of a country that was becoming of greater interest and which was undoubtedly home to some strikingly different animals and plants. Joseph Banks had brought back much plant material from Botany Bay during his and Daniel Solander's time on James Cook's first

6/7 'The nonpareil parrot': Sowerby's illustration in Shaw's *Zoology of New Holland*. Shaw wrote that as far as the details were concerned: 'These are so accurately expressed on the plate, as well as mentioned in the specific character, that it becomes unnecessary to particularise them here.'

6/8 A long-tailed opossum illustrated in Shaw's *Zoology of New Holland*. It is not clear whether Sowerby's engraving was based on a specimen or someone else's drawing.

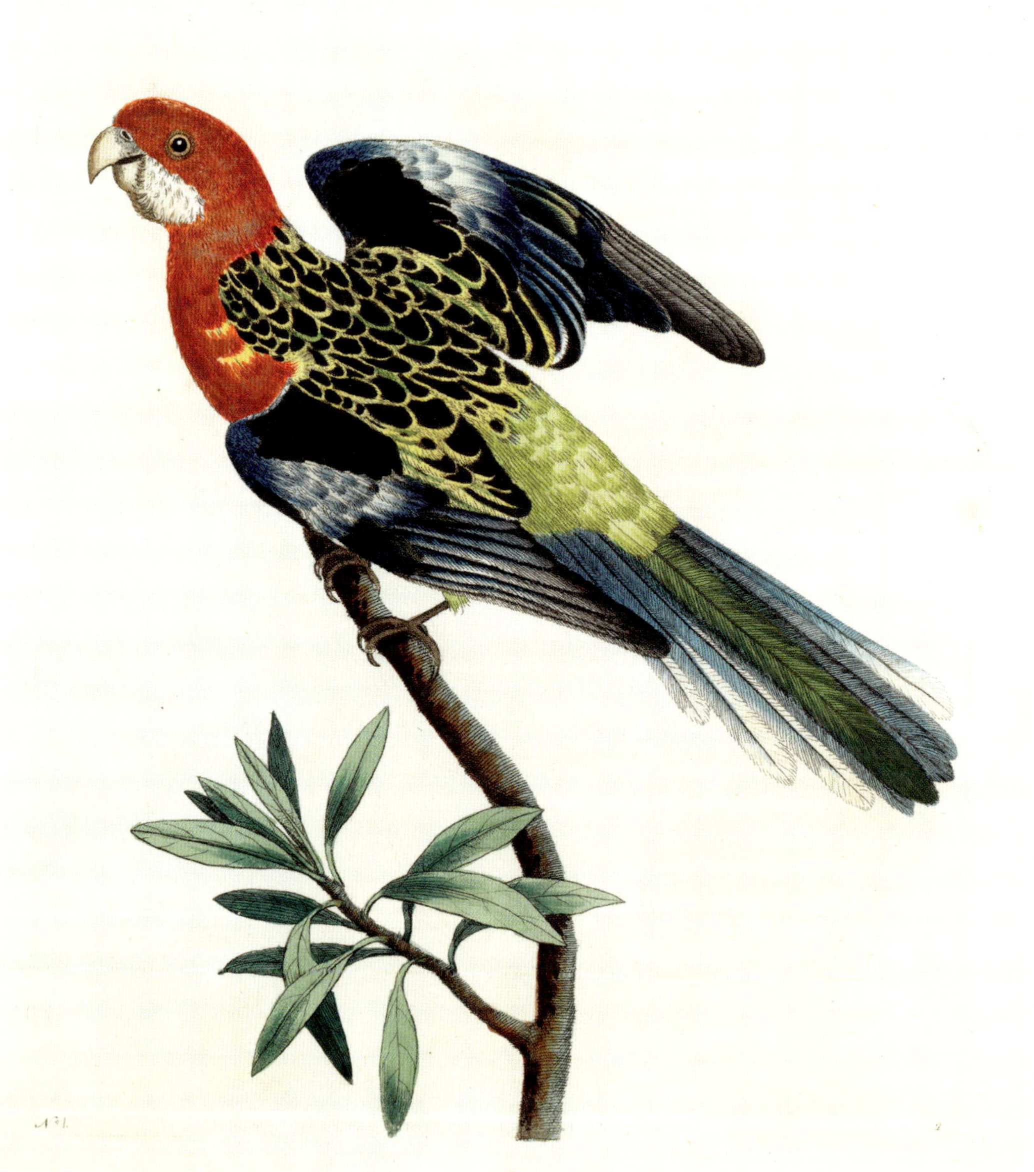

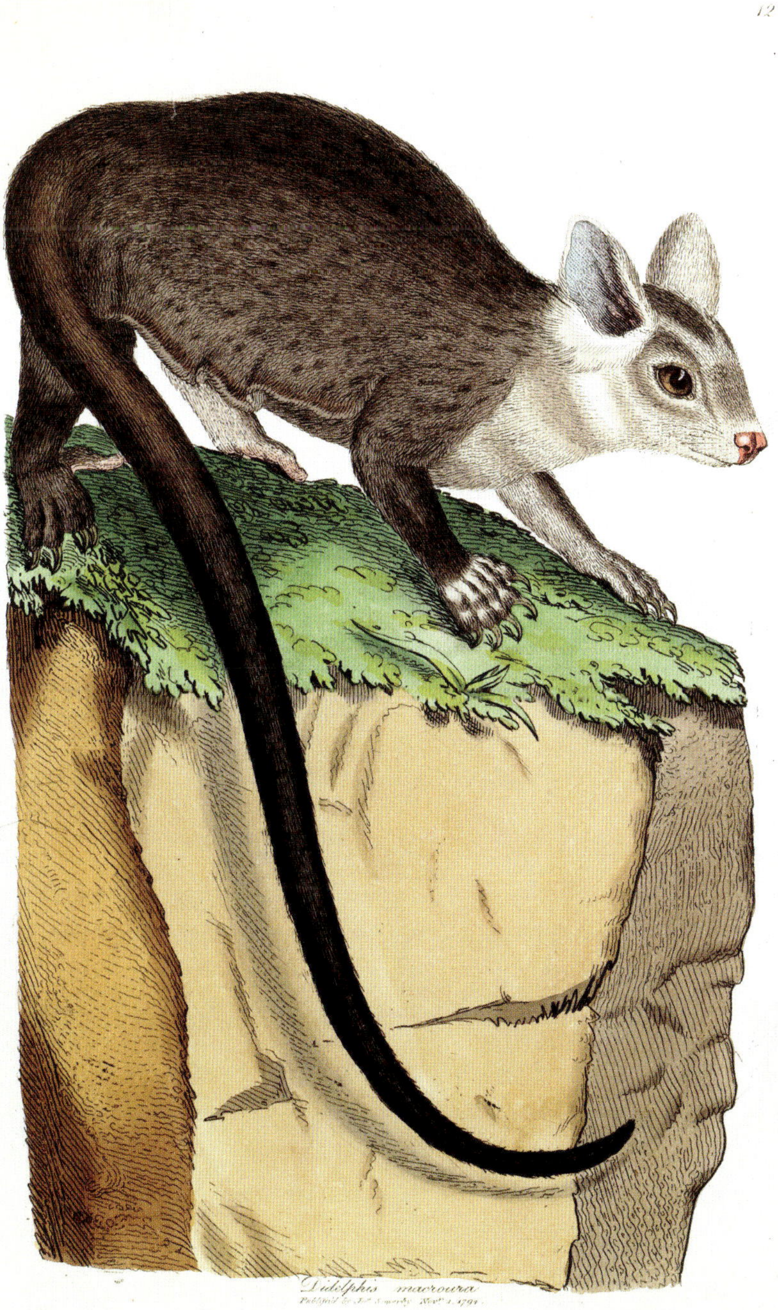
12
Didelphis macroura

6/9 *Banksia spinulosa* or prickly-leaved Banksia. In James Smith's *Botany of New Holland.*

Pacific voyage in 1770, including the genus *Banksia*, later to be named after him. Britain had chosen Botany Bay as the first site for its penal colony in 1778, partly with the intent to use the labour in developing the potential riches of this continent. People wanted to know about Australia. Shaw expressed, within his first few pages, a rather partisan and particular view:

> The vast island or rather continent of Australia, Australasia, or New Holland, which has so lately attracted the particular attention of European navigators and naturalists, seems to abound in scenes of peculiar wildness and sterility; while the wretched natives of many of those dreary districts seem less elevated above the inferior animals than in any other part of the known world... It is these savage regions however that Nature seems to have poured forth many of her most highly ornamented products with unusual liberality: where, in particular, she appears to have stationed birds, superior perhaps in elegance to those of most other climes; and which display their richly-varied plumage in scenes unknown to mankind in a civilised state, and where their beauties can only be contemplated by the eyes of barbarians.

The two volumes give an interesting perspective on how natural history was occasionally communicated at that time. Their importance is perhaps that they are examples of early illustrated accounts detailing the fascinating discoveries of nature's diversity in distant lands, although they contain a somewhat arbitrary mixture of species. They were not, nor were they meant to be, rigorous taxonomic treatises, although several new species were formally described for the first time, including Smith's account of *Banksia spinulosa* or prickly-leaved Banksia, with a beautiful Sowerby drawing. They were intended, as Smith said 'for the use of his countrymen and countrywomen ... written in their own language – a language every day growing more universal'. The illustrations are important, certainly in the case of Shaw's, for they impart much of the necessary taxonomic detail. These volumes were a popularising 'taster' about a land that was new to many and a chance for the authors to describe new species, some also of interest to horticulturists. Sowerby's plates are generally very fine, especially those of *Embothrium* trees. Without illustrations of this quality the volumes might have been of comparatively little value or interest.

These contributions were some of the last, at least for a while, that Sowerby did as a commissioned illustrator, for he was now intent on pursuing his own products and interests. He helped provide the plates for Charles Abbot's *Flora Bedfordiensis* published in 1798, but that stands out as an unusual event during these early years of interest in fungi and in his museum.

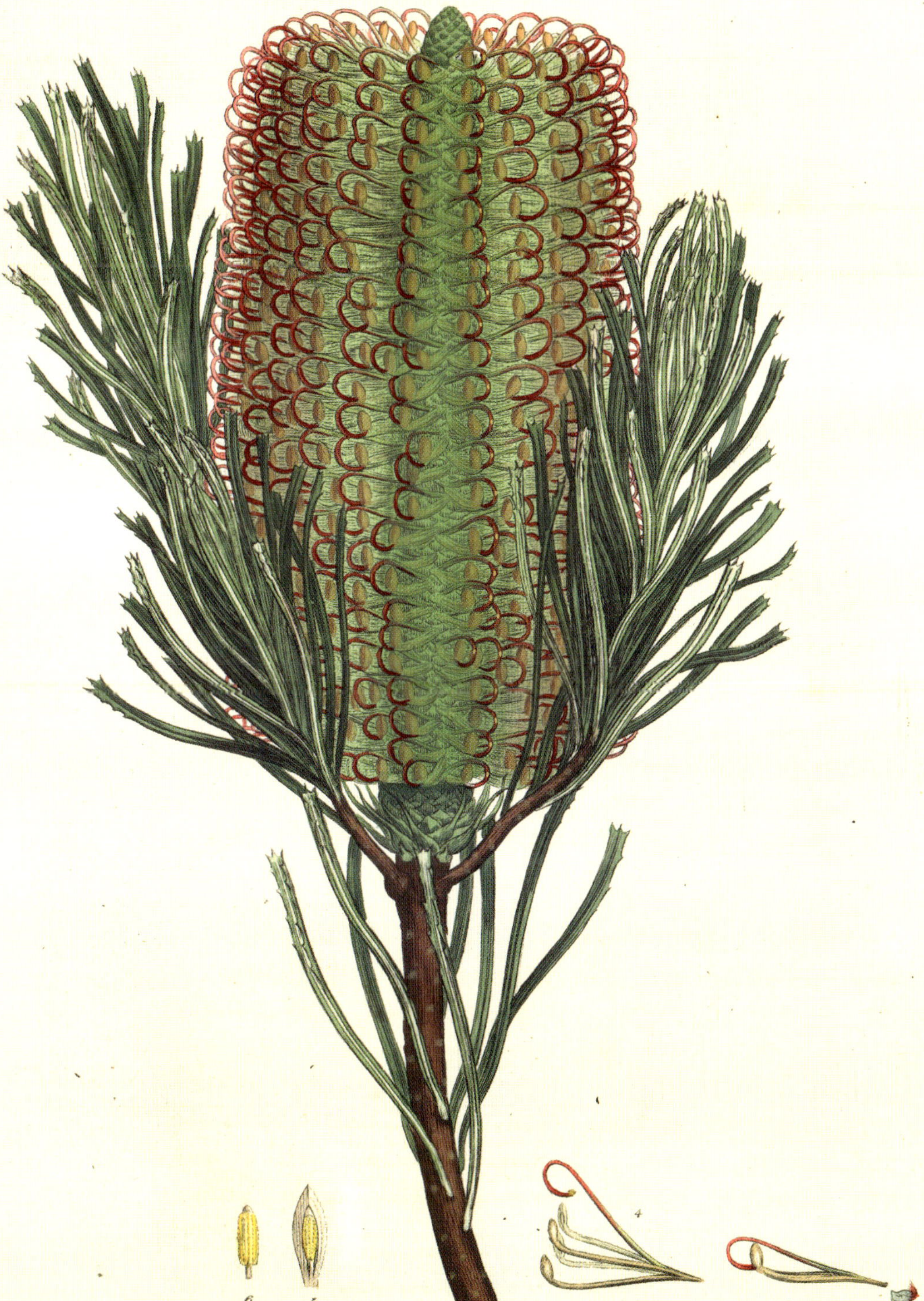

Banksia spinulosa.

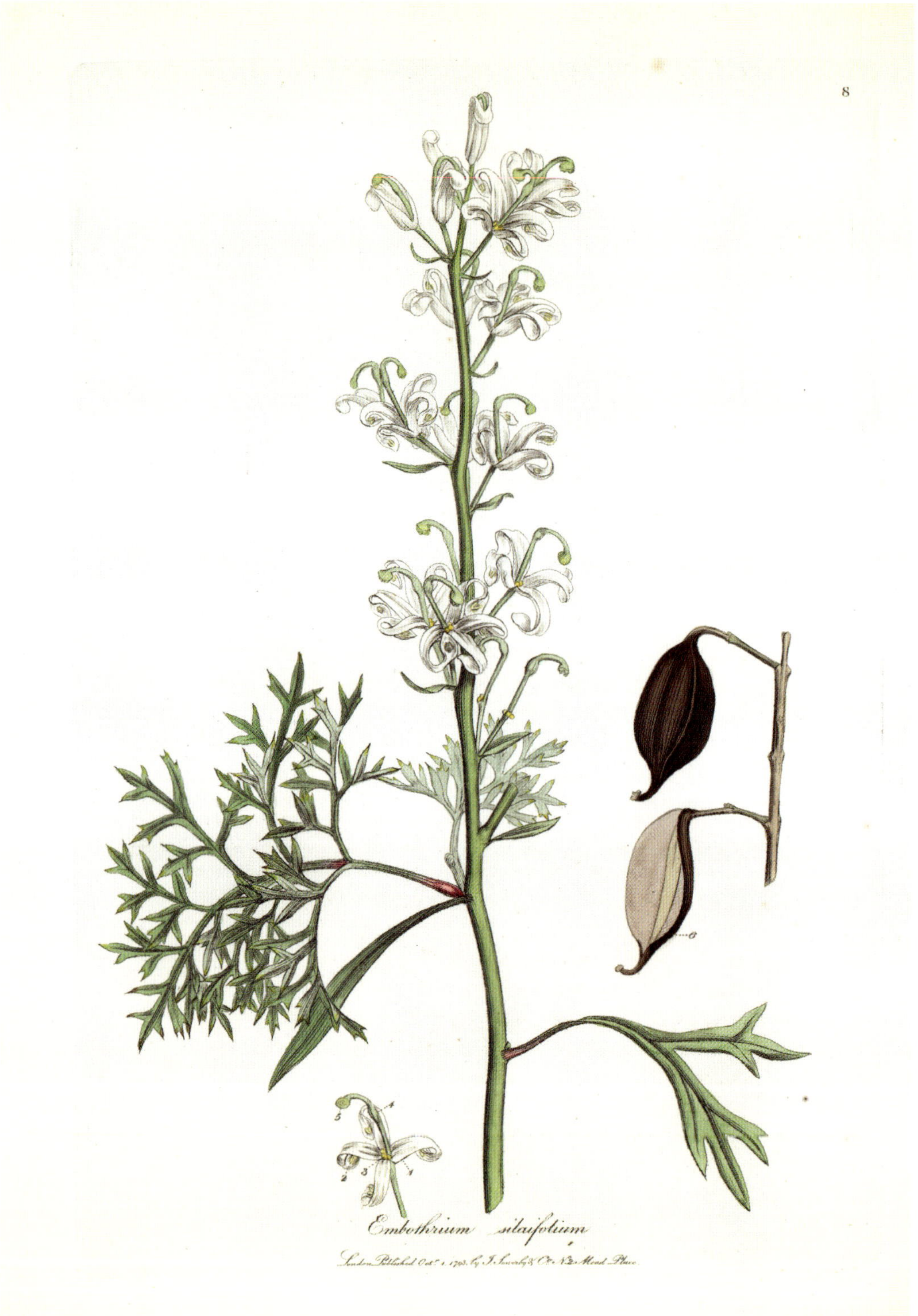
8
Embothrium silaifolium
London Published Octr 1 1793, by J. Sowerby & Co. No 2 Mead Place

6/10 *Embothrium silaifolium.* A small shrub of eastern Australia. Smith wrote: 'Of this new and very singular species of *Embothrium* a plant brought from New Holland flowered last summer, for the first time, at Messrs. Grimwood's at Kensington, from which our figure was drawn.' (*E. silaifolium* is a synonym of *Lomatia silaifolia*.)

He was self-driven to inform and enlighten the public, as well as to use his pencil and graver. No one else in England was doing what he thought must be done for England, and what was being done for France and elsewhere in Europe. The country needed, perhaps even urgently, a genuinely useful account of its fungi and to give this group the same attention given to botany. So, with years of collecting and observing behind him and with the sense that his friends would help in the ways they had always done, he started his single-author work on the *Coloured Figures of English Fungi or Mushrooms.* There is no question that he was beginning something that would become a major achievement. He worked on it diligently for eight years, then, after an interval, for several more years to produce supplements. He showed confidence in tackling a difficult domain while being aware of his own limitations. But he also knew that his experience, gained through specimen collecting and keen observation, gave a sound basis for his venture.

As was common practice, the work appeared in monthly numbers, with a blue paper wrapper, beginning in January 1796. Each number contained five plates and initially cost 2s.6d. but later (in September 1796) changed to twenty plates at 10s.d. published once every four months, so as to conform to the regulations of the Government's Stamp Office. It was printed as a small folio, a size that allowed comfortably for life-size illustrations, even though Sowerby

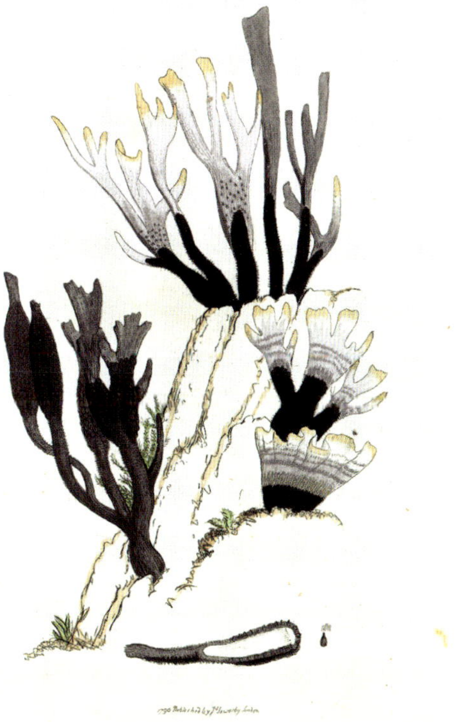

6/11 Sowerby's elegant depiction in *English Fungi* of candle-snuff fungus (*Sphaeria hypoxylon*, a synonym of *Xylaria hypoxylon*).

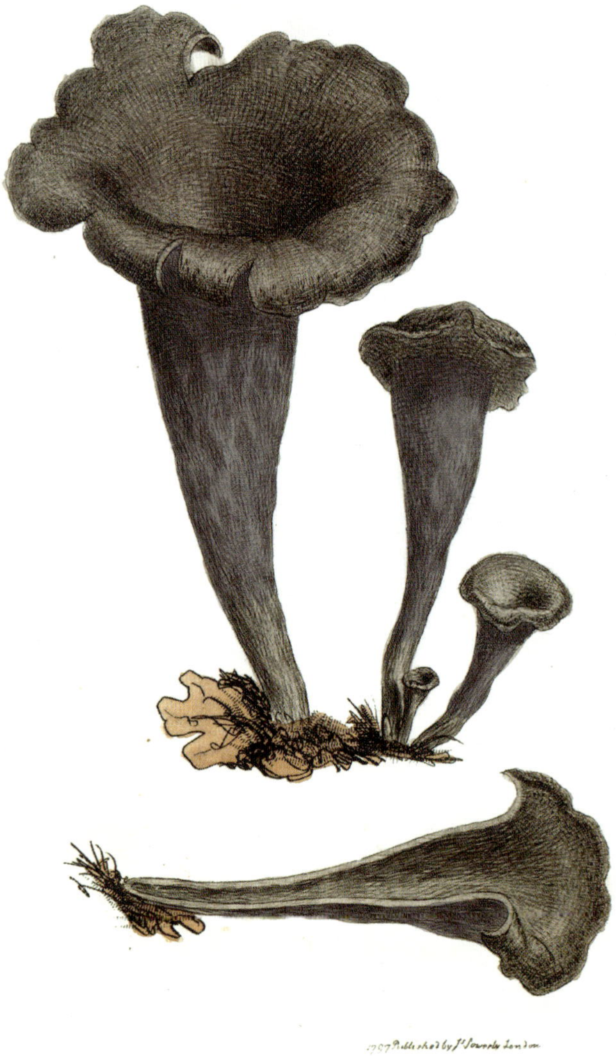

6/12 The 'horn of plenty', an edible fungus which, Sowerby said, was common in the woods around London. In *English Fungi* as *Peziza cornucopoides*, now *Craterellus cornucopioides.*

occasionally adapted his drawings so as to meet his engraving or printing needs. Early numbers of the work were accompanied by an introduction, in which Sowerby made clear that the modelling had become somewhat burdensome:

> It was not the Author's intention to have published this work, till he had nearly finished a laborious undertaking in which he has for some time been engaged, the making coloured models of English fungi in their different states, and thus as much as possible ascertaining and fixing the limits of each species: But that project being very extensive, it has been thought proper to begin publishing such as are least doubtful, in hope that in the mean time a more perfect knowledge may be daily acquiring of such as are to be given to the world hereafter.

He explained that he would try to make the names as correct as possible in 'so difficult a tribe' and that the emphasis would be on the illustrations, the text being concise. Later, the dedication was made jointly to Sir Joseph Banks and James Edward Smith, adding for the latter that 'as the writer of English Botany, I feel peculiar respect and esteem'. It is noteworthy that he did not use the word author.

Fungi are illustrated from numerous British localities and circumstances, thanks to the indefatigable cooperation of his fellow collectors and friends. Sowerby himself collected from Sir William Jerningham's plantations at Costessy near Norwich, from Earl Spencer's park at Wimbledon, from the hothouses in the gardens of the Duke of Portland at Bulstrode (on the western side of London). Sometimes he was accompanied by others such as Thomas Woodward, James Crowe or Benjamin Forster, especially when in places near Norwich. Lady Arden sent specimens from Nork Park in Surrey, Forster from a wine cellar among sawdust and bottles, and one found on a plastered wall of a ballroom. Another came from a beehive. Sowerby did not lack material or enthusiastic support.

Nor did he, of course, lack experience in drawing them. He had created several fine fungi illustrations for Curtis's *Flora Londinensis*,[21] and produced plates and drawings of them for himself and others, including for John Lettsom of Grove Hill, over several years. Perhaps inevitably these plates often lack the vitality and finesse found in many of his other works, as fungi can be more difficult to portray. The colouring, reflecting the nature of the material, is more restricted and sometimes only given to part of the whole illustration. Details are provided in separate small illustrations forming part of a plate, as with plants in *English Botany*. By the end of the first work phase he had made four hundred species illustrations; a further forty were produced for the supplements. Some of the original watercolours for the plates are at the Natural History Museum,

6/13 'Shaggy ink cap'. Sowerby's fine drawing in *English Fungi* of this common fungus, which can grow to over thirty centimetres high, in different stages of its development. *Agaricus cylindricus* is a synonym of *Coprinus comatus*.

London; they are good examples of Sowerby's artistic sensitivity and accuracy. The published plates do not always have the same finesse as the watercolours, some detail being lost, but they must rank as some of the best in Europe at the time, alongside Pierre Bulliard's fine work on French fungi (1791).

The text accompanying each plate varies significantly; sometimes it is just one line, other times nearly a whole page. The named illustrations are therefore the most important feature. Sowerby was concerned to give sufficient information, either by illustration or text, for the reader's use in identifying and observing varieties within a species. He admitted from the outset that his species assignments were not necessarily always correct. Part of the reason for publishing the work was that the subject was very much in progress. But he was confident enough to make comparisons, some critical, with the findings of other workers, quoting authorities such as William Withering, the German scientist Jacob Christian Schäffer and the French physician and botanist Pierre Bulliard, whose works were available to him.[22] He studied species as they aged or on different substrates and he described and named several new ones. He writes that *Agaricus hybridus* 'partakes a little of the characters of some other Agarics; I have therefore called it hybridus'.[23] *Agaricus cepaestipes*, which he collected at Sir Abraham Hume's estate at Wormleybury, Hertfordshire, in 1793, is named by him 'from the form of the stalk (like that of an onion) which will always characterize this species'.[24]

Sowerby did not adopt a systematic approach in his descriptions, far from it. His style is frequently casual and the content often limited. He was in contact with other botanists (particularly the mycologist Benjamin Forster) and discussed identifications and characters with them. He also used James Smith as a working critic, regularly sending him his drafts. Fungi were not Smith's strong point; he did what he could but that was rather little.[25] Smith wrote comments on some of the original drawings and was encouraging, sometimes praising Sowerby for his efforts. Smith's own lack of confidence about fungi is shown in one way by his request to Forster to provide a review of Sowerby's figures.[26] Forster's comments are what might have been expected from any conscientious reviewer. He praised and criticised, he agreed with the naming of most species but disagreed about some, such as *Helvella esculenta* (plate 51): 'Sowerby has very properly separated this species from the genus Phallus, but whether he is justified in ranking it with the Helvellas may be doubted – it certainly does not agree with the Linnean character of Helvella.' (This species is now named *Morcella esculenta*; see illustration 6/2b of the model.) But overall it is clear that he respected the quality of a work that was contributing much to both the knowledge and understanding of fungi. Sowerby

189

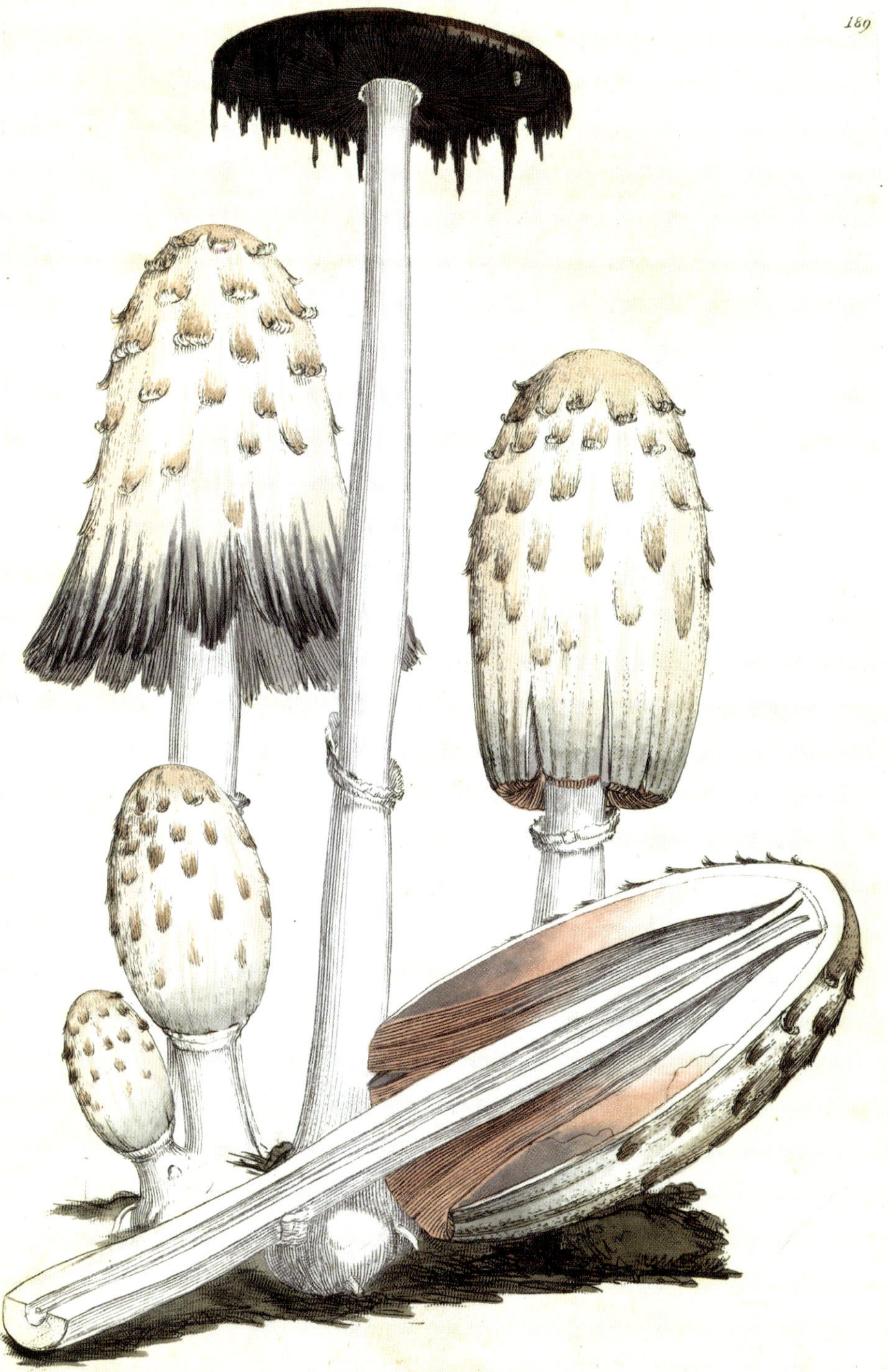

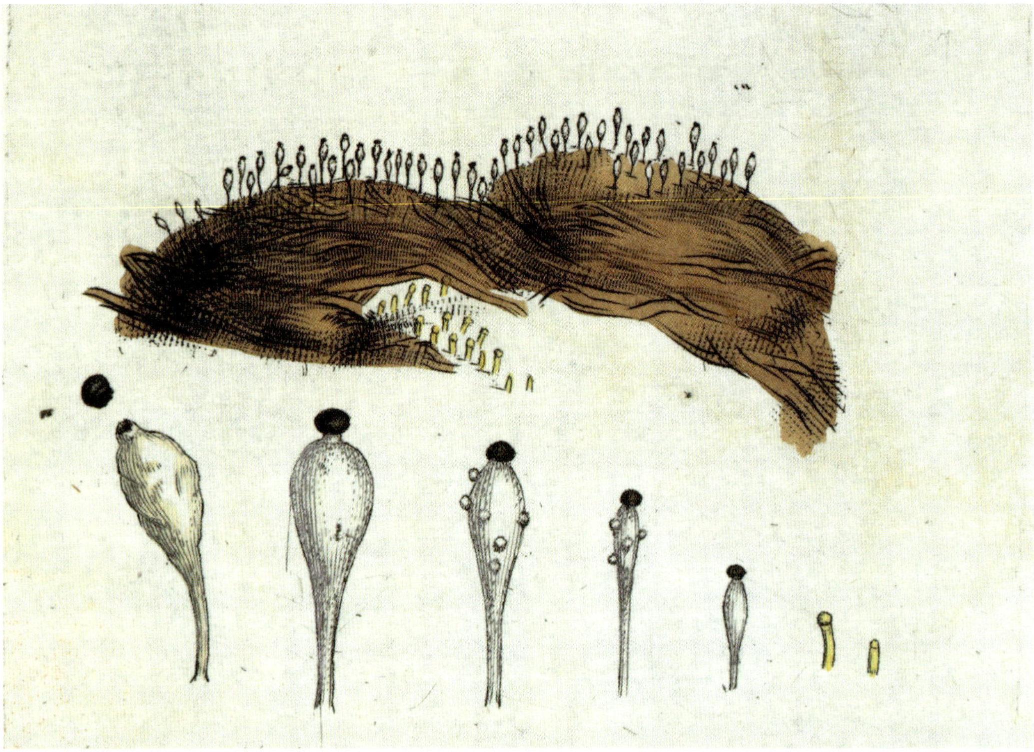

could be judged as one of the main English mycologists of his time.

Sowerby experimented and recorded some interesting snippets. He discovered, for example, that when in a ripe state, lycoperdons produce brown powders fit for immediate drawing when mixed with a little gum Arabic in water.[27] He ate some fungi with his friends and recorded the effects: 'Aikens at first merry as if in high spirits from liquors, afterwards sick. Eaten at dinner time.'[28] During specimen examination he found that 'many Agarica contain a loose pith but so tender it often slides along with the blade of the instrument used to divide the stem'.[29]

The Supplements, started in 1809, reveal his anxiety over people's ignorance of poisonous species, especially following news of a fatal event in Surrey, and how he hoped to enlighten them by illustrating the variety of shape, form and colour even within just one species. The Supplements' introduction ends with the words:

> I intend to finish models of the more particularly poisonous Fungi; and of those which are edible, to prevent, as far as possible, future mistakes, for the use of the Public and request those who have opportunities to help in their investigation, to favour me with information. I hope also that such persons will exert themselves to make the poisonous ones known as much as possible. Every Clergyman, Gentleman, Physician, and Apothecary, or public character looked up to for information, throughout the kingdom, should get acquainted with them for this philanthropic purpose; for which, and for other useful purposes, I have long kept my Collection open for public inspection every first and third

6/14 The intriguing small 'dung cannon' fungus which Sowerby explained occurs on horse dung, grows rapidly, and finally shoots its sporangium a considerable distance. His account shows that he was a keen observer of the phenomenon. In *English Fungi* as *Mucor urceolatus*, now *Pilobolus crystallinus*.

Tuesday in each month, from Eleven until Three o'Clock.

He also recorded how during his walks he might advise his fellow citizens about certain species:

> I gathered this sort in Kensington Gardens, in June 1795, and August, 1796: and I observed some boys gathering for Champignons... but advised them to throw them away, lest they should eat them. In the Green Park, about the same time, a nursery-maid, seeing me a little curious, asked me about some of those figured in this plate. I gave her, and two little ones that were with her, the same advice; she was, however, confident in her own opinion, that they were good to eat; but I had the satisfaction to see her throw them away when at a distance from me.[30]

On another occasion he shared the pleasure he experienced during his collecting of an *Agaricus*:

> I gathered these on the delightful Mount Edgecumbe in August 1812, where I had the most elevated pleasure I ever felt in my life, in contemplating the sublime ways of Providence, every aspect, hill and dale, bespoke its contrivance, a charming lesson for mortals, that 'Nought was made in vain, or not for admirable ends!'[31]

The Supplements appeared spasmodically until as late as 1815 and included varieties of species, especially the well-known ones. There was still no proper published competitor to his work and its popularity was maintained. Sowerby produced about five hundred copies of each Supplement – a healthy number for the time and subject.[32] Today, mycologists recognise that Sowerby had a lasting influence on fungus taxonomy; he described many species for the first time, and over two hundred new species names were given. One professional mycologist has recorded that 'in some genera (e.g. Sphaeria, Aecidium, etc.) several species were described for the first time, but in the fleshy fungi his illustrations were of such high quality that they permitted other workers to identify the taxa involved'.[33] Sowerby's work has stood the test of time, and not just for its artistic quality.

Sowerby collected specimens himself but was not much of a traveller. He was more comfortable working diligently on his specimens and art in his Lambeth home. He journeyed to nearby locations to satisfy commissions or for collecting, and occasionally to Norfolk and Suffolk for visits to friends, colleagues and members of his wife's family. But one exception to this pattern was caused by the enthusiasm of Mr Dawson Turner, the banker at Yarmouth, who persuaded Sowerby to join him on a month's collecting trip in the summer of 1799 to south-west England, a region for which no recent flora had been

6/15 A fungus that can be difficult to identify because, as Sowerby said, it can have many different appearances. Probably the conical wax cap *Hygrocybe conica* (synonym of *Agaricus aurantius*, as given in *English Fungi*).

written. Sowerby's willingness to go is somewhat remarkable, for his work on *English Botany, English Fungi* and his models was continuing unabated and there was always a hefty amount of correspondence and visitors to his museum to be dealt with. Such travel was somewhat adventurous and often laborious.

They set off in June, accompanied by Sowerby's eleven-year-old son, George, travelling more than a thousand miles and collecting from numerous locations both on their way, such as in Oxford, and then right down to Land's End in Cornwall.[34] George received some training, new specimens were gathered for *English Botany*, and later a report was submitted to the Linnean Society with a catalogue of more than one hundred rare species, including lichens and seaweeds.[35] In April of the following year, Turner wrote enthusiastically to Sowerby: 'Are you aware that in our tour thro' Cornwall we found Lichen leucophaeus n^r Camelford & L. composities at the Lands End? They are both great rarities and I did not know we had gathered them until a few days ago.'[36] The tour was an undoubted success and a good relationship between the two men was established, sufficient for Turner to suggest another 'jaunt' to the Isle of Wight two years later.[37]

With James and George away, business at 2 Mead Place continued with Anne in charge, assisted by the family. Steady production of the publications, especially *English Botany*, was essential. Any concerns Sowerby may have had were allayed by Anne's letter, sent to him in Bath:

> My Dear James,
> I once more take up pen at a meantime, I trust the Letter I wrote to you before you did not receive, its not of much consequence if you did not. by this you will learn that all at home are well & hearty, there is not any thing very pressing... to want you at present. I have received parcels and Letters from different correspondents all of which I have I think put by safe. Lord Seyforth and a Lady this Moment call'd to see the Birds, say he will call another time when you return. have recev'd your Box you sent first but no more, you did not say from whence you sent the other, if you do we may enquire after it, we are preparing the next number of E.B. which will be ready in good time. When you next write be so good as to say how George is in your last letter you did not mention him. James, Markham & Edward are well and tollerable good boys.....
> My love with this to you & George and believe me your Affectionate Wife
> Ann Sowerby[38]

In a postscript she tells him his son James has asked that should he bring anything home for him, he hopes he will not forget to take the plants up with the roots. The twelve-year-old son was instructing the father and master!

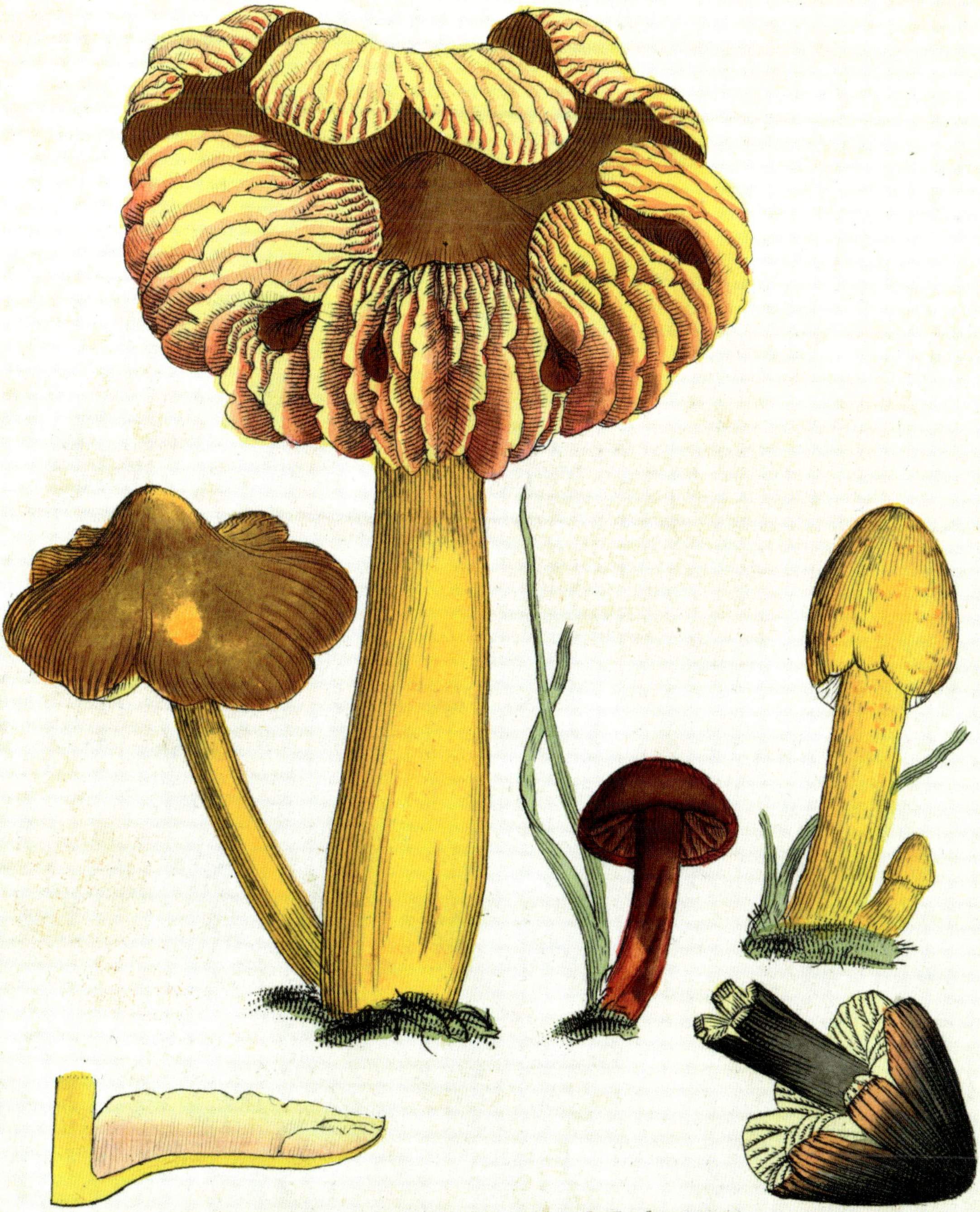

Janʸ. 1. 1803. Published by Jaˢ. Sowerby. London.

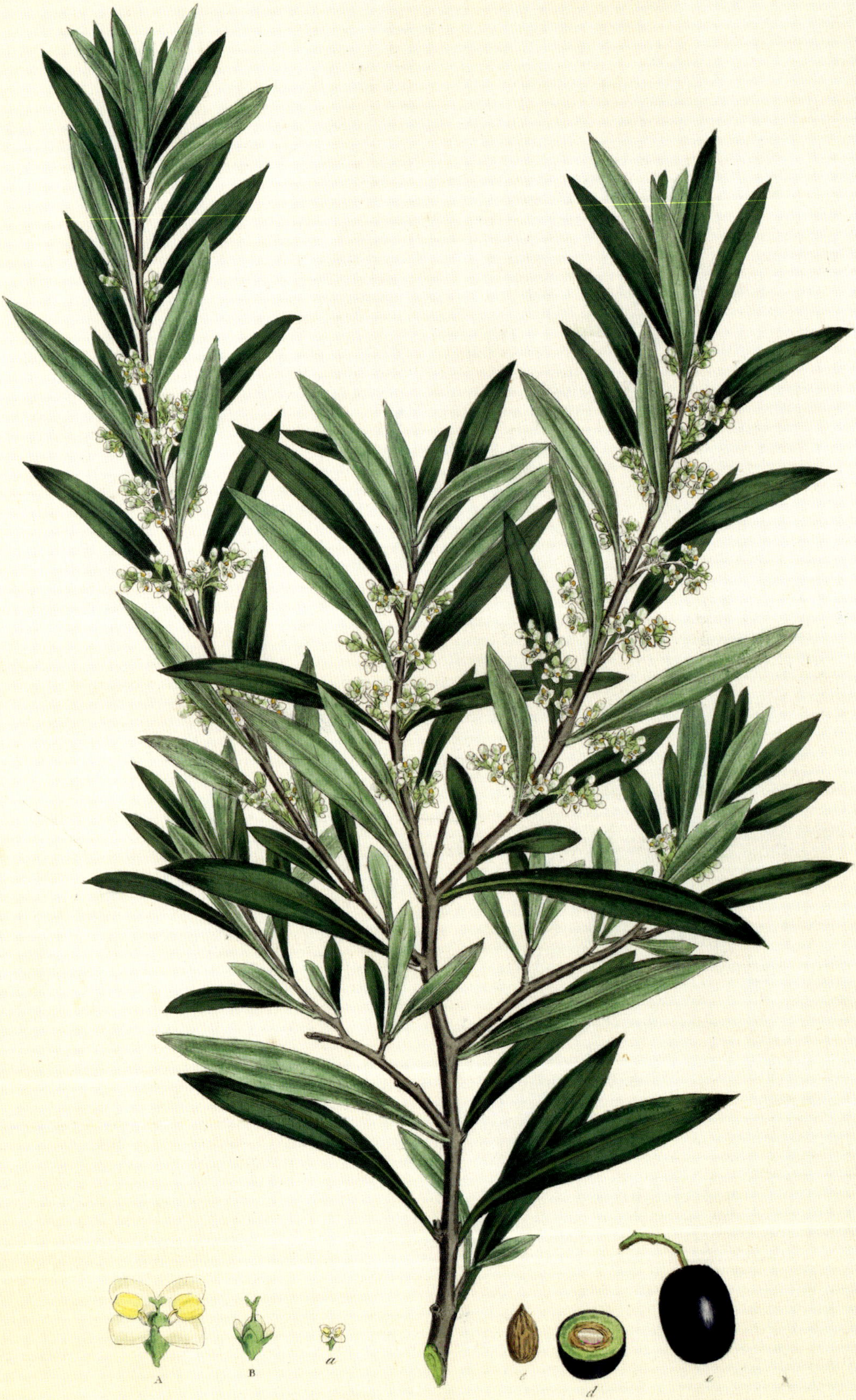
A
B
a
c
d
e

CHAPTER SEVEN

Troublesome exotic floras

7/1 *Olea europaea*.
The evergreen olive tree associated much with Greece and other Mediterranean countries. Sowerby's engraving for the third plate in the first volume of *Flora Graeca*.

Mead Place in Lambeth was the centre of Sowerby's life. Here were the tools of his trade, here people came to see him and his museum, here was his family and the family business, and here was his security. The outside world was present in the numerous specimens about him, in his herbarium, in his cabinet drawers and in the various packages arriving in the post every day. He seldom travelled far. Tours such as he made with Dawson Turner to the south-west of England and later to the Isle of Wight were rare indeed.[1] Even social visits became few and far between. James Smith, Dawson Turner and others invited him in vain to come and stay so they could botanise together. 'How can we get him out of Mead Place?' Dawson Turner once complained. The wider environment from which a specimen was collected seemed of small consequence whereas the immediate habitat, such as leaf litter, was noteworthy, especially if it could be drawn, modelled or at least easily described. Yet, as we have seen, Sowerby was moved by the beauty of the countryside and was always a keen observer in the field.[2] It was an almost obsessive passion for his work that kept him at home; other activities such as music, socialising or dancing were unwanted distractions. Even attendance at learned societies, the Linnean Society included, was not a regular feature of his daily life; people hoping to see him at a meeting were sometimes disappointed. Natural history, especially as could be represented on paper, was at the core of his being.

The last few years had gone well. *English Botany* continued with good sales and no end in sight. His publishing business was keeping the family secure and he was sufficiently occupied to be able to turn away commissions to illustrate other authors' works. But now came an unaccustomed period with setbacks that also affected his ambitions for Mead Place. Two works on plants from foreign lands caused him grief both financially and emotionally, as well as straining the working relationship with James Smith. Each of the two major projects started with their own difficulties. One, a flora of Greek plants – the *Flora Graeca* – was driven by others and later came to involve Sowerby, who desperately wanted to be involved but was initially thwarted by those in charge. It ultimately became a major publishing achievement, deemed to be Oxford's finest botanical treasure, but with a remarkably limited circulation of just twenty-five copies and at a formidable expense. The other work – *Exotic Botany* – was started by Sowerby but ran into the same kind of problems encountered years earlier with his *Flora Luxurians* in that illustrated books with a horticultural purpose were often not financially secure. It was moderately priced and accessible but its lack of support and Sowerby's own failing commitment led to its early demise. James Smith was a key person for both.

Sowerby had been overstretching himself, driven by his own enthusiasm for all parts of natural history and a near boundless energy to communicate while still making a living sufficient to support an expanding family. The two exotic floras, in their various ways, were bubbling away at the same time as he was busy with his existing publications (*English Botany* and *English Fungi*), starting new ones (works on British minerals and animals) and making contributions to the Linnean Society. Other stresses came into play. His seventh child, Felicia, was born in 1800, then Charlotte in 1802. Just a short while later his son Markham, aged only eleven, suffered a horrible death through being crushed by a wagon against the wall of a narrow city courtyard;[3] Anne would have been traumatised by this. His expanding business and range of activities placed demanding burdens on his management of his assistants (such as colourists) and no doubt on the family as working contributors, especially his wife and his two sons James and George. Pressure on habitable and working space was eased by taking a lease on the next door property (number 3 Mead Place), which was later to be occupied by James De Carle's family and George. All the while his museum continued to develop and to be open to visitors. Energy could not have been in short supply but satisfaction and contentment sometimes were.

Flora Graeca holds a particular fascination. It is a work without parallel, created through a call from the grave and involving the dedication of a small team of people over several decades.

Sowerby's involvement was essential, yet sometimes fractious; the outcome was magnificent but the initial impact was marginal. It is a subject that inevitably has attracted the attention of several authors, resulting in the publication of two fine books.[4]

The story can be summarily told of the wealthy and young Oxford professor of botany, John Sibthorp,[5] travelling in 1786 and 1787 to study and collect the little-known flora and fauna of Greece, Crete, Cyprus and Turkey – lands of rich natural history treasures, awaiting discovery. He was accompanied by the already renowned illustrator Ferdinand Bauer (brother of Franz, or Francis, Bauer) and, for some of the journey, by John Hawkins, who had a wide and knowledgeable interest in geology, mineralogy and topography, especially of the Balkans.[6] Bauer was well-treated by Hawkins but not so by Sibthorp, who seems to have failed in giving him the respect he deserved or expected. Nonetheless, he returned with excellent pencil sketches of numerous species, many of them numerically colour-coded so that back at home he could correctly colour his full-scale drawings.

Travel stimulates the mind and so it was with Sibthorp; he set out again early in 1794, joined for some of the time by Hawkins, but not by Bauer, who had written to Hawkins complaining of Sibthorp's cavalier attitude in planning for the journey and excusing himself.[7] But calamity struck; Sibthorp returned earlier than planned, seriously ill, in the summer of 1795. He died in Bath the following February, leaving a detailed will with instructions on what should be done with the vast amount of his collected material – material of much consequence and interest – and dedicating at least some of his considerable wealth for the purpose. This led to a unique and famous botanical outcome. Two of his three executors, John Hawkins, and Thomas Platt (a prominent solicitor in London), were well chosen; they fulfilled their obligations with tenacity and dedication. The third, Francis Wenman (Professor of Civil Law at Oxford) died soon after Sibthorp.

Their daunting task was to arrange the production of a *Flora Graeca* to be published in a folio edition of twenty parts each of fifty coloured plates, using the specimens and drawings of Sibthorp's travels. It was a daunting task not only because of the large quantity of material but also because of the inadequate field records made by Sibthorp. Nor was it necessarily easy to find the right people to do the job. James Smith was recognised as the only person capable of bringing scientific sense to the disorder and of writing the *Flora*, and it must have been a relief to the executors when he accepted their commission in 1799. Specimens, documents and drawings were soon despatched to his home in Norwich and from then on Smith worked diligently until his death, exercising his now remarkable scholarship and seeking the help of

James Sowerby. He seemed indefatigable; his other works, including *English Botany*, suffered little or no delay.

The other task for the executors was to find someone to engrave and colour the plates; this was also a massive undertaking involving nearly one thousand detailed, large plates for a planned production of fifty copies. Ferdinand Bauer showed significant interest, claiming he would do the engraving better and cheaper than another potential artist Daniel MacKenzie. On the one hand there was talk of a trial between the two men and on the other there were concerns over the amount of time that Bauer would need if he worked alone. The executor, John Hawkins, remained uncomfortable with any of the proposals, suggesting even that if MacKenzie were to be employed then the more simple plants should be selected for him to work on.[8] He was also much vexed by the knowledge that many of the drawings were in a half-finished state and that only Bauer could complete them. By the end of 1799, Hawkins was out of his depth and sought Smith's advice. 'I am moreover of opinion that the immensity of the work requires us to enlist in it other artists of merit such as Sowerby.'[9] No other artists' names were mentioned in these exchanges. High-quality engravers and colourists were far from common. Sowerby was in a small top league even if not, perhaps, quite as high as Bauer. Smith had undoubtedly been pushing for Sowerby to be involved although correspondence suggests that he was being, at times, a little unwilling.[10]

The passing of time resolved the issue. Bauer became cool about the project, MacKenzie died (in February 1800) and the task fell to Sowerby early in 1800. Bauer was probably well out of it, for by mid-1801 he was sailing in the *Investigator* with Flinders on the circumnavigation of Australia. By the summer Smith was already getting impatient for the first plates from Sowerby; by the autumn he was receiving and criticising his efforts:

> am highly pleased with the greek plants – You know I shall not flatter, but on the contrary it is my business in this stage of the business between you & I to find all the fault I can – I think the outline on the Veronica particularly, too strongly engraved or etched, & as the impression will be but few, I wish you to guard against that imperfection in future.[11]

Sowerby did not, could not, do all the work himself. He employed assistants and colourists as well as involving some members of his own family. The correspondence shows that he even employed some engravers – even though much of this specific task must have been his. But all was not well. Hawkins visited Mead Place to check on production and ensure good quality but he was far from happy and reported to Smith:

I have carefully examined the coloured Plates lately executed and was sorry to find some essential faults which I pointed out to Mr Sowerby & trust he will in future avoid. I lamented to see the miserable state in which Bauers beautiful drawings came out of Engravers hands. Surely there is no occasion for all this negligence and it may be avoided! Does Mr S think that when the Flora is published the drawings will be of no further use? Were they even not so superior as they really are to the coloured engravings they ought to be carefully preserved as public record belonging to the University of Oxford. But I will venture to assert that as works of Art they are superior to any thing of the kind existing and will constitute one of the most valuable treasures of the University.[12]

Smith, in telling Sowerby of the news, put a more positive spin on matters:

I am very glad Mr Hawkins & you understand one another so well. It is a great satisfaction to have him approve the plates. He laments bitterly that the drawings are so dirtied – cannot this be in some degree prevented? The engravers should be told they must keep them clean.[13]

The difficulty was resolved and plate production continued regularly, with them going to Norwich for checking. Smith wrote in praise, saying they were 'excellent' and sometimes even 'very excellent'. Payment was occasionally a problem especially when the executors had to apply to the University for release of some of Sibthorp's legacy, a process exacerbated when a new Vice Chancellor was in office. Smith expressed his views in no uncertain terms to Sowerby:

Mr Platt, though a man of business, is, as you and I should be, very impatient when he meets with the insolence of office in any one, and it must be galling to him who has so much trouble at this affair, for no profit, to meet with petty delays & mistrusts in these Oxford people, who are great only on their own dunghill – I mean Vice-Chancellors & such people.[14].

If matters were then harmonious it would not be for long. The time had come for publication of the first fascicle, which meant distribution and selling, and who better for this task than Sowerby himself – a publisher who well understood the nature of, and the market for, this special publication. But the choice was not his. The hard work on the plates did not give him any right to the selling of the *Flora*, which might be financially advantageous to the seller given the high price of the work. But Sowerby was passionate about this work, it was no ordinary commission and he

7/2 *Morina persica.* Commonly known as prickly whorlflower. An evergreen perennial used as an ornamental plant. *Flora Graeca.*

wanted as much association with it as possible. Early on he had told the executor, Mr Platt, of his keenness and that he would lease the house next door (No. 3 Mead Place) so that he would have space for the work as well as for its sale. He was sufficiently encouraged by Platt to go ahead and by the next summer the deal was done.[15] With the investment being made he was anxious for it to yield fruit but Platt and Hawkins had other plans and other interpretations of events. Claims and counterclaims went between them, Smith and Sowerby.

Letters from Smith to Sowerby reveal much about how the confusion arose and was exacerbated by the involvement of several people, often at some distance from one another:

> I could not mistake the words Mr Platt said, & they are now fresh in my memory. But I am more & more convinced that he often speaks from the impulse of the moment, when hurried or fretted perhaps about other matters, & I am certain he is perfectly well meaning & honourable. I would advise you therefore not to mind his occasional hurries, but to go on as you have done. We certainly depend upon you to continue the work, & I promise you I will withhold my part in case you are unjustly used. I can have no objection to your selling the work, & as you say Mr Platt now agrees to it, there will be no difficulty.[16]

Sowerby firmly held the belief that Platt had agreed to him selling the *Flora Graeca*. Confronted with the opposite he became seriously distressed. He may have railed against Platt himself for his inconsistency and for now starting to say that it all depended on Hawkins's wish. Sowerby could turn to no one to vent his anger except, of course, James Smith.

Now matters took an unpleasant and personal turn. Sowerby in his extreme anxiety had his son George (aged 16 and starting to act as an amanuensis to his father) write a bitterly complaining letter to Smith, which seemed to blame him for his troubles. He went so far as to say that he 'must bid adieu to all appearances of friendship in future to any concerned in the work', and spoke of 'the baseness of mankind' such that he would 'give everything up' and 'can transact no other business'. Smith responded rapidly, with astonishment and dismay, and explained he had always wanted Sowerby to have 'the sole management' of the *Flora* and that he would most assuredly mention him in his preface to the *Flora* with high commendation. He asked Sowerby to reflect a little and to tell him what was vexing him, what was making him angry with Smith, and to do him justice in his turn. He concluded: 'Pray give me a line – more like your usual & long-lived friendship for whatever you may think, I am as I always have been Y^r faithful friend J E Smith.'[17]

Just three days later Smith received more explanatory and calming words from Sowerby.

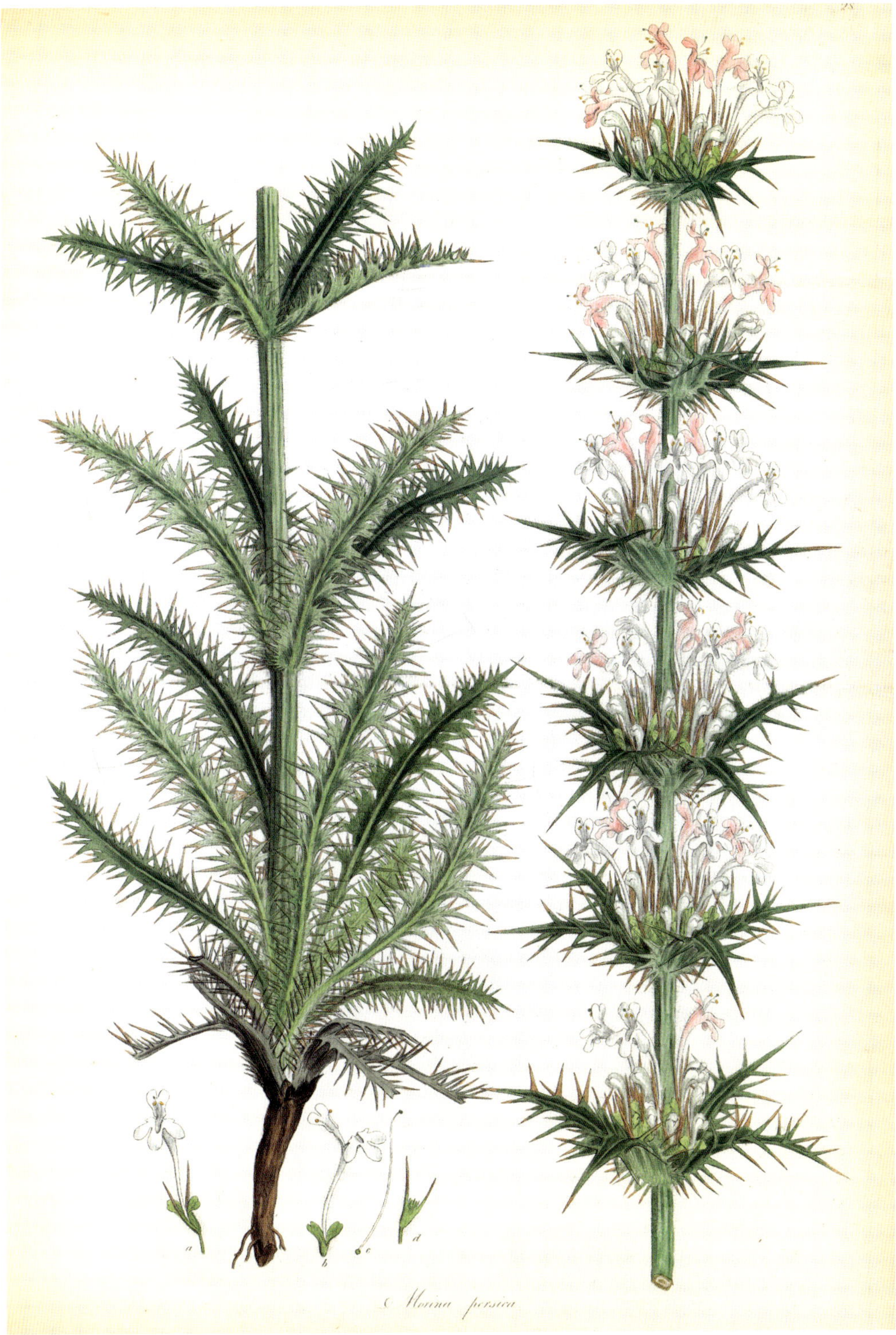
a
b
c
d
Morina persica

He explained how he had been sorely mistreated, especially in being so misled by Platt, writing that 'all I have done may be wrong altho done with the most pure intention of doing right'.[18] He claimed that if he were not to be the distributor then, as Hawkins had purportedly said, he should be indemnified for his expenditure on leasing a house. But Hawkins protested this was untrue and quickly put on record to Smith how he had discussed matters with Sowerby on a visit with his wife to Mead Place:

> ... when I went with Mrs H last summer to his house on purpose to see the plates which he had prepared and spent nearly the whole time I staid there in diligently comparing them with the original drawings. My language upon this occasion to Mr S- was, You must consider me as a person who being the Executor of the will of a friend & a Trustee for the publick has a paramount duty to fullfill which will not permit him to overlook any faults in your execution which are capable of being corrected. I shall therefore take upon me to criticize your work and shall have this apperance of being severe which you will have the candour not to take amiss.
> I accordingly pointed out several things for correction which were admitted by him. The same language I held when he talked of his disappointment in not being commissioned to deliver the work which, he said, he had been led to expect & of the loss incurred thereby for he told me had even hired a house on purpose'.................'I must therefore conclude with repeating my surprise at this unwarranted assertion of Mr Sowerby, a conduct which together with his arrogant and illibiral treatment of Mr Platt last summer, renders it necessary for me to avoid in future any communication with him.[19]

It would be difficult for an executor to put himself into the position of not communicating with someone already heavily involved in *Flora Graeca* but Hawkins was showing his irritation – and his lack of support for Sowerby – certainly over anything to do with the publishing. Smith, true to his word, continued to support Sowerby while Platt vacillated only to become finally swayed by Hawkins who then informed Smith:

> I have said so much in my former letter upon the subject of Mr Sowerbys cause of dissatisfaction that I trust there is no further need of disculpation on my side, nor is there any. I believe on Mr Platts in spite of Mr Sowerbys persevering complaints. That this gentlmwas sanguine enough to hope that his interest with the Trustees would procure him a Commission so desirable and lucrative as this is, I do not much wonder at; but every day confirms me in the opinion that we did right in rejecting him & upon this occasion I cannot

help feeling some concern at our differing from you. Many persons of consequence may go as you say, to Mr Sowerbys, but do they go there twice? Whereas these same persons are in the habit of going to Mr White. The one an Engraver lives in a distant part of the suburbs, the other a Bookseller in the most frequented part of the town and is acknowledged to have the largest stock of books of any person in the trade besides dealing particularly in works on Nat^l History. Can we then hesitate in our decision?[20]

They did not hesitate much longer. Hawkins had his way and White the bookseller was commissioned to do the sale over the coming many years. But why had Sowerby become so upset and over quite a long period? His purchase of the lease of number 3 Mead Place seems premature if it were solely for *Flora Graeca*. But there were other factors. The lease had become available and could be taken by someone else at any time. Sowerby's family was growing and needed space – there were about seven children in the house at this time with two of them very young[21] – and his interests in natural history were expanding into mineralogy and zoology, as were his collections and publications. The house next door seemed an obvious solution to several problems provided there was sufficient money to make it happen. Sowerby had sometimes written of wishing to get out of debt at certain times, but that was not an unusual matter in those days. The scale of any debt is unlikely to have been very large and his son George had acquired one or more properties mostly through inheritance; the rental helped maintain the Sowerby coffers. His publications were doing at least reasonably well, especially *English Botany*, and he was undertaking paid work such as teaching and preparing material for the *Transactions of the Linnean Society*. His publishing business, for that is what it was, continued successfully. The taking of the lease seems reasonable on many grounds.

Flora Graeca is a most remarkable and beautiful accomplishment and it is no wonder that commentators and critics have acclaimed it. From a somewhat inauspicious beginning, the dedication and persistence of just a few well-chosen people resulted in a work of considerable scientific and artistic integrity and balance. Its scientific value rests not just in its botanical descriptions and accurate renditions of the many species but also in its demonstration of the pursuit of scholarly natural history at the beginning of the nineteenth century. It probably could not have been achieved without Smith's dedication and knowledge, nor without the skills of Bauer and of Sowerby. Yet, it is ironic that while Hawkins had written of Bauer's drawings as being a future treasure of the University of Oxford, it is necessary for a reader to search quite hard for his name in the published volumes; similarly for Sowerby. Sibthorp and Smith are rightly

7/3 *Cyperus comosus* (now *C. rotundus*). A handsome starchy sedge that can grow to well over a metre. *Flora Graeca*.

7/4 *Salvia argentea*. A Mediterranean plant grown for its large silvery leaves. *Flora Graeca*.

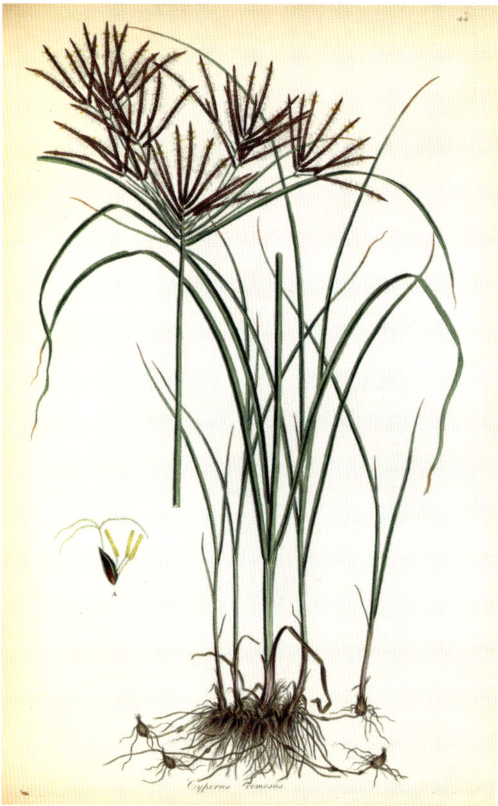

prominent but rather as with so many Gothic cathedrals, the architects are known while the identities of the master craftsmen remain obscure. Bauer's and Sowerby's names are to be found as brief acknowledgements by Smith in his long Latin preface to the work. For Sowerby, he wrote:

> My first task was to find someone capable of engraving illustrations accurately from the original drawings. This work is committed to Mr James Sowerby, long celebrated in this art. Before the illustrations were engraved on copper, I took care to compare each one with the original, and then I checked diligently that the colour and appearance of each coloured illustration were correct.

But despite Sowerby's initial desires to be so involved with the *Flora*, its result was something of an antithesis of his own ideals; ultimately it was only for the rich, it was almost inaccessible since only twenty-five copies of the first issue were finally made, and it did not truly reveal Sowerby's contribution. As a significant product of the Enlightenment it failed to have a proper initial impact despite its revelations of the botany of an intriguing part of the world (including illustrations of new species) because these findings were not effectively shared. There was little sense of urgency in getting the scientific results out to the wider world. Those who really wanted to communicate their science did so by publishing in the transactions of the established and new learned societies, including the Linnean, or by a relatively rapid publication of floras. The executors had, however, fulfilled Sibthorp's wish and now, after a further one hundred and seventy years, the artistic merits of this monumental work are really accessible through the University of Oxford's Bodleian Library's 'Digital *Flora Graeca*', with the complete series of printed

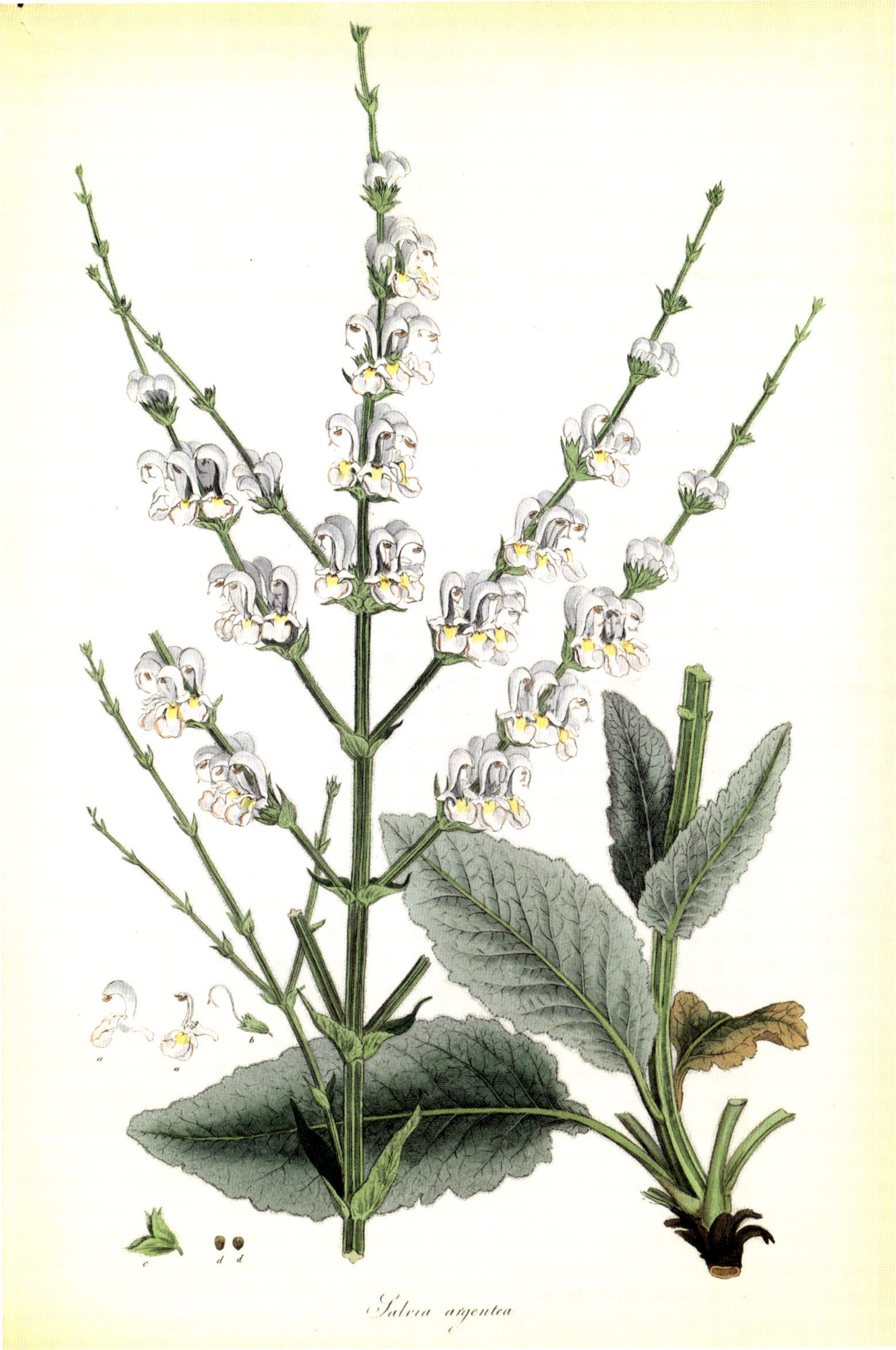

Salvia argentea

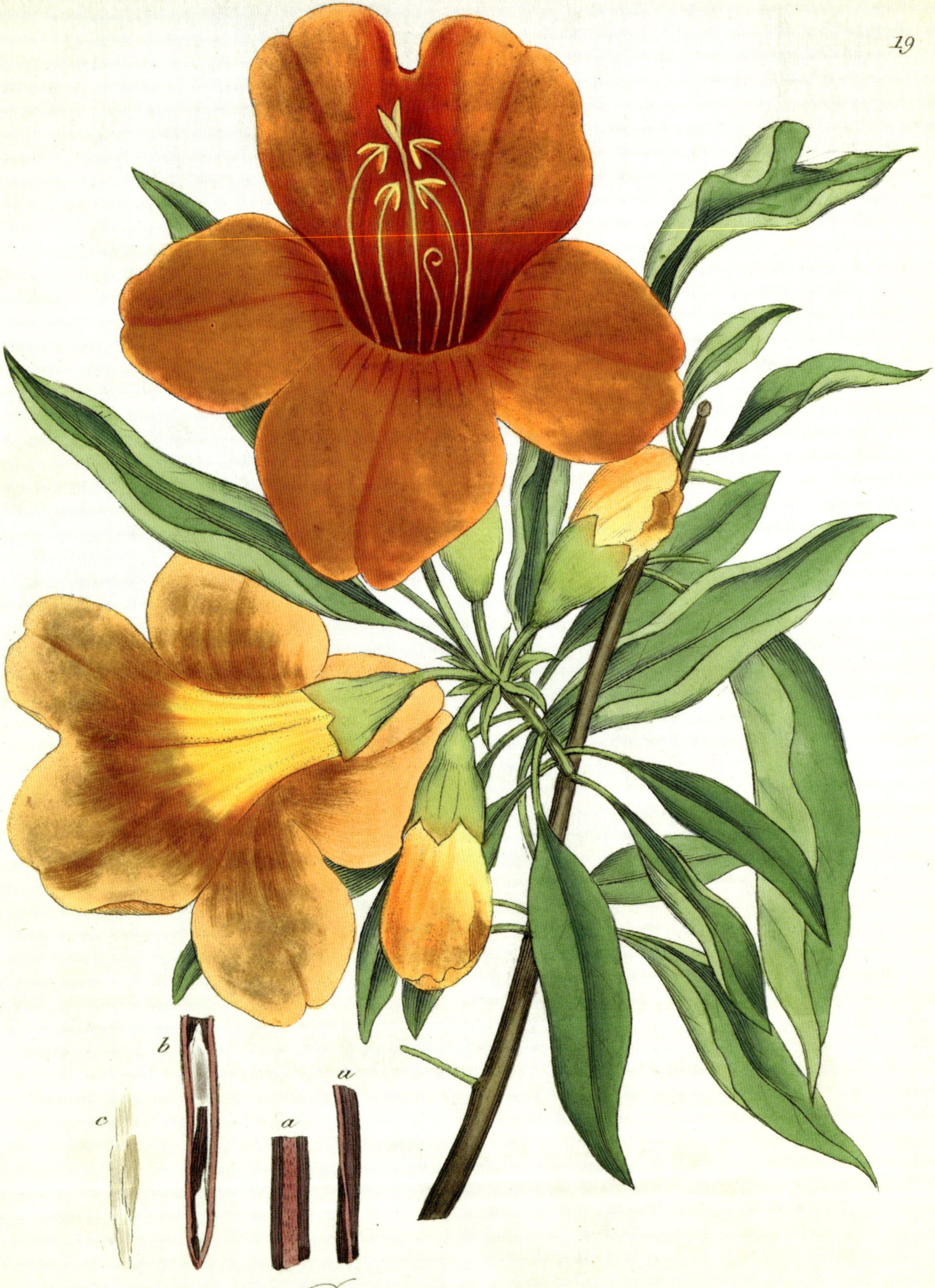

Bignonia undulata.

March 1. 1805. Published by Jas. Sowerby. London.

7/5 Wave-leaved orange trumpet-flower (*Bignonia undulata*). A large tree discovered 'in the northern forests of Hindustan'. The emphasis in the illustration of the flower is a characteristic of many of the plates in *Exotic Botany*.

volumes, original drawings and accompanying unpublished papers.[22]

Flora Graeca continued on its way as a serial publication for many years. It was a useful source of income for the Sowerbys, even after James's death in 1822, but it was a problematic project through its cost and slow rate of production, not finishing until 1840, more than ten years after Smith's death and more than forty after Sibthorp's. But as Stephen Harris has said in his finely illustrated book about the *Flora*: 'the romance of John Sibthorp's journeys through the Greek Islands and the little known Ottoman Empire, his untimely death, the astounding costs, workmanship, and intellectual effort involved in its completion have all added to its mystique'.[23]

Sowerby had probably been overworking and was vulnerable. He was forever wishing to push forward and keen to start new projects; he did not like being thwarted. During the stresses of *Flora Graeca* he was starting his own *British Mineralogy* and *British Miscellany*, and working with Smith on *Exotic Botany*. Social demands and the Napoleonic wars, especially the declaration of war between Britain and France, added to Sowerby's real and perceived burdens. He was able to pay his way out of the summons to take an oath for serving in the Militia of the County (in 1803) but a little later he ended up as a Special Constable, to act under the direction of the Civil Magistrates 'in case of the Enemy attempting to invade this Kingdom', and also as a parish officer in Lambeth.[24] Life was demanding.

Smith was to avoid any risk of not being the recognised author of the planned collaborative work on exotic plants, *Exotic Botany*. There was to be no repeat of his mistake over the authorship of *English Botany*. His sensitivity on the topic even surfaced over the financial arrangements when he suggested that should sales be high (into the low thousands) he should receive a share of the profits on the grounds that the success of the work might 'partly' be owing to his writing it (his underlining).[25] The work was planned to be of rare and beautiful plants that could well be cultivated in British gardens and would, like *English Botany*, have hand-coloured plates and scientific descriptions. Just as with *English Botany*, Smith wrote the text and made remarks on each plant's qualities, history and modes of treatment. Sowerby produced the figures and the whole thing was published by him. The work was dedicated by Smith to his friend, the banker, historian and patron of the arts William Roscoe, who among his many activities founded a botanical garden in Liverpool. He had invited Smith to give a series of botanical lectures in Liverpool; these were a success and the two men became long-standing friends. Smith's wordy preface includes reference to the artist and natural historian:

> Of Mr. Sowerby's abilities for his share of the task it would be improper, as well as

7/6 Red-gum eucalyptus (*Eucalyptus resinifera*). This can grow to become a large tree. A native of New South Wales said to exude a red-coloured resin. Its first English flowering (1804) was in a greenhouse at Paddington, London, and the specimen was used by Sowerby to illustrate it elegantly for *Exotic Botany* in 1806.

> superfluous, to express any thing here; but it would be ungrateful in me not to say that, after so many years' experience, I enter on a new work, assisted by his pencil, with the most perfect confidence and satisfaction.

This rather implied the work was Smith's initiative whereas it was really a shared one, and with its future determined by Sowerby. But Smith had high hopes and having agreed to receive money for his efforts rather than copies to sell, the terms were set such that if more than 500 copies were sold Sowerby 'would allow' Smith to receive '2 guineas and half addition and so on for every 500 in future'.[26] But sales never reached such levels, despite Smith advertising the work in London newspapers and during his lectures and visits to Liverpool (where booksellers had orders for about eight or ten sets!). Any thoughts of the work selling thousands of copies were soon forgotten. Within just over a year Smith sensed that Sowerby no longer wished to continue but he nevertheless sent plants for drawing while tentatively checking whether or not he should continue with writing his descriptions. When he learned of Sowerby's plan to stop the publication, even after a slight increase in sales, he offered to forgo his earnings for a period and suggested that sales would increase because of its good reputation, finally telling Sowerby 'pray cheer up, & do not distress yourself nor me by any groundless fears'.[27]

But the trials and tribulations were not just those concerning sales. While Sowerby was still much out of heart about *Flora Graeca*, a new difficulty came through Richard Salisbury's possible criticism of a plate in *Exotic Botany*. Salisbury, a Vice-President of the Linnean Society, had earlier dismayed Smith with his published criticism of the text of *English Botany* and was now possibly repeating the process over an inaccuracy in the drawings of a plant. Smith expressed his deep anxiety to Sowerby and asked how such an unfortunate fault could have occurred, mindful of Salisbury's vindictiveness.[28] We do not have Sowerby's reply, although he appears to have been upset by such happenings especially since accuracy of representation was his continuing mission. Smith, on the other hand, calmed down quickly and was soon asking Sowerby to run errands in London (such as buying a mahogany writing desk for him) and agreeing that the plate would have to be published as it was. A little later Salisbury resigned from the Vice-Presidency, which was a relief to the membership of the Society.[29]

Even without Salisbury, *Exotic Botany* was never going to be a success. It probably lacked a sufficiently clear focus at a time when other floras contained more scientific information. Competition was also strong and included Curtis's continuing *Botanical Magazine*. The plates are, however, of a high standard as expected of Sowerby – accurate and often

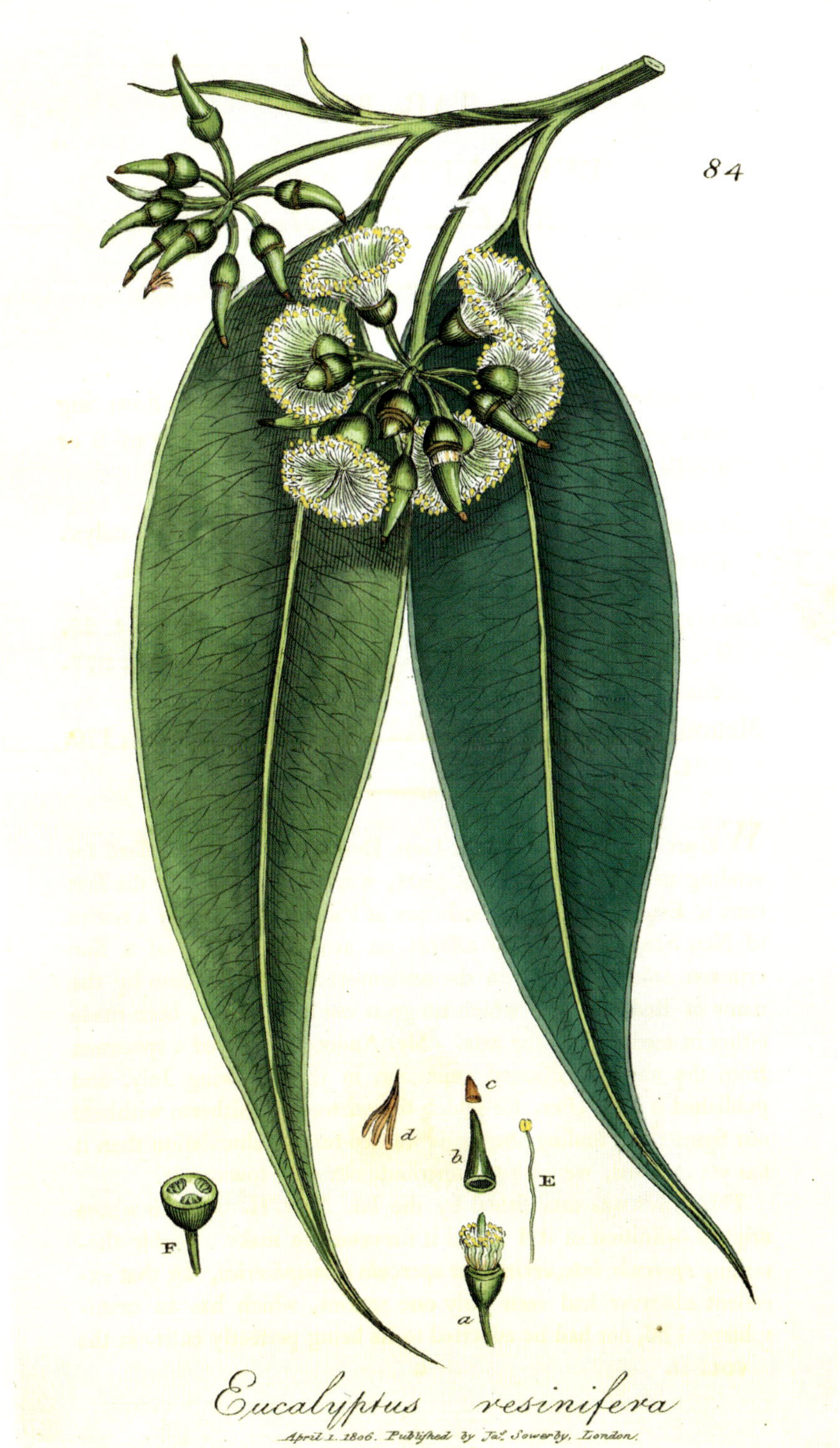
84
d
c
b
E
F
a
Eucalyptus resinifera
April 1. 1806. Published by Jas. Sowerby, London.

7/7 Blue dotted diuris (*Diuris punctata*). A beautiful plant growing two feet tall and a native of New South Wales. The genus was first described in the *Transactions of the Linnean Society* and at that time none had been brought alive to Europe. *Exotic Botany.*

7/8 Purple roscoea (*Roscoea purpurea*). A fine engraving for *Exotic Botany* of a Nepalese plant, based on drawings and specimens owned by the botanist Dr Francis Buchanan.

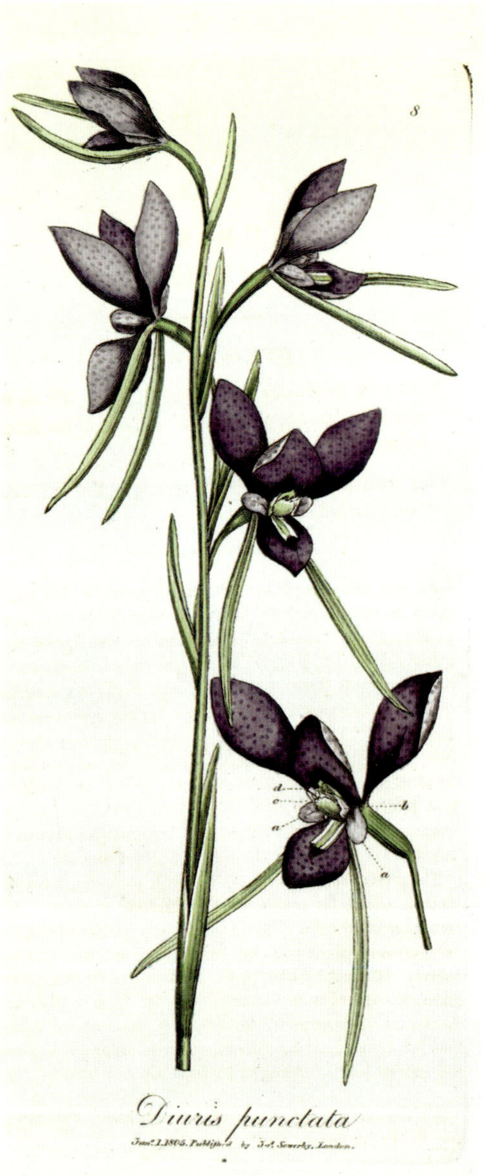

beautiful, some delicate and refined, others blousy and ostentatious as are the plants themselves. A great variety of plants feature: some with attractive foliage such as *Canna*; or with exotic flowers such as *Hedychium* and *Alpinia*; or with unusual and complicated flower structures such as *Roscoea*; or even with remarkable properties such as the aromatic *Amomum* and the carnivorous *Utricularia*. There are also fine examples of *Epidendrum* species from the orchid family. There was no shortage of potential specimens. Smith had told his dedicatee, Roscoe, that it 'it will abound with rarities & novelties',[30] and his preface explained that many plants would come from both Sowerby's and his own collections, many of which had not been published. Indeed, Smith cultivated some himself. Several also came from Sir Joseph Banks's substantial herbarium and other contributors such as the mineralogist and horticulturist Charles Greville, who had a garden in Paddington Green, Middlesex. Specimens originated from South Africa, Jamaica, Australia, regions of India, Nepal, and other exotic places, whether the tropics or the mountains. Sowerby also took the opportunity to use drawings he had made much earlier, from the greenhouses of Sion House, the stove of Messieurs Lee and Kennedy (whose nursery was at Hammersmith), and the Duke of Portland's hot-house at Bulstrode. Some, such as those from Nepal, were redrawn or engraved by Sowerby from original drawings

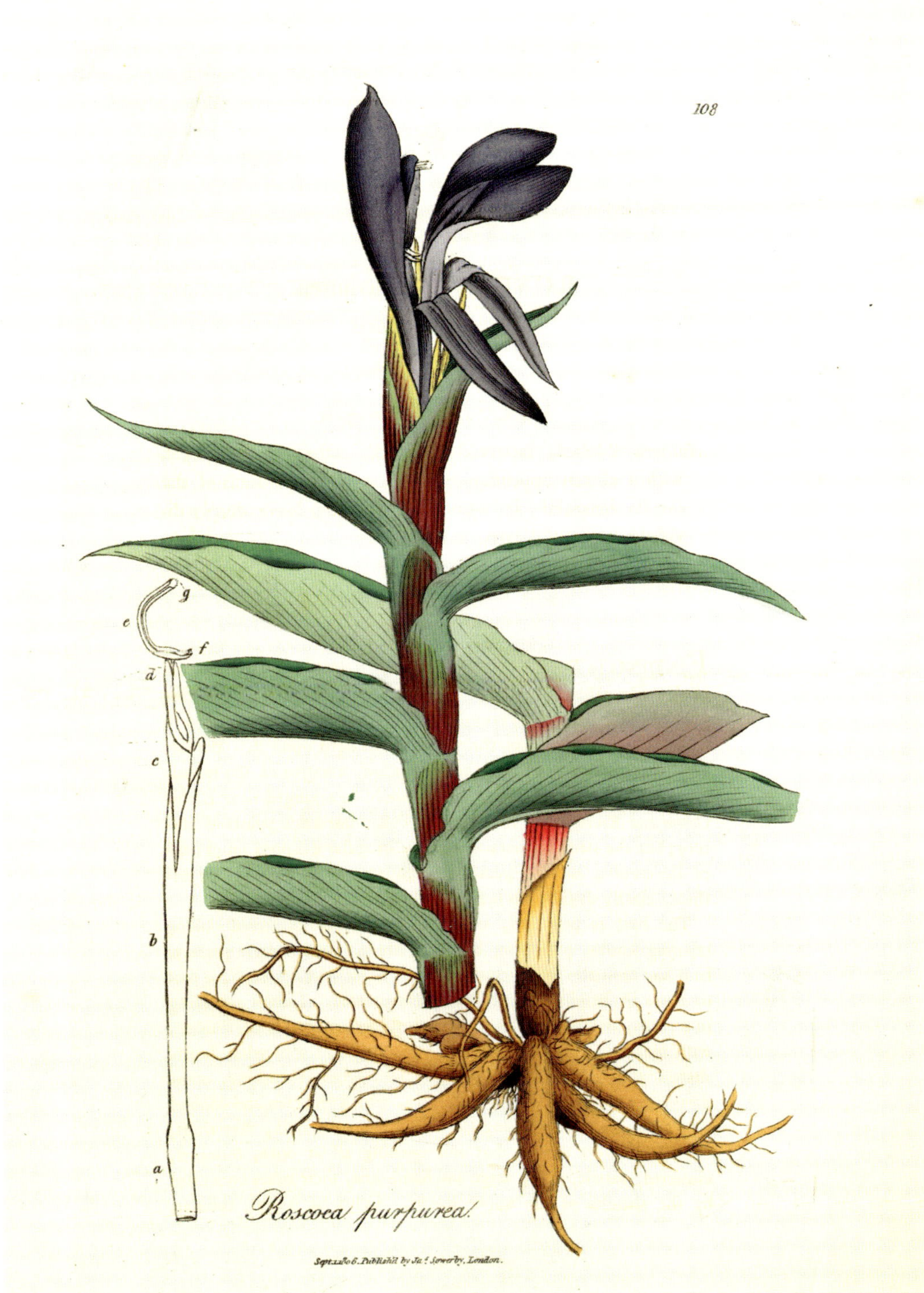
108
g
e
f
d
c
b
a
Roscoea purpurea.
Sept. 1. 1806. Publish'd by Ja.s Sowerby, London.

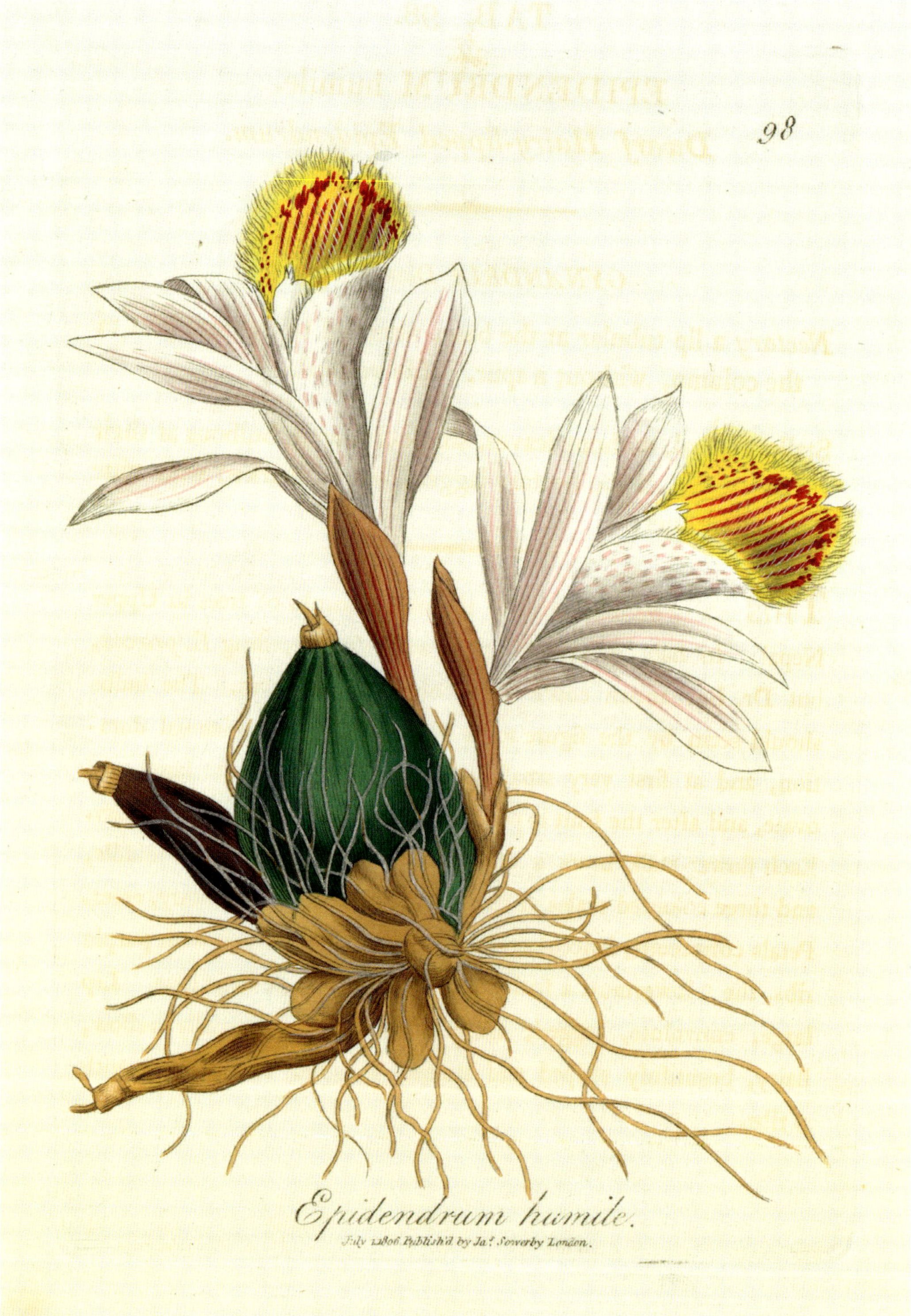
98
Epidendrum humile.
July 1.1806 Publish'd by Jas. Sowerby London.

7/9 Dwarf hairy-lipped epidendrum (*Epidendrum humile*). Grows on the mossy trunks of trees in Nepal. The engraving by Sowerby was based on one of the drawings owned by the botanist Dr Francis Buchanan.

by others.[31] The short text accompanying each plate is straightforward, often relatively low on scientific information (despite the claim of the authors) and pitched for the horticulturist, with some new species described for the first time.

Smith had a strong commitment to the work and although he knew of the poor sales he tried repeatedly to persuade Sowerby to carry on, suggesting again that he would forgo his money, but to no avail. He complained bitterly to Sowerby, asking for an explanation over delays in issuing some of the numbers and becoming exasperated when his letters went unanswered. In a rash moment he even suggested that he would work with someone else for he did not wish his material to lie idle, but then quickly said that he would not work with anyone all the time that a Sowerby carried on with what he called the 'executive' part.[32] Smith informed Roscoe, the dedicatee, of his dismay over Sowerby and his knowledge that booksellers were not making enough sales.[33] Their relationship was strained again, even though Smith sensed the inevitable. *Exotic Botany* ran to just two volumes with one hundred and twenty plates, with several of these being published rather late, nearly four years after the start. Yet, despite all this, Smith always expressed his esteem of Sowerby, even if he was sometimes critical of the way he managed his affairs. When Roscoe had intimated he wished to buy more copies of *Exotic Botany* Smith assured him that Sowerby was a most honest and honourable man albeit bad at accounts, and warned him to make sure he received good copies.[34]

Sowerby was now in his late forties. His way of life and business had reached a new phase. Some of it was not satisfactory, such as the problems with the two *Floras*, and some of it was rather routine, such as with *English Botany*, but much of it involved new stimuli and a need to fulfil his driving ambition to learn and communicate the natural history of Britain and even beyond. Mineralogy and zoology were just as much a part of this history as botany, from which he now partly diverged. A look at the nature of publications over time shows a distinct shift – those before 1804, whether his own works or as the artist for others, are almost exclusively botanical; those after are predominantly, but not exclusively, his own and are more about minerals, crystals, fossils, animals, and even the physics of light. He was an inquisitive observer – his passion to tell people what was 'out there' continued unabated. Unlike Smith and many others, he was not content to devote his life to just one field of study. This meant, of course, that he would be unlikely to reach new and deeper levels of scholarship, but that was not his way – all of natural history was his arena.

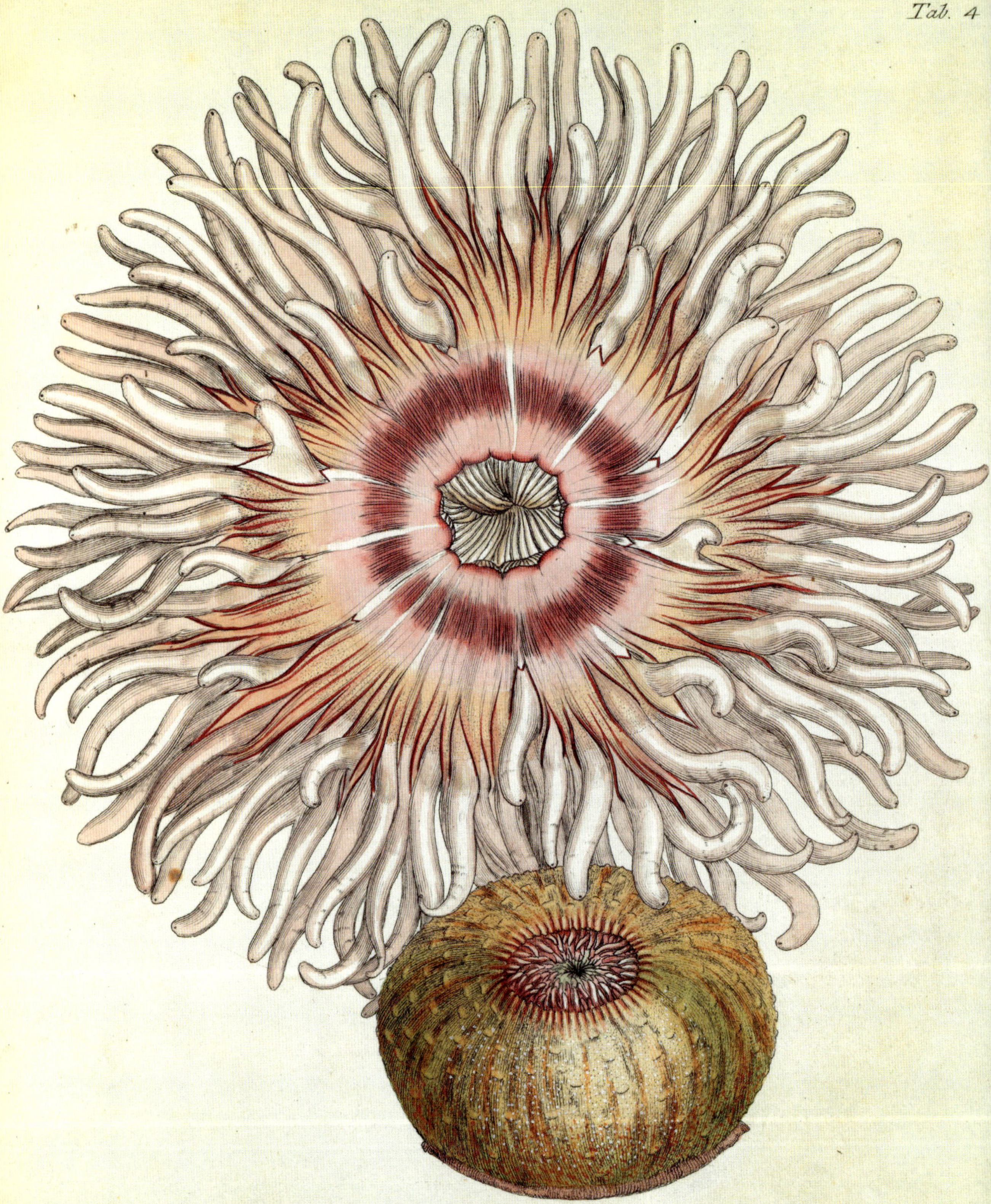

Oct.r 1. 1804. Published by Ja.s Sowerby, London.

CHAPTER EIGHT

Sowerby's whale

8/1 The common sea anemone, *Actinia equina*. From Sowerby's *British Miscellany* (1804).

8/2 (overleaf) 'Sowerby's whale'. The illustration, which accompanied the first description of *Physeter bidens* (now *Mesoplodon bidens*), in Sowerby's *British Miscellany* (1804).

If one thing alone can give a person lasting fame then for Sowerby it is a species of whale. A fine but unusual specimen was stranded on the coast of Elgin, Scotland, in 1800 on the estate of James Brodie, who was not only member of Parliament for Elginshire, but more importantly keenly interested in natural history and a Fellow of the Royal and Linnean Societies. Brodie immediately put himself to work, observing and measuring the large animal (some sixteen feet long), and soon arranging for its head to be sent to Sowerby at Mead Place. Sowerby and Brodie exchanged a lively correspondence and shared a mutual enthusiasm. This was fortunate for Sowerby; he was not an established zoologist but he was judged to be a good and keen observer, well capable of rising to the challenge of making an accurate description and of taking the pains to identify it properly.

Identification was both difficult and easy. Difficult because this was an unusual whale not matching any known species; easy because it had distinguishing features strongly suggesting a new species. It was a beaked whale, that is one with a jaw showing some similarities to a bird's beak, and, most unusually, with two short teeth positioned approximately in the middle of the lower jaw. Luck was on Sowerby's side for this was a male and it is only the males that properly reveal the teeth. James Brodie provided important information on the size, colour and other characteristics. Sowerby himself made careful

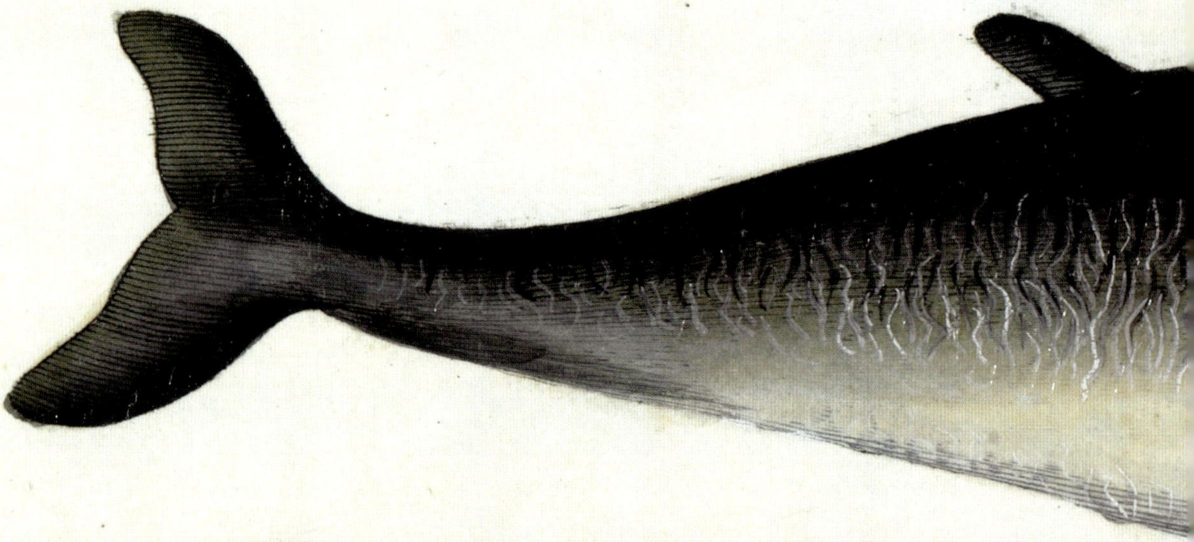

Oct. 1. 1804. Pub

1

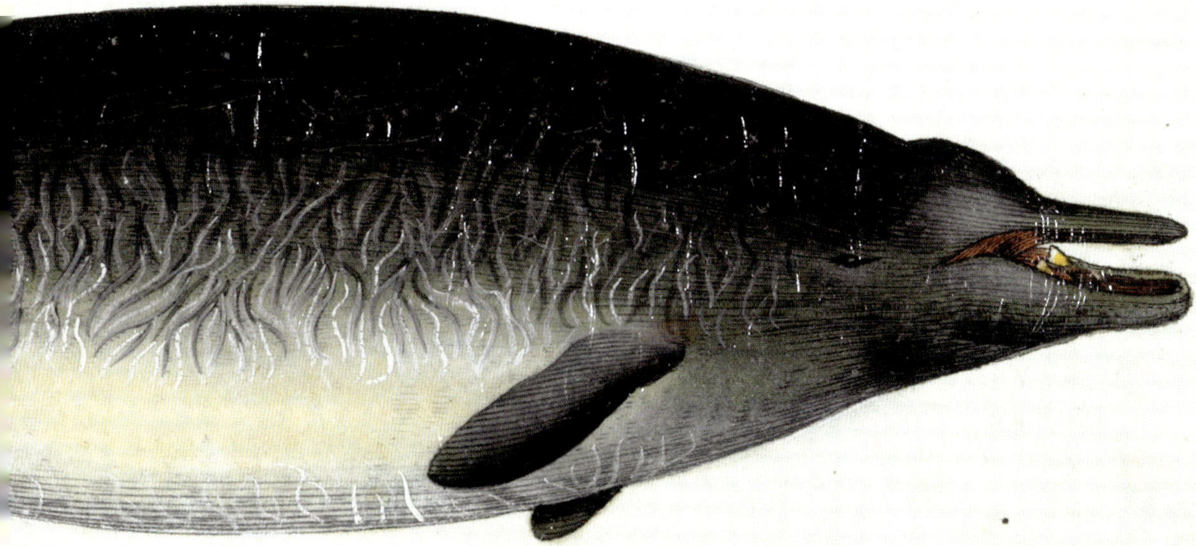

d by Jas. Sowerby, London.

8/3 The skull of Sowerby's whale over which there was discussion at Joseph Banks's house. Beached at Elgin in Scotland, 1800. The specimen (OUM 6998) is now in the Oxford University Museum of Natural History. Credit: Oxford University Museum of Natural History.

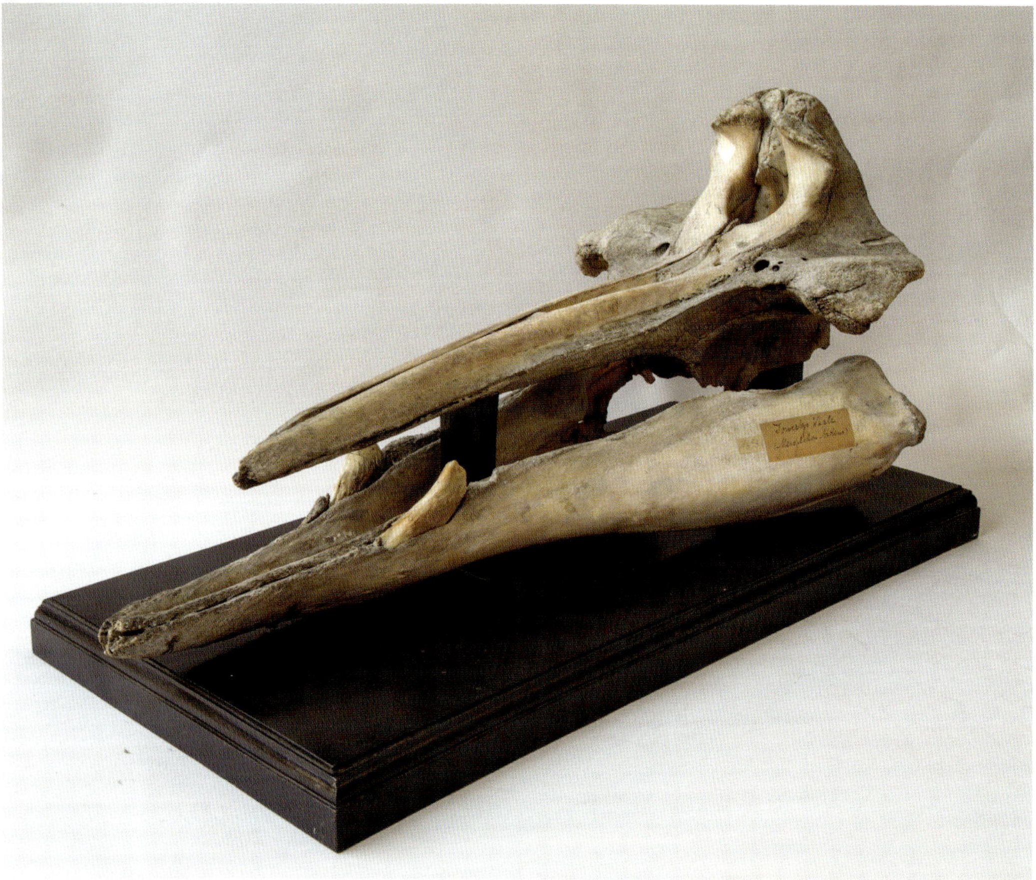

observations of the head, drew sketches and also discussed his specimen with fellow naturalists, including a session with others at Sir Joseph Banks's residence in Soho Square, probably at one of Banks's breakfasts. He presented a sketch of the animal to the Linnean Society in March 1804, describing it as a new whale, and was confident enough to make it the first entry in his new work on animals, *British Miscellany*, published later the same year.[1] He named it *Physeter bidens*, the word *bidens* from the Latin for two teeth. No one argued or disagreed; Sowerby now had credentials as a competent zoologist. Later authorities on whales, such as Henry Dewhurst,

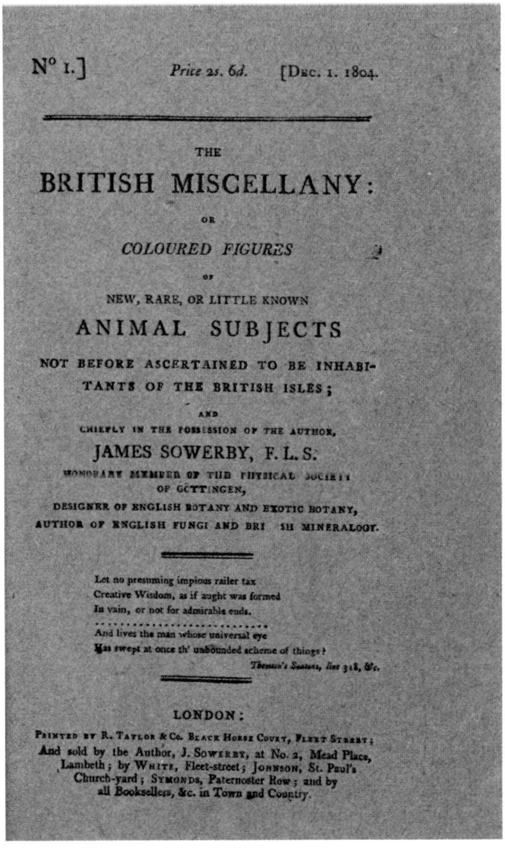

N° 1.] *Price 2s. 6d.* [Dec. 1. 1804.

THE

BRITISH MISCELLANY:

OR

COLOURED FIGURES

OF

NEW, RARE, OR LITTLE KNOWN

ANIMAL SUBJECTS

NOT BEFORE ASCERTAINED TO BE INHABITANTS OF THE BRITISH ISLES;

AND

CHIEFLY IN THE POSSESSION OF THE AUTHOR,

JAMES SOWERBY, F. L. S.

HONORARY MEMBER OF THE PHYSICAL SOCIETY OF GÖTTINGEN,

DESIGNER OF ENGLISH BOTANY AND EXOTIC BOTANY,

AUTHOR OF ENGLISH FUNGI AND BRI SH MINERALOGY.

Let no presuming impious railer tax
Creative Wisdom, as if aught was formed
In vain, or not for admirable ends.
. . . .
And lives the man whose universal eye
Has swept at once th' unbounded scheme of things?

Thomson's Seasons, line 318, &c.

LONDON:

Printed by R. Taylor & Co. Black Horse Court, Fleet Street;
And sold by the Author, J. Sowerby, at No. 2, Mead Place, Lambeth; by White, Fleet-street; Johnson, St. Paul's Church-yard; Symonds, Paternoster Row; and by all Booksellers, &c. in Town and Country.

8/4 The blue cover, or wrapper, for the first issue of *British Miscellany*, 1 December 1804.

supported the observations and naming.[2] James Smith, however, bemoaned the fact that Sowerby was no longer devoting his life to botany alone.

Physeter bidens (now renamed *Mesoplodon bidens)* is not a common whale. It is rarely observed and only occasionally found stranded on British or other shores; when it is, it makes the news and is still referred to as 'Sowerby's whale'. Circumstances had been kind to Sowerby: the pursuit of natural history was enjoyable when it was like this rather than battling with executors over the production and selling of a foreign flora. He was in control and enjoying the sense of discovery through observing and drawing, what was for him, a whole new world of animals. Furthermore, a fine specimen of a whale skull was added to his museum for all to see. There was every encouragement to follow this new path.

Many popularising naturalists such as Gilbert White, Thomas Beswick and Thomas Pennant were revealing more about the variety of nature and using fine illustrations to aid their purpose. The sheer scale of animal life, as well as plants, was becoming better appreciated. The optical microscope was an important tool for the natural historian, not only in studying plant and animal components such as the cell, but in observing morphological details, especially those relevant in seeking relationships between different organisms, either living or fossil. The Sowerby household was a place where new techniques were welcome; George especially was an adept user of the microscope, especially in his study of insects, which would feature in the new work.

British Miscellany first appeared at the end of 1804 as a monthly issue or part wrapped in the usual blue paper, and containing four coloured plates with a short text of description and classification. Its enticing title indicated a publication about really new things and ones

8/5 An opah or kingfish (genus *Lampris*), found near Weymouth and recorded as a very rare occurrence in British waters. *British Miscellany.*

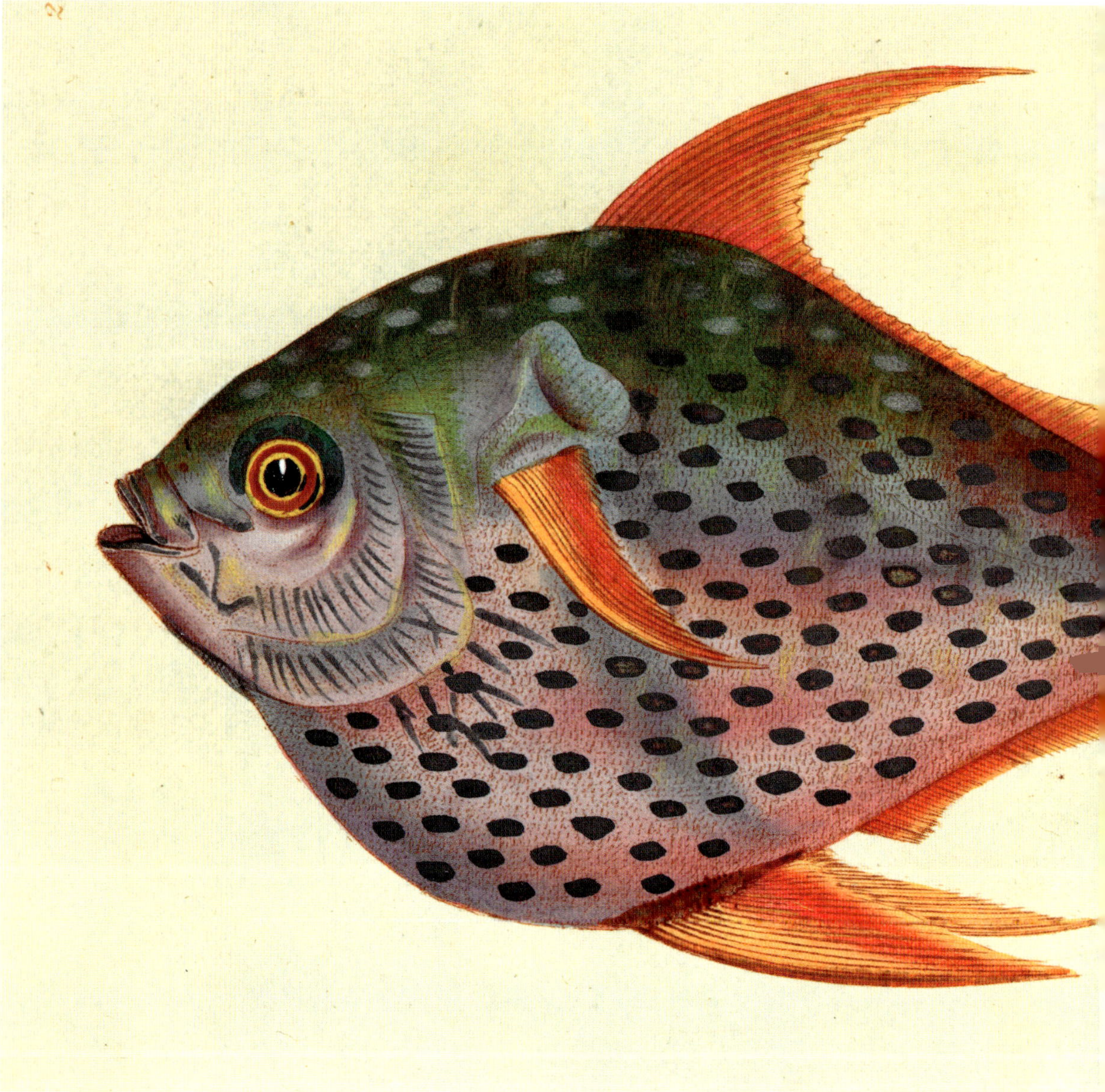

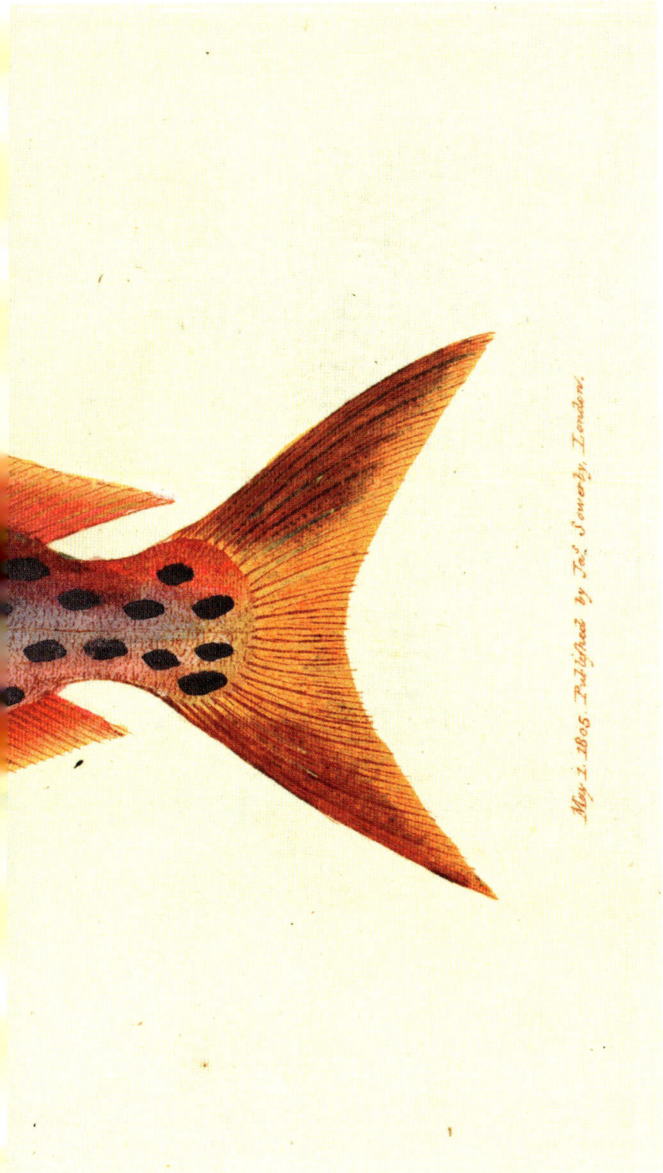

which could be seen at the author's own museum if the reader cared to visit: *The British Miscellany or coloured figures of new, rare, or little known animal subjects many not before ascertained to be inhabitants of the British Isles; and chiefly in the possession of the author.* Insects, birds, worms, mammals, crustaceans, coelenterates, sponges, reptiles and others are included. Insects make up nearly half the work. It is a kind of potpourri of life with no reason for its existence other than that the animals were considered rare or new, and also, of course, that Sowerby had acquired some of the specimens. He described several species new to Britain for the first time, such as the colourful starfish *Asterias equestris* (provided by his friend James Brodie) and the bryozoan *Millepora compressa*.[3] Some are still taxonomically valid and so have the descriptions attributed to him. One such, a fish parasite, is a fine example of his observational powers, even regarding a very small animal which would have been unfamiliar at the time.[4] He told of how in 1799 and again in 1804 he had observed a 'strange and curious animal fixed to small sprats' and notes how this 'grievous parasite, like a harpoon inflicts a serious wound' and that its 'neck is not only long so as to penetrate deep but is notched to secure it in addition to the barbs on the sides of the head'. The parasite embeds itself in the eye of the fish and sometimes in the dorsal musculature – both situations shown in his illustration. In this example his recorded observations are

8/6 A parasitic copepod (in the genus *Lernaea*) first described by Sowerby in his *British Miscellany*. The parasite is attached to the fish's eye and back of the head.

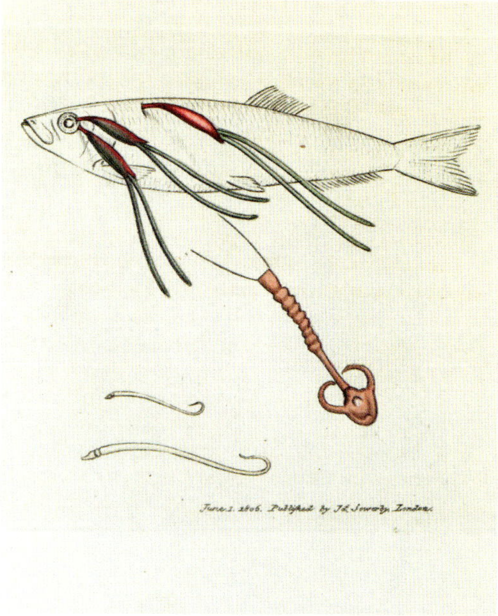

sufficient to distinguish this species from other closely related ones but for some animals his descriptions are often brief, sometimes almost negligible, while his illustrations can be detailed and accurate. Occasionally a full account would be given, especially if there were doubts over the naming, as in one case 'so that our readers may be better able to judge of its claim to be considered as belonging to a distinct genus'.[5] Then reference to other published works was made and different examples examined.

Sowerby's choice of specimens reflects his enthusiasm for every part of natural history whether large or small: a leech collected by him from the side of the River Thames, a new species of insect collected by his son in Swansea and a small larva of a lacewing covered with lichen are just some examples. *British Miscellany* was appropriately dedicated to Lord Seaforth (who had provided Sowerby with a collection of birds)[6] and James Brodie (who supplied the whale's head and other specimens).

Numerous naturalists, including William Hooker, James Brodie, Aylmer Lambert and Charles Lyell, who had been sending him plants for *English Botany*, now sent him worms, butterflies, wasps, beetles, shrimps, fish and much besides. The botanist Aylmer Lambert sent a fine gillaroo trout collected from Lake Carra in Western Ireland, together with notes on his observations of it.[7] Later, Sowerby worked for Lambert by illustrating some of the species in his fine work on the *Pinus*. The often-called 'father of British ornithology' George Montagu drew attention to an unusually long marine worm, the black line-worm, *Lineus longissimus* (now known as the bootlace worm; it can reach lengths of a few tens of metres). Montagu was unable to send a specimen because it was so prone to decompose, and so when Sowerby was finally able to illustrate it he reported on the difficulty of obtaining specimens:

> It seems to have been long and well known to the fishermen of the coast; but after they have told one that it is many fathoms in

8/7 Sowerby's precise illustration of a dragonfly (*Libellula conspurcata,* synonym of *L. fulva*). *British Miscellany.*

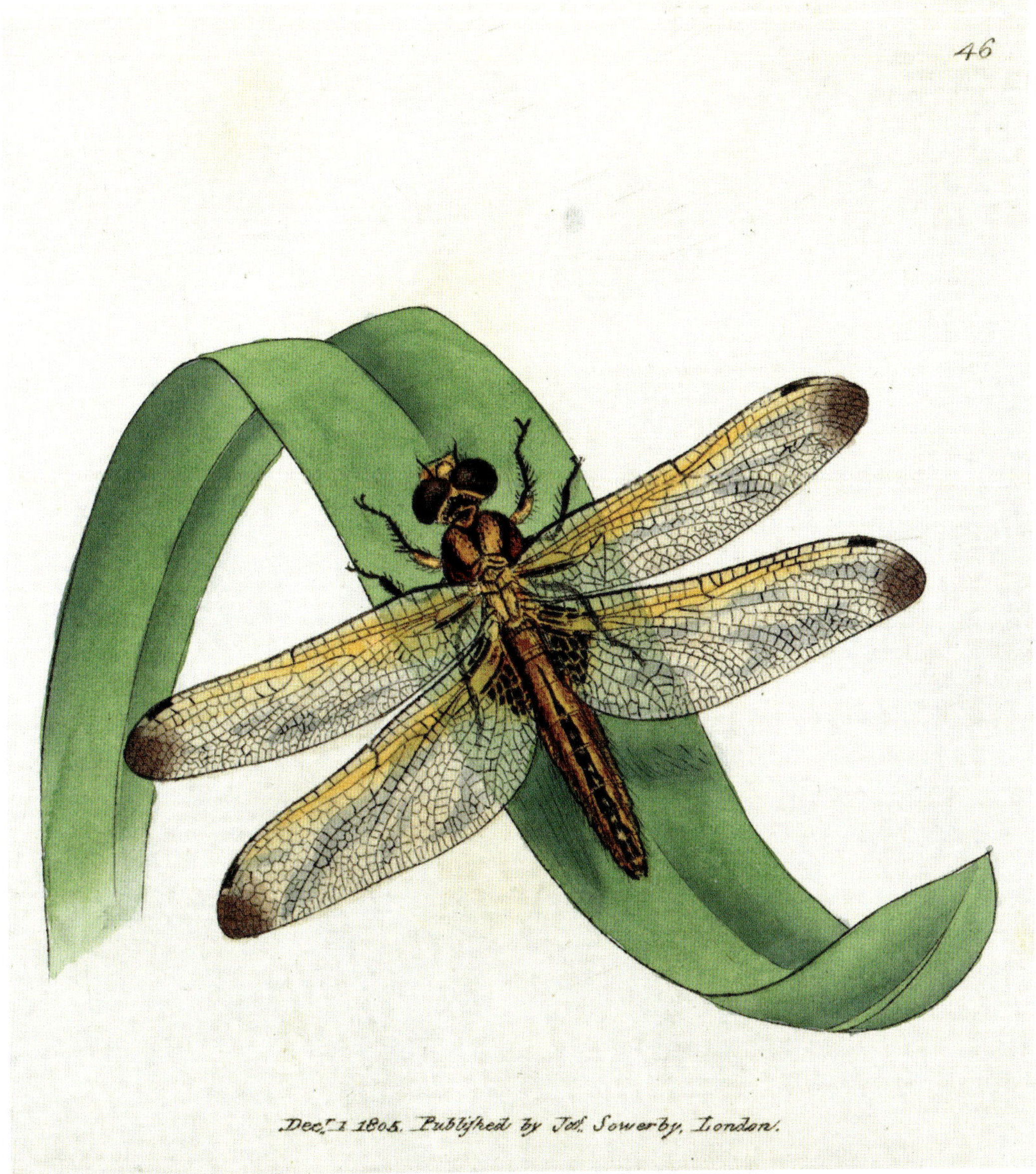

8/8 The cirl bunting. Sowerby's coloured engraving for the frontispiece of Montagu's *Ornithological Dictionary* (1802).

> length, and that though they are continually hauling them in as they would a rope, they never find the extremity, they are then sufficiently satisfied that one knows enough of the matter; and persuasion or money will scarcely procure specimens from them.[8]

Montagu often offered specimens of various kinds and expected in return that Sowerby would help him with guidance and materials for producing his own illustrations for his publications on birds and shells.[9] This Sowerby indeed did, in his usual generous way, also engraving the illustration of a cirl bunting for the frontispiece of his ornithological dictionary.[10] This was a time when Sowerby was almost obliged to fulfil people's wishes over their specimens; Montagu was far from being the only one and instructions could be demanding. The Reverend Bale sent an owl accompanied with the strong request:

> I shall be much obliged to you to have the Strix Otus well stuffed & set up for me – no time should be lost – there is a man in St Martin's Lane ... but any one who shall do it well may be employed... the bird itself is not only very rare but very interesting in all its parts... it was shot on Monday – & no time should be lost.[11]

Some specimens for *British Miscellany* were collected by his sons. George, aged fifteen, also went on a small excursion to south Wales in 1803 with the architect and natural historian Joseph Woods, where he collected insects, mosses and lichens and sent fish home to Mead Place.[12] The Reverend William Kirby, a keen and scholarly entomologist,[13] sent specimens from Suffolk and helped with descriptions,[14] sometimes accepting his errors when corrected: 'Mr Sowerby suggested to me that what I took for larvae of this insect were really pupae:- To this ingenious conjecture I readily accede.'[15] But he would also be directly critical to Sowerby, especially over the costs of sending his descriptions and appraisals of the drawings and on the treatment of specimens:

> – in the first place you ought not to put me to the expence of carriage, in the passing to & fro of subjects for British Miscellany since it is for your benefit – in the next place as a friend I must advise you when Gentlemen entrust you with subjects for your work, to return the specimens as soon as you have done with them.

He added that others also complain of this habit and finishing: 'You see I take a friendly liberty with you – my notice is the desire I feel to see British Miscellany gain ground & to keep the Cabinets of the principal Entomologists open to you.'[16] Kirby's letter has the tone of someone who considers himself a 'gentleman' dealing with someone who is not 'one of them', despite

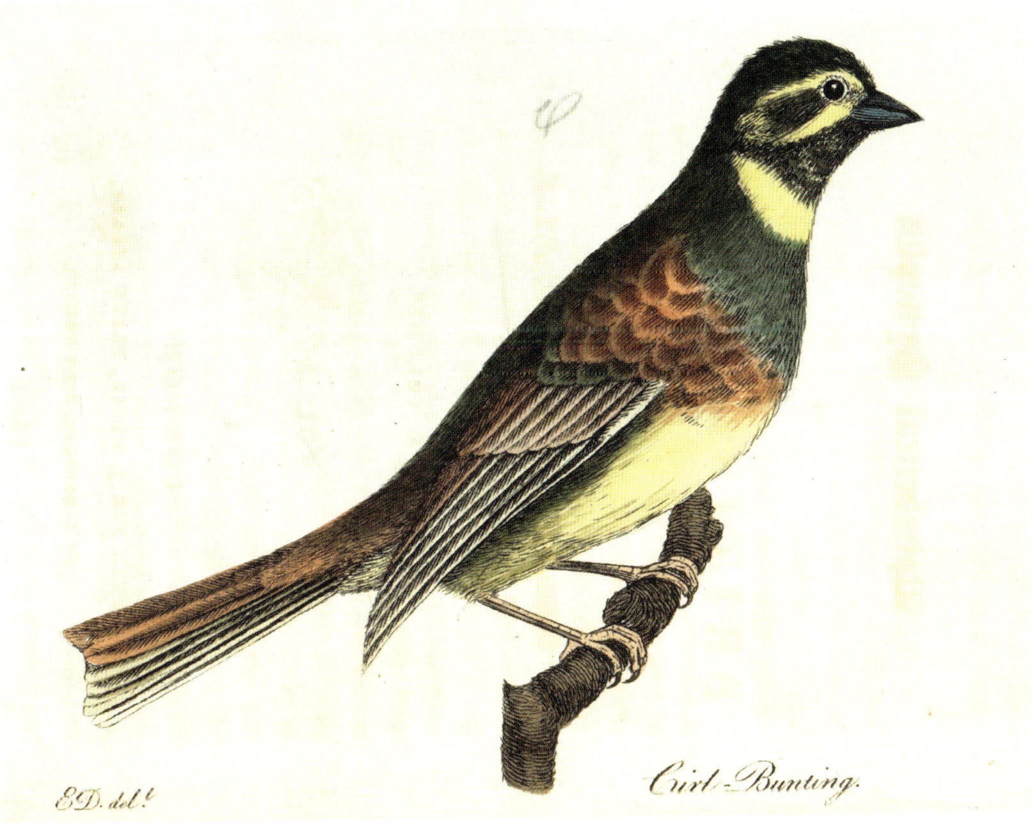

recognising Sowerby's knowledge and having commissioned him to do some illustrations for his work on bees a few years earlier.[17] Kirby was a rector and the son of a solicitor, although he had no special wealth or estate. For him Sowerby was in a different class. The Enlightenment had its social distinctions even when people were working for the same cause.

British Miscellany is a rather charming and unusual publication demonstrating Sowerby's inextinguishable urge to observe as well as to draw and describe nature. Most, but not all, of the illustrations are of good quality and are accompanied by a short, accessible text with citations to other authorities. A sense of excitement is present in descriptions of some new genera and species, which were not at this time so elusive as to cause Sowerby much difficulty. Even the pressure of all his other work was not allowed to prevent him from this initiative; there was to be no improvement in the management of his affairs nor in keeping sound financial accounts; productivity was a key objective. His family kept the business in order with his son George developing his roles as his father's amanuensis and as an entomologist. His son James was now a skilled artist and a knowledgeable botanist, especially on mosses; he was already able to fulfil many of his father's tasks. With his pleasant personality he worked successfully with other naturalists on their research and publications, acting as a family ambassador in enhancing the Sowerby business. Both sons were involved in the production of

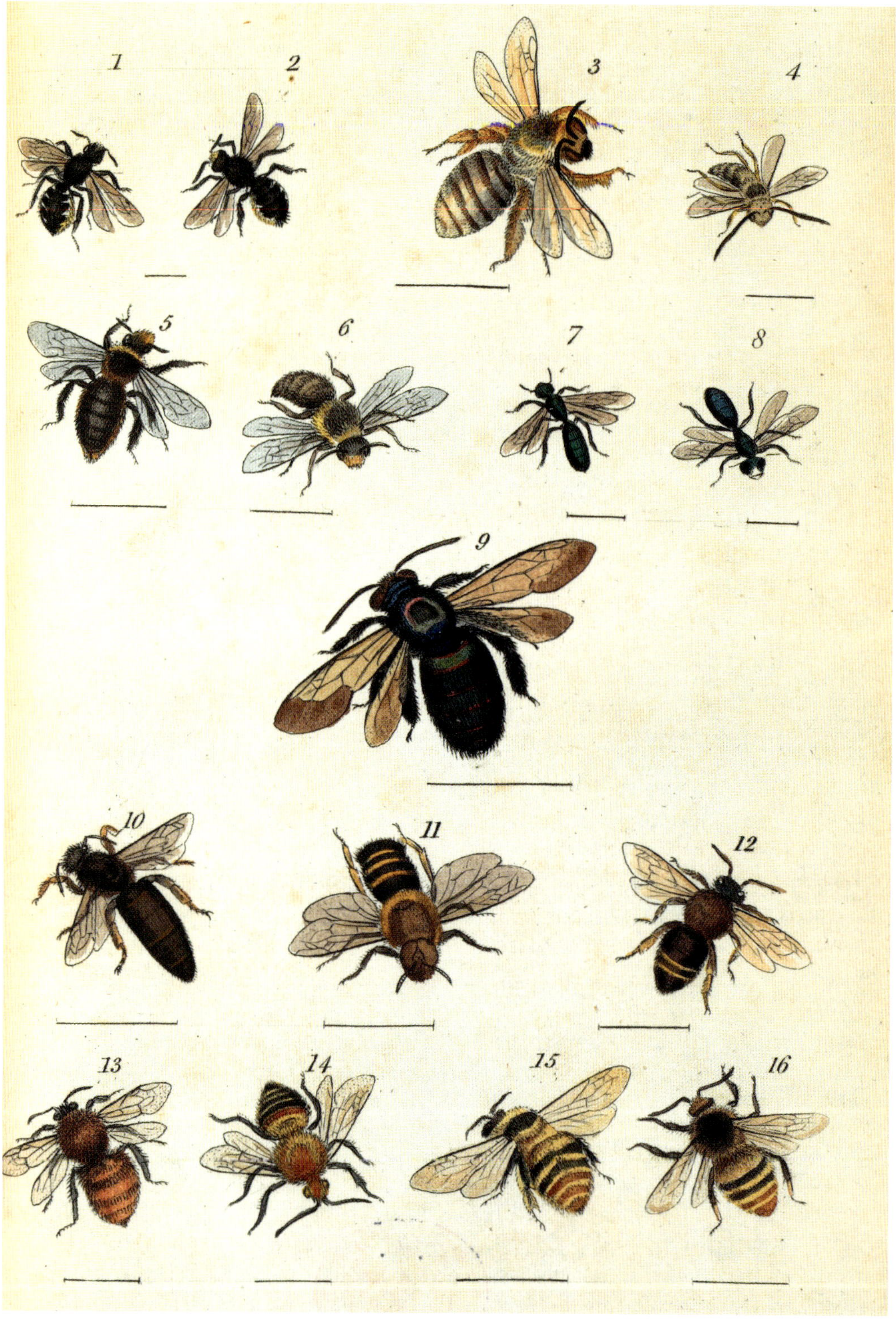
1
2
3
4
5
6
7
8
9
10
11
12
13
14
15
16

8/9 Sowerby's plate of several species of bee (*Apis*) for Kirby's *Monographia apum Angliae* (1802). Sowerby produced four coloured plates for volume two, which also showed that Sowerby had numerous bee specimens in his museum.

British Miscellany. George would study the insects with the microscope and he and James would draw some of them. Anne would assist with the production, collation of plates and text, and administer the orders.

The young James was not, however, always living at Mead Place around this time. He was becoming involved in the gentlemen's world of natural history, partly through visiting his relatives in Norwich as well as James Smith and his friends or associates, and partly through his father's desire for him to assist others such as Dawson Turner, the banker in Yarmouth, with their investigations and publications. These experiences were grist to James De Carle's own mill.

In the autumn of 1802, now aged fifteen, he stayed a short while with Turner's family. He kept in contact with Mead Place by sending specimens such as three sharks he had acquired on one of his coastal walks, and he put his mother's mind to rest by saying he had no need of shirts or stockings as they had all been washed in Norwich.[18] A longer visit followed during the next summer, to help with drawings for Turner's work on mosses, to accompany Turner's young daughters, to receive some schooling, and to collect specimens. Regular letters to his father assured him that he was doing some drawings for him, and told of how Turner was teaching him to scan verses which would help with his pronunciation, and how he might be drilled with a musket in the light of a possible invasion by the French. The sense of an invasion was high; those who were wealthy enough sent their wives and children away from the coastal towns such as Yarmouth.[19]

The visit was a success and Turner pleaded with Sowerby to be able to keep him a little longer: 'we get on extremely well together in Latin and arithmetic... I have indeed great pleasure in his company, independently of the use he is to me but I fear you are dreadfully in want of him.'[20] Sowerby agreed and soon after James De Carle wrote:

> have fired a Gun three times & rode six miles on horseback but have not been drilled, because Government having refused to take all the volunteers who have offered their services none of Mr Turner clerks have been exercised. I shall bring your spaniel to London which Mr T has had the goodness to give me. The drawings I have done will be sent to Norwich next week.[21]

He later wrote to say how sorry he was to learn of his father's skin infection but also, as a dutiful son, to remind him to do the colouring for the first three fascicles of James Dickson's work on cryptogams.

James De Carle's artistic work for what Turner called 'my little book on the Irish mosses' (*Muscologiae Hibernicae Spicilegium*, 1804) is superb.[22] He captured the intricate detail while sensing the overall appearance of the plant. He may have worked hard and long on each

8/10 James De Carle Sowerby's watercolour of *Dicranum flexuosum* for Dawson Turner's book on mosses (1804).

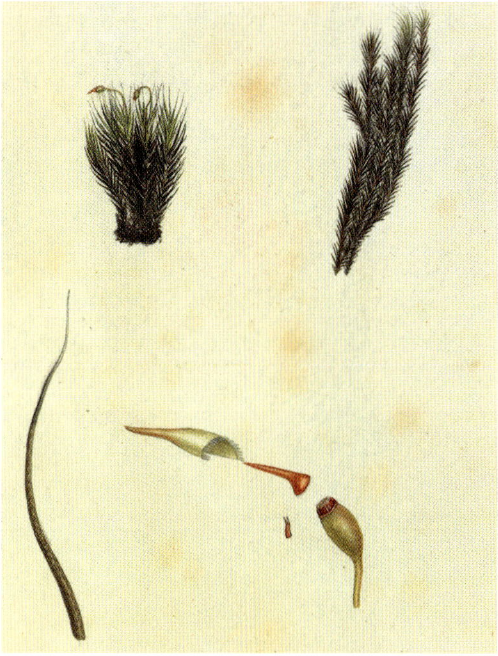

8/11 Engraving, by James Sowerby, of his son's drawing of the moss *Dicranum flexuosum*, in Dawson Turner's book on mosses (1804).

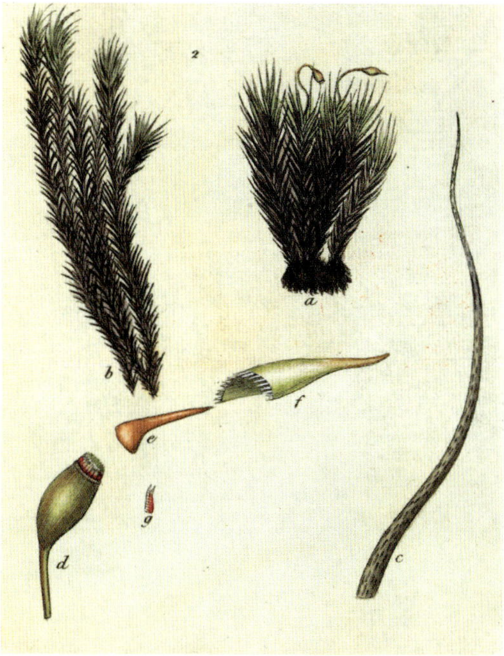

illustration but the result shows a consummate skill in one so young and with limited experience. His father was certainly involved for he gave advice to Turner to reduce the plates to octavo size, and he was certainly involved in much, if not all, of the engraving, and possibly the printing and colouring, for Turner showed his impatience with Sowerby's output from Mead Place when preparation of the work was slowed down.[23] As was often the regrettable practice, no acknowledgement was given to either Sowerby in the published volume but Turner had treated James De Carle well, almost as though he were his own son. Typical was the gift of a puppy spaniel to the young man and father, which was soon despatched to Mead Place to swell the hubbub in the busy household.

British Miscellany ran to just twelve numbers, comprising two volumes, the second of which is incomplete. Sowerby had been encouraged to continue with its publication by many, including Dawson Turner and William Kirby, but his interests and his time were increasingly directed towards minerals, rocks, crystals and the science of colour. *British Miscellany* had been enjoyable and fulfilling and it could now be laid aside, perhaps

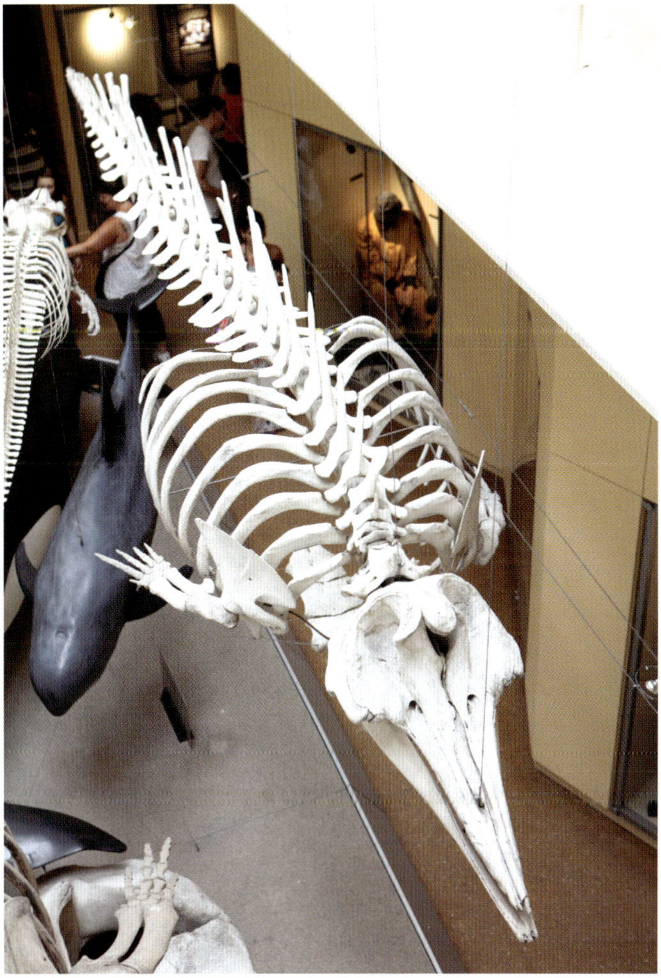

8/12 Sowerby's whale – skeleton at the Sea Museum Aquarium, Biarritz, France. Beached near there in 1951. Photo credit: Michael Venning.

the intention was for a while, but in fact it was never returned to. It was in any event an unusual publication. 'Miscellanies' did not meet the same need for knowledge as did more specialised publications such as floras. Buyers might purchase the parts for their novelty but interest could fade if the longer-term objectives were vague. The smallish size of the plates (octavo) also did little for those who might wish to collect natural history artwork for display. It is possible any financial success may have seemed very uncertain.

The whale which was stranded in Elgin in 1800 is now firmly linked to Sowerby's name. Strandings are rare but when they do occur Sowerby's name is usually given. In September 2013 the British Broadcasting Corporation reported on a 'rare whale' beached on an island off the west coast of Scotland. On this occasion it was accompanied by its young calf, still alive in the shallow water. The mother died and the calf swam out into open water with an uncertain life ahead of it. The report told of Sowerby's discovery and also that most of what we know has come from stranded specimens – something that James Brodie and James Sowerby initiated just over two hundred years ago.[24]

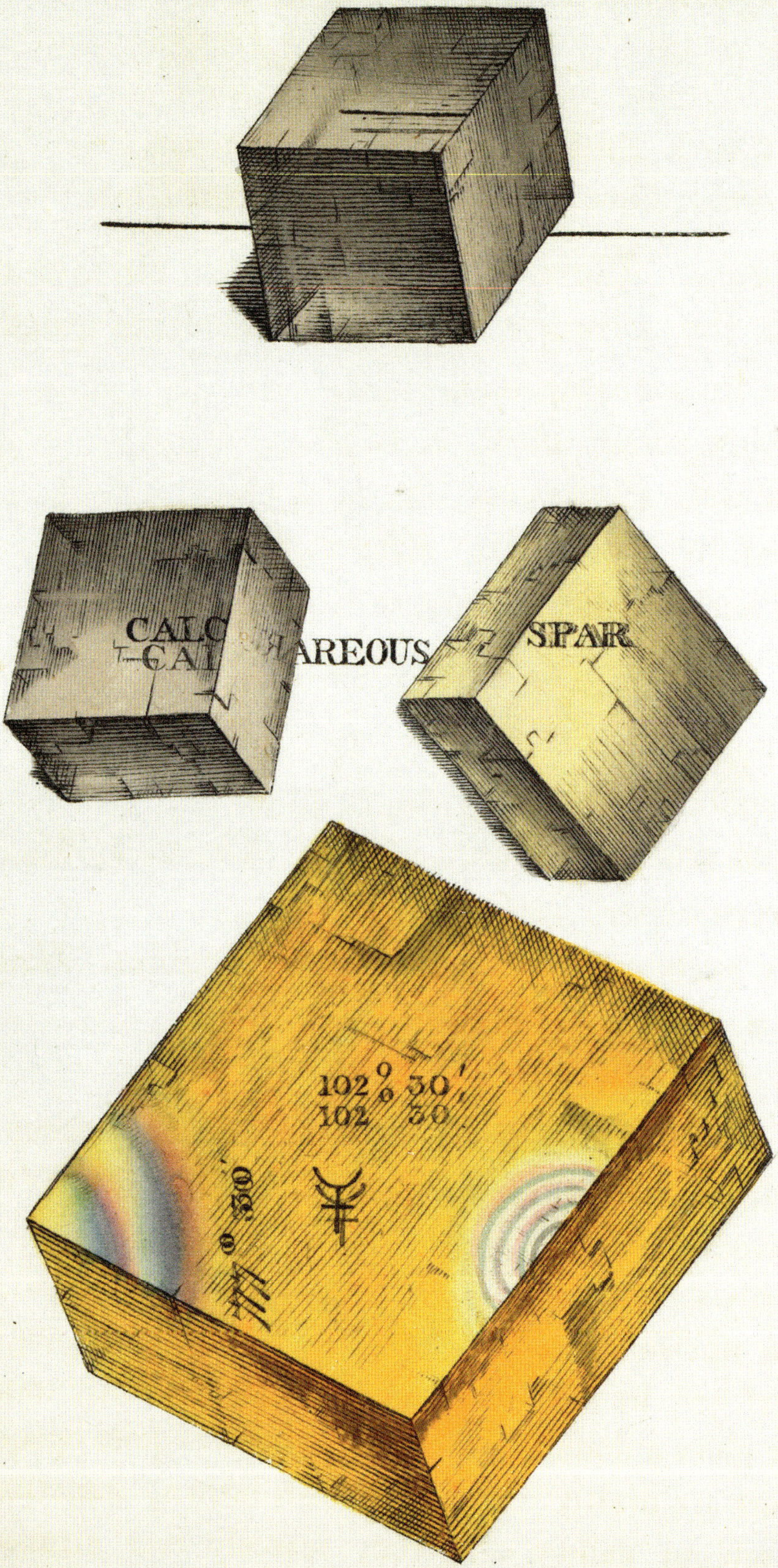

Nov. 1. 1802. Published by Jas. Sowerby, London.

CHAPTER NINE

Minerals, crystals and colours

9/1 Crystals of calcite. The second plate in *British Mineralogy*, illustrating the crystal form, interference colours and double refraction shown by this mineral.

Botany was in rather good shape in Britain at the end of the eighteenth century. Many local and regional flora had been published during the last few decades; universities had botanical posts and gardens; Kew Gardens was thriving; the President of the Royal Society, Sir Joseph Banks, was influential and an active botanist; the Linnaean classification system for plants and Linnaeus's collections had been adopted and bought; there was a good network of botanists; apothecaries helped to advance knowledge and continued to write books on the medicinal value of species, and the voyages of discovery had returned with fine illustrations and examples of exotic specimens.

Mineralogy was not in the same condition. British mines such as those for tin and copper in Devon and Cornwall, or for lead in Derbyshire, were important economically and attracted scientific enquiry but there were few university posts in mineralogy or geology. Sir Joseph Banks showed an interest but much less so than in botany, the equivalent of floras existed in only a minor way, and there were no institutions of mining or geology as were found in parts of the Continent.[1] To compound matters, there was no easily usable and agreed system of mineral classification, despite numerous attempts to devise one.

Linnaeus with his methodical and organising mind had attempted a classification using a 'sexual' analogy, rather as he had done for plants but this time assigning male and female roles to the 'salts' and 'earths' respectively. These came together to give groups (or 'classes') of minerals each of which contained genera and species. Distinguishing the various genera or species from each other was done using both chemical and surface (mainly crystallographical) features. However, other people thought differently. Some, such as the Swedish chemist and mineralogist Torbern Bergman, preferred chemically based methods; others, especially the French school through the distinguished crystallographer René-Just Haüy, favoured ones based on outward or crystalline form. The influential German geologist Abraham Gottlob Werner well understood that mineral identification was different from mineral classification and pragmatically used external characteristics for the former and chemical properties for the latter. Among all this, Sowerby adopted in part a Linnaean approach, although he was anxious later on to give his own thoughts on the matter.

Despite these circumstances there were plenty of mineral and rock collectors, some with very fine cabinets. It would often be sufficient for them just to identify the mineral, which became easier the more that minerals passed through their hands. But since many were not of species showing fine crystal faces but of the much more common irregularly shaped form, it was necessary to rely on other physical properties such as colour and hardness, and sometimes on chemical tests and analysis. Some collectors would know of the crystallographic work by continental, especially French, mineralogists such as the renowned Jean-Baptiste Romé de l'Isle and his *Cristallographie* (1783) with its insights into the 'structure' of minerals (including the constancy of the angle between adjacent similar faces of crystals of the same substance), and René-Just Haüy with his *Traité de Minéralogie* (1801) and the concept of the crystallographic unit cell. But if these advances were known about they were rarely debated or used by most British collectors. People, with some notable exceptions, did not keep their cabinets for the purpose of scientific research. Interest in mineralogy could be encouraged by many, including mineral dealers such as John Mawe in London, with their fine specimens available for display or study, even if at a price. The Linnean Society sponsored mineralogical lectures by the German apothecary and mineralogist Johann Schmeisser,[2] but unlike the Geological Society, founded in 1807, did not have geological collections.

Collectors were from several sectors of society but especially the upper and parts of the middle classes, including the clergy. It was an age in which many people could become famous just for their collections. Names which resonate still with mineralogists and mineral collectors include the Cornishmen William Borlase, the Reverend William Gregor (the discoverer of titanium), John Hawkins and Sir John St Aubyn, and others such as John Woodward (whose collection is at the

9/2 Witherite (barium carbonate) crystals on galena. Sowerby said this was the first time that it had been seen with hexagonal elongated spicules. *British Mineralogy*.

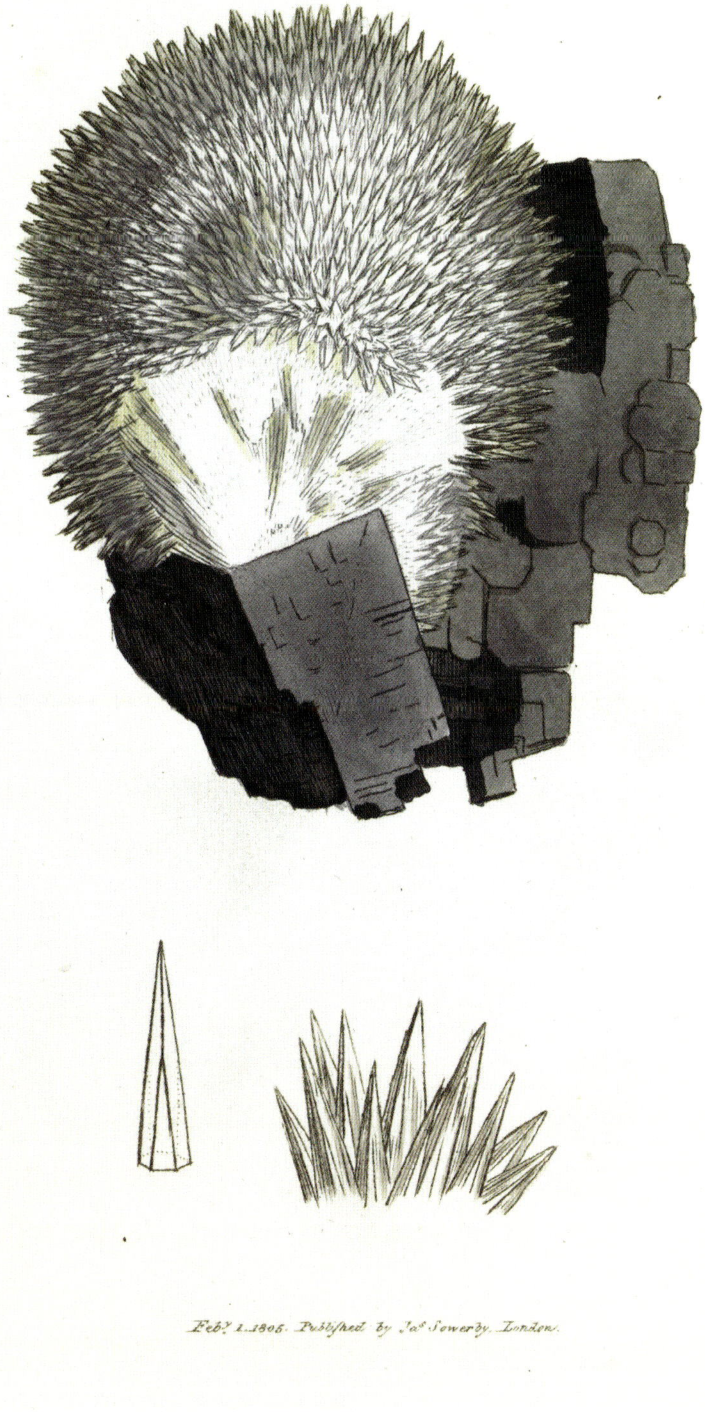

9/3 Fine specimen of copper arsenate (the rare mineral lammerite). The engraving shows the fibrous crystal form in this specimen lent to Sowerby by Philip Rashleigh. It came from Gwennap in Cornwall, once an important mining area. *British Mineralogy.*

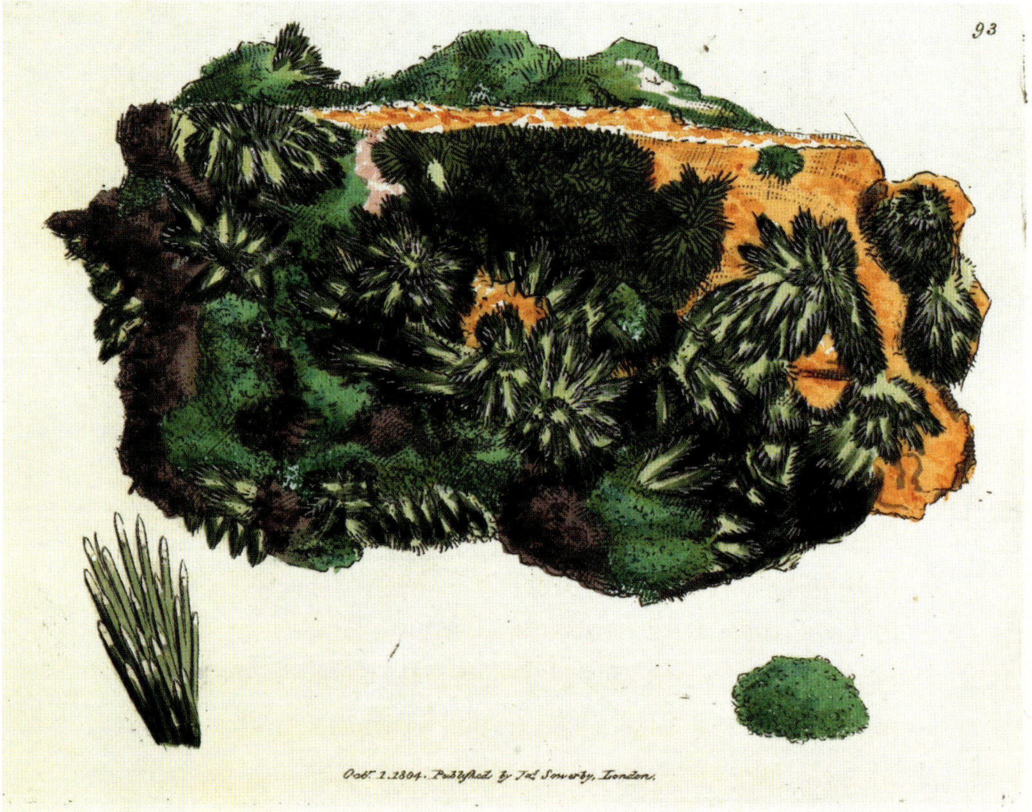

Sedgwick Museum, Cambridge), the Earl of Bute, Thomas Pennant, Charles Hatchett (the discoverer of the element niobium) and Sir Abraham Hume and William Babington (both founding members of the Geological Society, 1807). Babington and St Aubyn employed the Frenchman Jacques-Louis Comte de Bournon, who had fled France and the Revolution, to help organise their collections. He was also a founder of the Geological Society and deemed to be particularly knowledgeable about crystallography. Collections would change hands, sometimes for considerable sums of money; Sir John St Aubyn, for example, bought a collection from Babington who in turn had bought much of Lord Bute's collection soon after his death in 1792.[3]

Sowerby dealt with all these people. He was as much a collector as anyone, with minerals and rocks forming an important part of his museum.

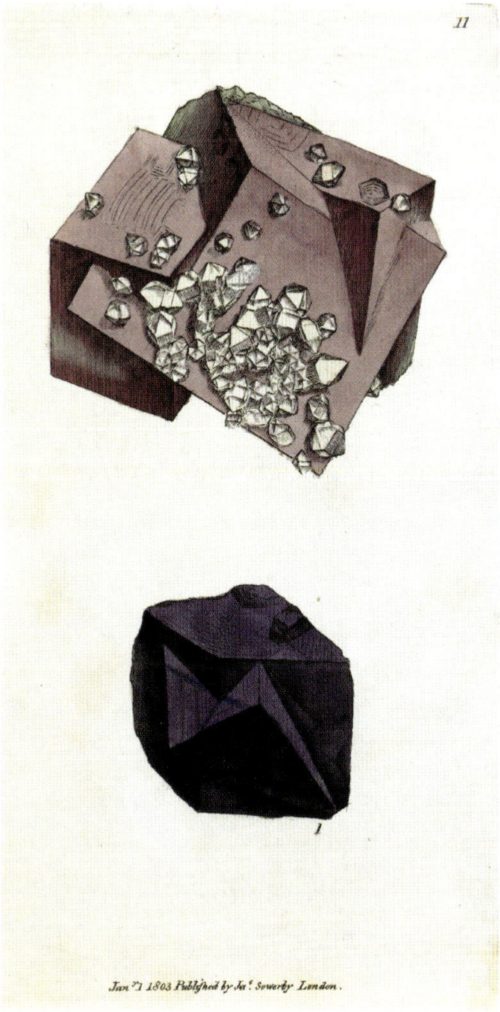

9/4 Crystal of fluorite with interpenetrant twins. The upper one has small crystals of quartz attached. *British Mineralogy.*

He traded, exchanged, borrowed and received numerous specimens for several years before and after buying, in 1800, a significant and organised collection from Mr William Day, previously an active collector who sometimes went with a pick to examine the spoil heaps in mining areas.[4] Sowerby's network of correspondents was established long before he started publication; several were in the mining areas, especially of Cornwall and Devon, and they had specimens ready to hand. He also acquired material from mineral dealers, among whom were some renowned practitioners. So with good material, a keen eye and fresh enthusiasm he was well placed to use his observational and artistic skills in a new venture – one that would complement his major work *English Botany.* But unlike *English Botany* this was to be one he would do on his own – no James Smith or equivalent person to share the burden or even inform on the science. Despite his years of collecting minerals he must have been on a steep learning curve, acquiring knowledge through his own careful observations and by studying the various books on the subject, as well as from his many correspondents and visitors. He did ask James Smith to advise him and he was willing but said 'I am ignorant of fashionable mineralogy – even its terms'[5] and so his contribution was more to help with the style of Sowerby's writing.

British Mineralogy started in November 1802 for which subscribers, who paid 2s.6d, received

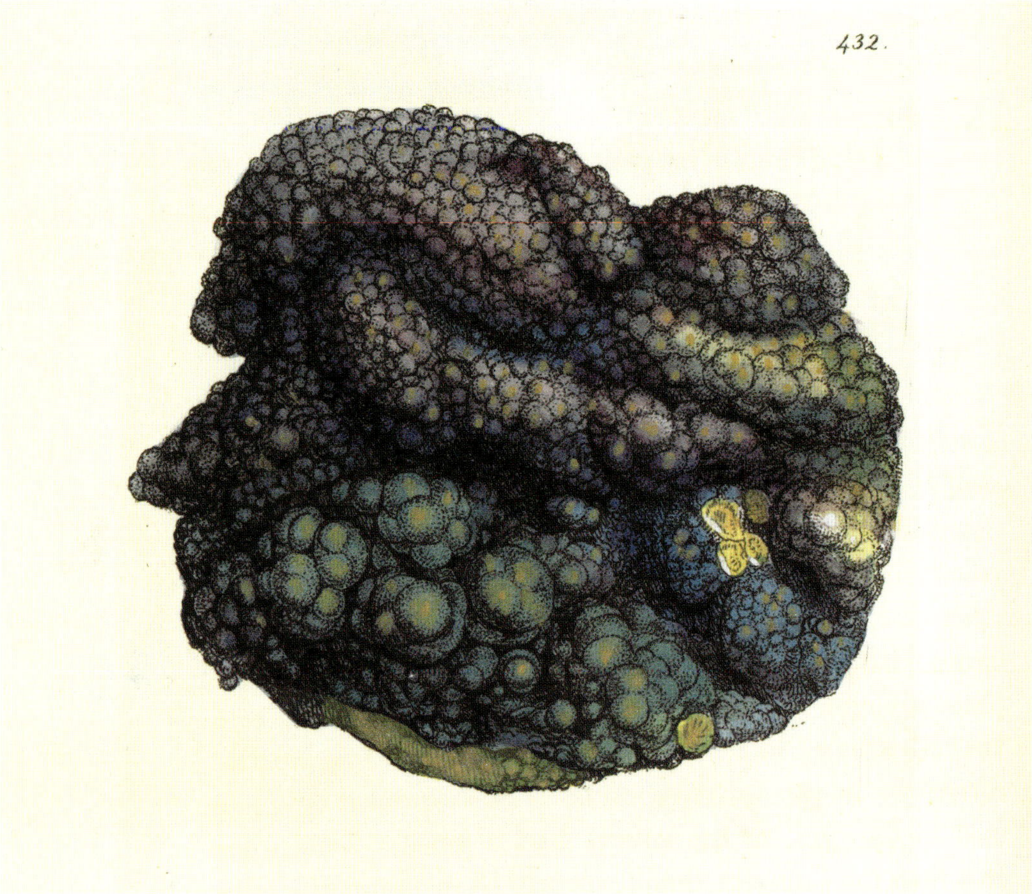

two sheets of text and two hand-coloured plates 'bound' with one long stitch in the customary blue wrapper.[6] It carried on monthly and later bimonthly for a further fourteen years to give a total of 550 coloured plates and text, all of which could be handsomely bound in five volumes with appendices. It is a remarkable work not only for its illustrations and originality but also for what it reveals about the sciences of mineralogy and chemistry at that time, and their development over the years of publication.

Classification remained a thorny issue, bemoaned by several mineralogists.[7] Sowerby wrote about and indeed included at the beginning of his work a simple system based on three divisions: combustibles (coal, peat, oxygen etc.), earths (clays, carbonates, quartz etc.) and metals (oxides and sulphides of metals etc.); but he only occasionally used it to advantage. He was also initially keen on specific gravity as a means of discriminating some minerals but he rarely quotes the values of the different specimens he drew and described. The reality was that no classification was easy to use on a routine basis; a convenient technology was not yet readily available. People identified specimens mostly on the basis of experience: there was nothing better than handling many specimens and 'getting one's eye in' using properties such as colour, friability, relative hardness and sometimes,

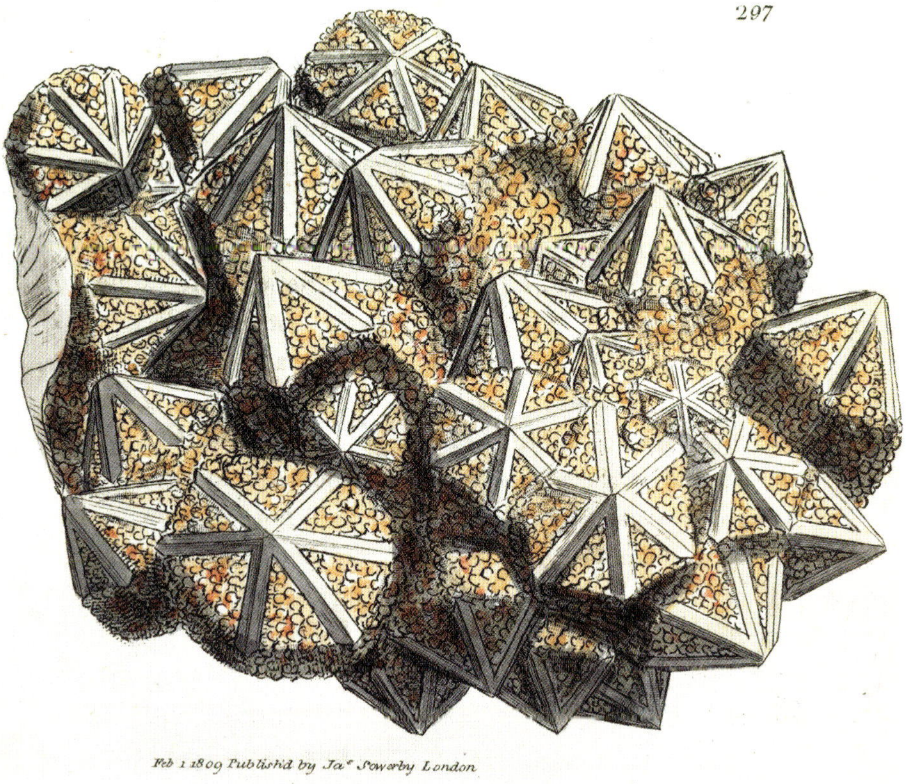

9/5 Colourful mass of copper sulphide (chalcopyrite), described by Sowerby as 'blistered' (now 'botryoidal') because of its form. The specimen came from Cornwall. *British Mineralogy.*

9/6 Unusual crystals of quartz, modified by growth of another mineral (possibly dolomite) on the faces of the pyramids. Sowerby was always keen to illustrate unusual occurrences. *British Mineralogy.*

if one was lucky enough to have the right sort of specimen, crystal shape. Sowerby himself summed up the problem and gave his answer: 'The different appearances that various substances assume is the greatest stumbling block in Mineralogy: it becomes therefore necessary to learn them.' Chemical composition is a key component of any meaningful classification but methods of analysis were at an early stage of development. It was, of course, also a time when some of the chemical elements were first discovered or isolated.

Sowerby was far from averse to doing chemical tests, joining with his sons James De Carle and George in performing various experiments. Sowerby had acquired practice in the use of the blowpipe, and often referred to it in some of his mineral descriptions. This instrument, simple in design but requiring particular skill in its use, was to play a significant role in descriptive mineralogy. It consisted of a curved (or bent) tube with a suitable mouthpiece used to blow a current of air through the flame of a candle onto a very small amount of the mineral, which was usually placed on a block of carbon or piece of platinum as a support. With a well-focused tube and a skilful practitioner the temperature could reach that of a smelter.[8] The behaviour of the mineral would be observed closely, on its own or with one of a small number of fluxes such as borax and soda. The nature of the residue was also important. The blowpipe was in its heyday as Sowerby was producing his *British Mineralogy.* It was used by several chemists including the Swede Jöns Jacob Berzelius, who discovered the element cerium in 1803, and by mineral collectors such as James Smithson, the founder of the Smithsonian Institute, Washington DC.

The instrument would have been in Sowerby's hands when he needed confirmation of an unusual species or to test a rare mineral. A rare white stone, *Silex granatus*, he confirmed as a garnet since: 'the trial of the species was chiefly strengthened by the aid of the blowpipe, under which it resembles the common Garnet'.[9] His account of an uncommon form of the well-known mineral galena (lead sulphide) is a good example of how he used the blowpipe:

> A piece being laid on Charcoal, and the flame gently applied by the blowpipe, it became red-hot, the Sulphur sublimed from the interior and melted with a greasy aspect, and a trifling blue flame, leaving the substance rather lighter in colour when cold. On further urging the flame, white Sulphate of Lead appeared on the surface which bubbled and blistered, and on cooling showed some signs of facets and spiculae. After a while it was reduced to a globule of Lead. On further examination we find it to be composed of sulphate of Lead, and about 10 per cent. of Sulphur.[10]

So much rested on visual observation. Some chemical tests such as solubility in alkalis or acids were made and compared with results reported in recognised texts of the time. Sowerby's possible favourite, for he quotes it often, was Kirwan's *Elements of Mineralogy*, which gave details of mineral properties, including chemical behaviour, and was probably best used in confirming the identity of a mineral rather than as a tool for identifying a complete unknown. Keeping well abreast of the latest chemical advances and other discoveries was a necessary part of any mineralogist's life. Sowerby reported on the results of Charles Hatchett's analyses and his discovery of niobium, while at the start of his fourth volume he highlighted Mr Humphry Davy's first isolation of the elements sodium and potassium.[11] These last achievements were by electrolysis, using the recently created voltaic cell, a device which gave a new buzz to the world of chemical discovery, hand in hand with mineralogy. Richard Kirwan, too, had noted the change in the state of mineralogy over the years between the first and second editions of his *Elements*. In 1784 he was bemoaning the lack of government and other interest in Britain, a situation in stark contrast to many other European countries. Ten years later he was delighted at the successful use of chemical analysis and that it now 'nearly reaches the precision of an algebraic formula'.

British Mineralogy thus appeared at a period of much activity in the world of chemistry and mineralogy. It acted as another catalyst to those collectors who also pursued research, often in their own homes. One only has to look at the *Philosophical Transactions of the Royal Society* to get some sense of the new vistas that were unfolding. Just the four years from 1802 to 1805 record the following: Charles Hatchett's discovery of niobium; more papers from Hatchett (four in 1804), including one on magnetic 'pyrite'; and papers on mineralogical topics by the now-famous names of Humphry Davy, James Smithson, Count de Bournon, Charles Greville, William Gregor, Smithson Tennant and William Wollaston – the last two on the discovery of the platinum group elements, rhodium, palladium, osmium and iridium. People sensed there was much still waiting to be found and that almost any unusual mineral could yield a significant discovery. Many of these scientists were correspondents with Sowerby.

This was also the time when geologists were in hot debate on the origin of rocks. The Neptunists (such as the mining professor in Saxony, Abraham Werner) argued strongly that all, including basalt, came from the agency of water, while the Plutonists or Vulcanists (supported by the Scot James Hutton, and so sometimes called Huttonians) had many rocks, including granite, originating in heat or fire.[12] Sowerby was well aware of the controversy but through the care of his observations was able to conclude that neither extreme was right. In describing a particular copper mineral he wrote:

> The vitreous appearance might give occasion for dispute between the Neptunians and Volcanians; but as I consider both modes of formation to be more or less concerned in many instances where it has been imputed to one only. I shall leave it to every philosopher to consider it in his own way, as such a specimen may give force to either theory when considered alone.[13]

He did not just describe specimens, he thought about their initial formation and possible subsequent alteration. He had seen enough specimens to wonder at the nature and causes of their variation. Having described, for example, the disposition of minerals in a specimen

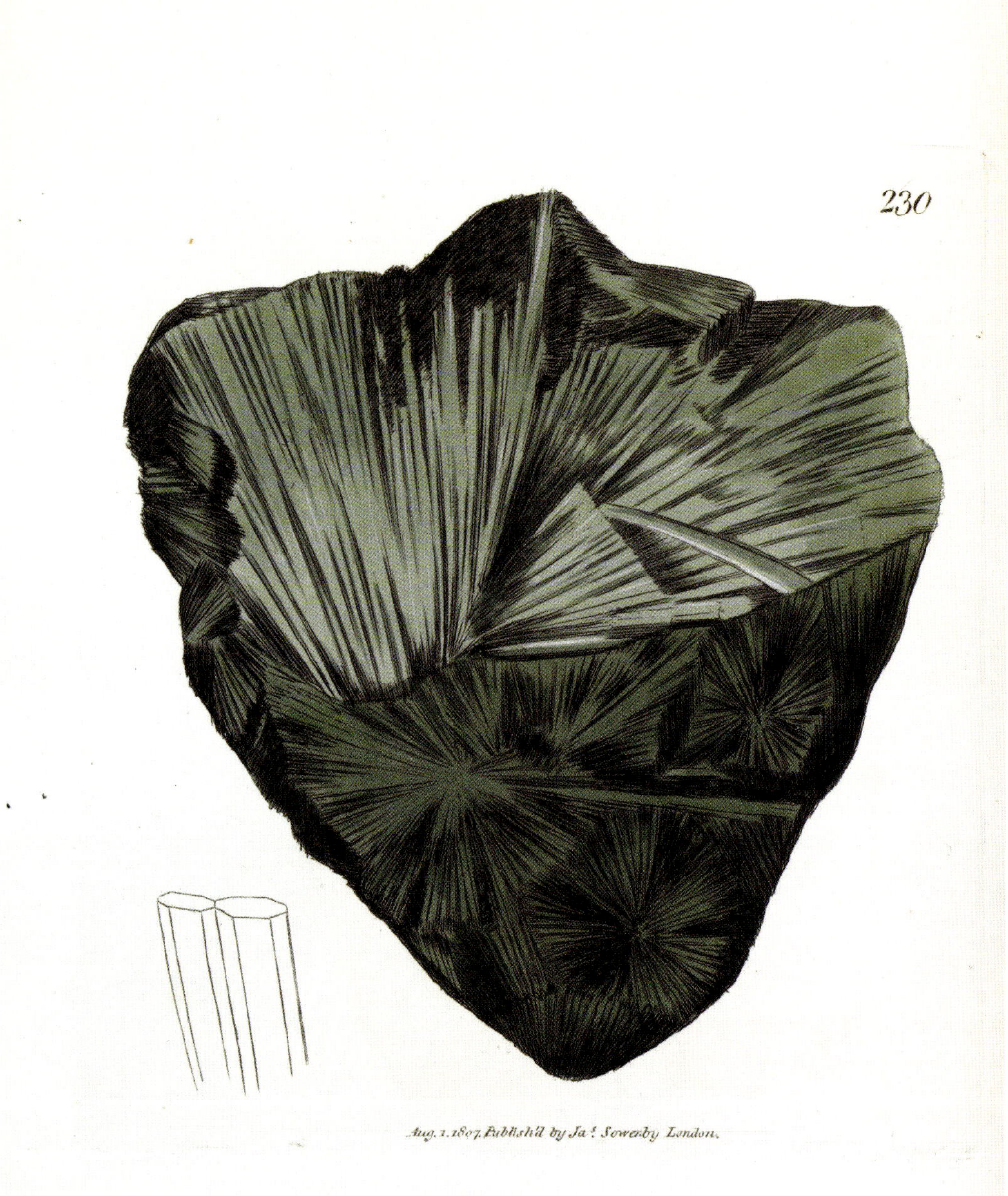
230
Aug. 1. 1807. Publish'd by Ja^s. Sowerby London.

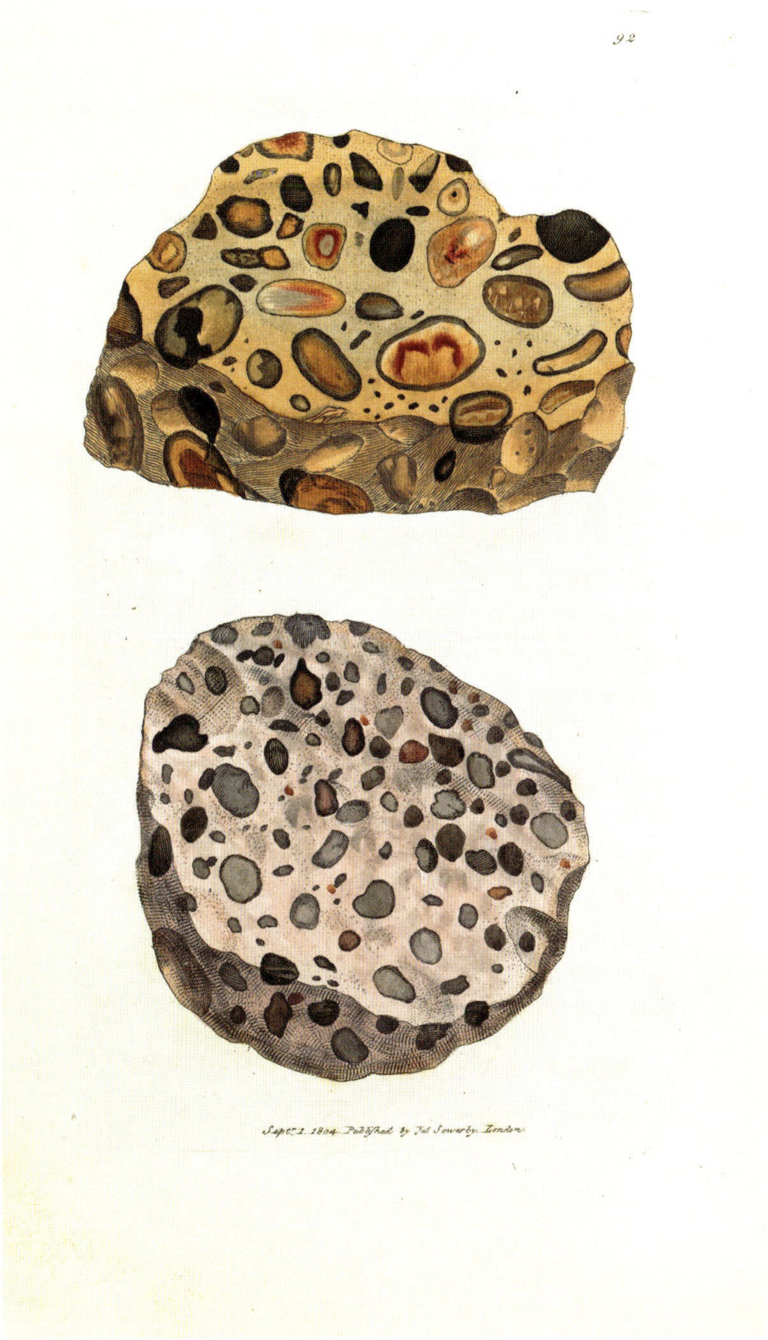

9/7 Beautiful radiating crystals of the mineral actinolite. A specimen from a vein in the Isle of Skye. *British Mineralogy.*

9/8 Puddingstone. Rounded pebbles set in a finer matrix. The upper specimen from Hertfordshire, the lower from Kent. Sowerby occasionally included rocks in his *British Mineralogy,* especially if they showed some unusual or interesting features.

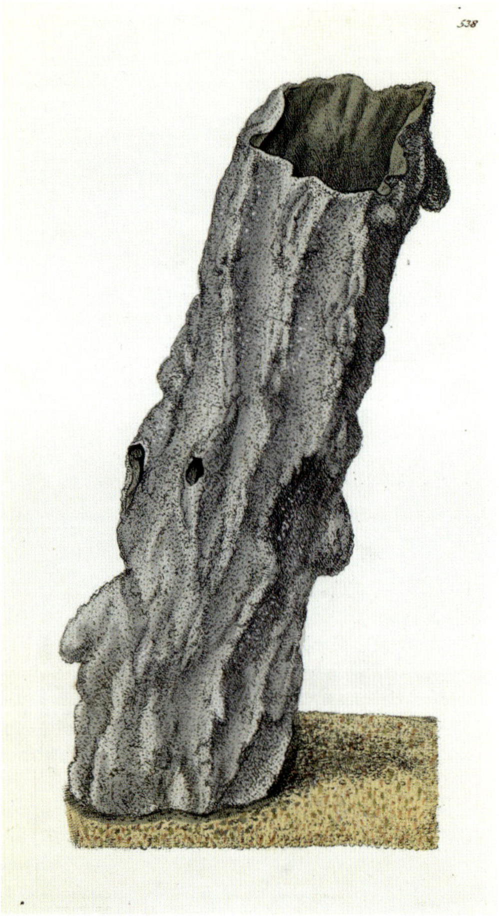

9/9 Vitrified sand tubes (nowadays called 'fulgurites') caused by fusion of sand grains from a lightning strike. The tubes are a few centimetres in diameter but can be many metres long. This illustration came near the end of *British Mineralogy* and Sowerby wrote: 'Ere I terminate this work I have the satisfaction to add the figure of another phenomenon almost as extraordinary as the Meteor stones.' *British Mineralogy.*

containing zinc sulphide he concluded that mineral formation and loss within the specimen had taken place over time and wrote, somewhat wordily but with insight:

> Thus may be seen the changes wrought underground by means of various agents; and thus we have continual proofs of chemical agency, which perhaps may lead us to inquiries that may be of considerable importance, while the subject thus comes to hand is an example too curious to be passed over at a time when investigations drawn from Nature are found to be the truest mode of enquiry and improvement.[14]

In another sentence he proclaimed positively on the formation and benefits of coal:

> In Coals we see an astonishing contrivance, not only to preserve ancient and vast woods and forests to provide for future contingencies, but, as it were, to the very produce of each successive term of the former vegetable surface; thus acres are piled upon acres and laid up in the grand storehouse of Nature, that nothing may be lost to all favoured Mankind!

He was just as enquiring about the processes of fossilisation when examining a shelly rock, saying: 'the animal parts, exclusive of the Carbonate of Lime, must pass away, while

Carbonate of Lime, or Flint, is filling up the vacancy', but he then questions whether it is right for a mineralogist to investigate this process: 'yet it is convenient to have such information, as it may be sometimes of great importance; and it must appear remarkable to all, that although the petrified remains of other animals are very universal, yet those of our own species are never found'.[15] His ability to embrace perceptively so many branches of natural history was a hallmark of his work.

In its scope and style, *British Mineralogy* was unlike any other work. Its standing rested also on the quality of the material Sowerby used, and through the people involved in providing it. Lords and Ladies were among the many who loaned or gave him specimens from their own cabinets,[16] but two collectors stand out as being particularly important – Charles Greville and Philip Rashleigh. Greville was a horticulturist and mineralogist living close to London. One-time lover of Lady Hamilton (mistress of Horatio Nelson), he used his wealth to acquire fine and rare specimens from numerous parts of the world to help satisfy his desire to own beautiful things. His mineral and gem collection became recognised as probably the finest in private hands in England, if not the world.[17] Access to this wonderful resource was freely given – a fact well acknowledged in *British Mineralogy*. But Greville did not live for many years more and Sowerby then had to forgo the privilege until most of the collection was bought soon after Greville's death by the British Museum and his access was thus restored.[18]

The mineral enthusiast, landowner and Member of Parliament Philip Rashleigh, who lived near Fowey, Cornwall, and thus in the tin and copper mining district, had amassed a notable collection, especially of Cornish specimens. He and Sowerby became frequent correspondents, sharing specimens and information. Rashleigh had already led the way with his own illustrated work on *Specimens of British Minerals*, based on his collection, the first volume appearing in 1787, and the second five years later. This was effectively the first successful book of its kind in Britain[19] but was not really a competitor to Sowerby's *British Mineralogy*, and neither did the two men see themselves as competitors. The fifty-three plates are by Henry Bone (enameller to George III) and Thomas Underwood, and, although good, are not generally in the same class as those by Sowerby. The focus of the work is, naturally enough, on minerals from Cornwall and Devon with many examples of tin and copper ores. Each illustration is accompanied by a short description of the specimen but with little or no scientific information or reference to other authors. Indeed, in his second volume Rashleigh admitted to a lack of chemical knowledge and content and hoped that accurate representations of specimens alone 'may contribute to the amusement of those who delight in such studies'. Rashleigh commissioned Sowerby to do some

9/10 *Cuprum carbonatum* (azurite) in an unusual form of many spicules. Both specimens formed on quartz crystals – the lower one right on the tip. *British Mineralogy.*

separate illustrations for some of his more 'difficult' specimens, telling him 'by what I have seen of your execution, nothing is beyond your performance'.[20]

Rashleigh did only a little research on his minerals for he was primarily a collector concerned with identification, locality and outward appearance, delighted when something new or very rare occurred, and always keen to share his enthusiasm. He was happy to help in Sowerby's quest for new specimens, some appearing in *British Mineralogy*, but warned him not to always expect success over acquisitions from the mines as: 'many things you wish for can scarcely be procured, some of the Mines being out of work, & what is a common case, things which are plenty at one time may never be seen again'.[21] They were generous in their exchanges as fellow enthusiasts often are. He thanked Sowerby warmly for the 'very handsome box that contained six capital specimens which will be an addition to my Cabinet. The Green Carbonate of Strontian is the finest Specimen I have seen on Account of the Crystals it contains, & the White Sulphate of Barytes is a great ornament to it', but also tells him that he has little interest in rocks: 'Some Gentlemen Collect Rocks & similar Stones, which I think encumber a House too much, therefore have never attempted it: unless they have something useful or ornamental belonging to them, they seem to me not worth a place in the Barn.'[22] Minerals were his love, and he was particularly pleased when his own specimens were figured by Sowerby. His love of minerals extended so far as to fear that should Napoleon successfully invade then his collection would fall into enemy hands. Many of Sowerby's fine Cornish specimens must have been supplied by Rashleigh. Fine crystals of cassiterite, green octahedral fluorites from St Agnes, botryoidal malachite from Wheal Unity, gold nuggets and lustrous jet-black sphalerite were just some of the Cornish minerals which, alongside others from many English and Scottish localities, filled the cabinets in his museum at Mead Place.[23] Rashleigh died in 1811 and Sowerby lost a good friend.

The willingness of people to cooperate freely in the *British Mineralogy* venture is a striking feature of the age. Some wrote of their experiments, chemical and physical, including from one correspondent on the sounds that different thicknesses of basalt make when struck by a hammer. Others were willing to despatch awkward specimens, such as a granite post containing large grains of feldspar all the way from Penzance, and many were anxious to ensure that the specimens provided were of good quality.

A cutler from Sheffield exchanged and gave specimens, including insects, shells, plants and fossils over many years and sometimes sent victuals for the Sowerby family table, including in one parcel a Yorkshire ham 'cured the beginning of this year by a skillful person… I have put in

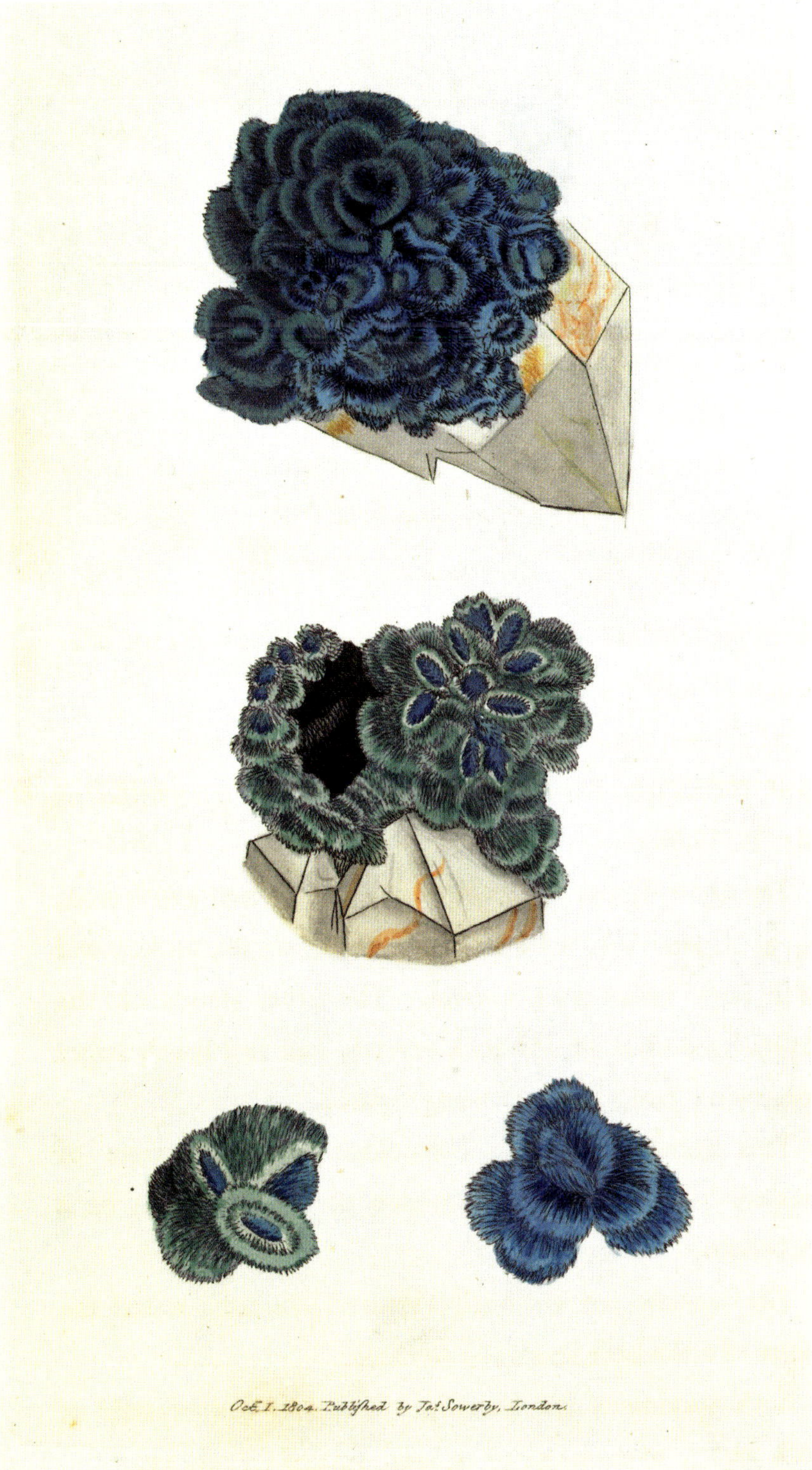
Oct. 1. 1804. Published by Jas. Sowerby, London.

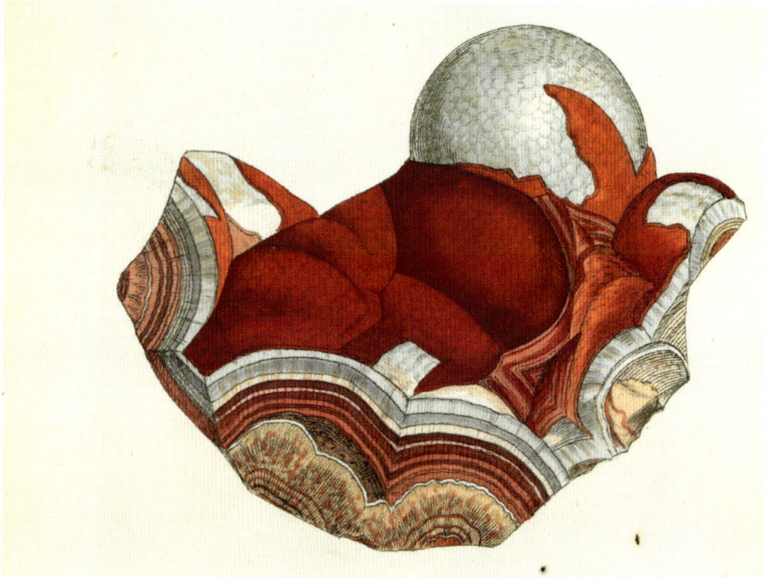

9/11 Agate. A beautiful specimen from Somerset. *British Mineralogy.*

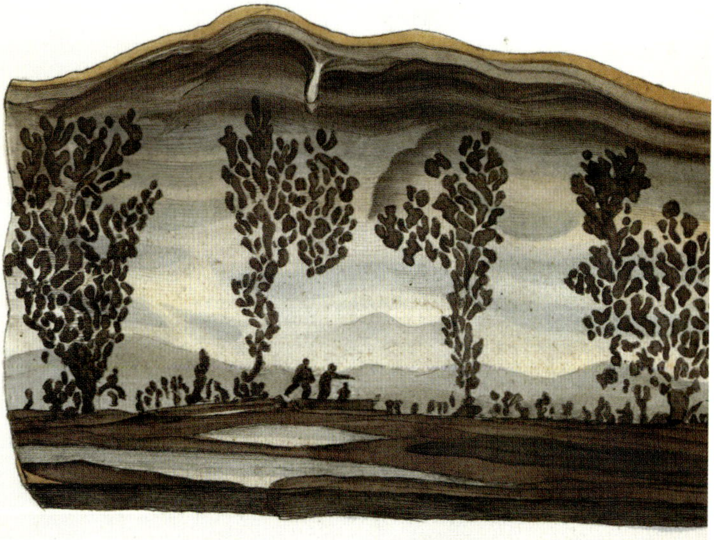

9/12 Cotham Marble. A limestone from Cotham, Bristol which shows a 'landscape' on its cut surfaces. Sowerby conjectured a little on its origin; we now know that it was caused by algal growth during formation. *British Mineralogy.*

a few fossils'. Like others he told Sowerby of the uses of some specimens, such as bloodstone, which 'is used by the Platers here for polishing their goods and sometime ago it was in great repute among Cutlers for Burnishing Wood knife handles but now the common flint is liked better for the purpose'.[24]

British Mineralogy had something of a mixed reception on its first appearance. One particularly sharp and early critique appeared in Aikin's *Annual Review* for 1804:

> We are sorry to witness in this work the injudicious misapplication of ingenuity and talents, which we have often admired, and from which we have derived both pleasure and information. The crystalline and other particular forms of minerals, together with their colour, in most instances may be represented by painting, and are so in the volume before us with considerable success; but the infinite gradations of lustre, the play of light, and the still more essential character of fracture, it is utterly impossible for the best artist to delineate. Where the objections to a work are radical, it is useless to point out minuter imperfections.[25]

This is the full length of this dismissive comment. It was perhaps made worse by being in the journal edited by the geologist and mineralogist, Arthur Aikin, who later became a founder member of the Geological Society and wrote the popular *Manual of Mineralogy*. Fortunately others thought it an admirable work, although the mycologist and friend of James Smith, Benjamin Forster, joked with Smith:

> I suppose you have seen Mr S's Mineralogy if he goes on as he has began, (8 plates carbonate of Lime I think) when will the work finish? & where is the Library that will hold the volumes? or where the purse that will pay for them? – I should suppose upon a very moderate computation he might pave a foot path with his pebbles ores etc (the Plates) from our house at Walthamstow to yours in Surry-street.[26]

The mineralogist and fellow of Jesus College Cambridge, Edward Clarke, was more encouraging to Sowerby: 'it is a most pleasing work, & will be very useful. I hope you will persevere and carry on with your usual accuracy & elegance', although he had also chided him for not using exotic minerals for then he would have had specimens showing good crystal forms.[27] But Sowerby declared that he wanted to depict 'moderate' samples rather than rare showy ones as it would help 'to overcome difficulties that we ought to be accustomed to'.[28] Clarke and numerous others lent or gave specimens to Sowerby in a genuine desire for success. Aikin also accepted the value of the work when he later wrote in his own *Manual*

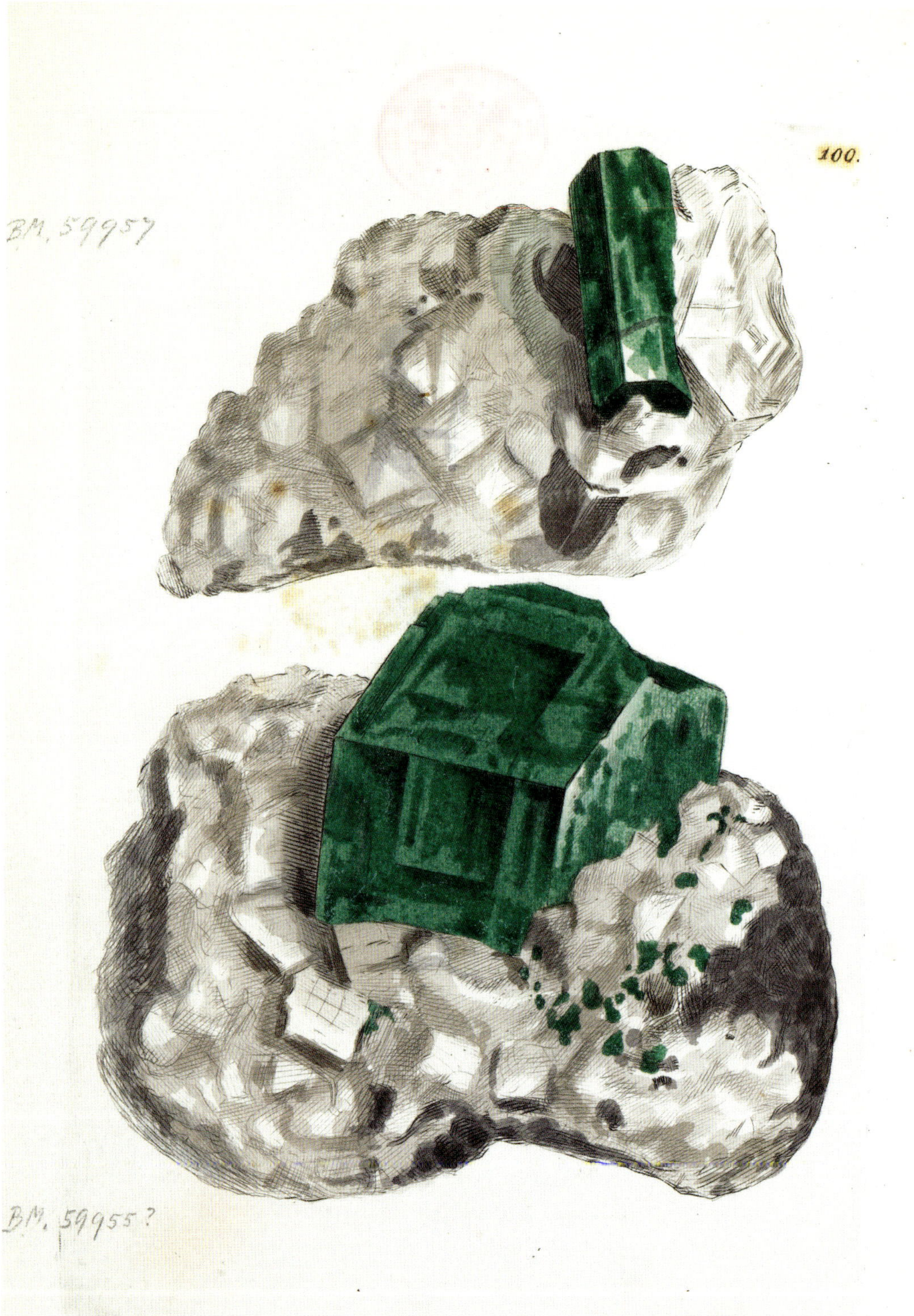
100.
B.M. 59957
B.M. 59955?

9/13 Emerald crystals (gem-quality beryl). Sowerby wrote: 'Of all the precious stones the emerald is acknowledged to be the most pleasing in colour.' *Exotic Mineralogy.*

of Mineralogy of 'Mr Sowerby's elegant work on 'British Mineralogy illustrated with coloured plates'.' The work's popularity was evidenced by good and continuing sales over the fourteen years of its production and the later assessment that Sowerby 'was, for many collectors, almost their sole source of information about British minerals and localities'.[29]

The success led Sowerby to start a companion publication *Exotic Mineralogy* about eight years after the start of *British Mineralogy.* This was encouraged by many mineralogists including, of course, the Cambridge professor Edward Clarke. It is an intriguing publication for what it shows about the development of mineralogy as well as of the knowledge of 'new' elements including some of the platinum group: platinum, palladium and iridium (with the chemist William Wollaston, discoverer of palladium, supplying Sowerby with specimens). It depicted minerals recently named and analysed such as gadolinite from Sweden, a species bearing the rare earth elements – elements which are nowadays of such significant economic importance. It had fine examples of specimens from many parts of the world, including a blue spinel from Sri Lanka, mercury ores from Spain, the beryllium mineral euclase from Peru (brought back by Dombey, presumably with his collection of plants), and ilvaite (formerly yenite) from the Isle of Elba. *Exotic Mineralogy* ran in parallel with *British Mineralogy* for a few years and then continued for about a further

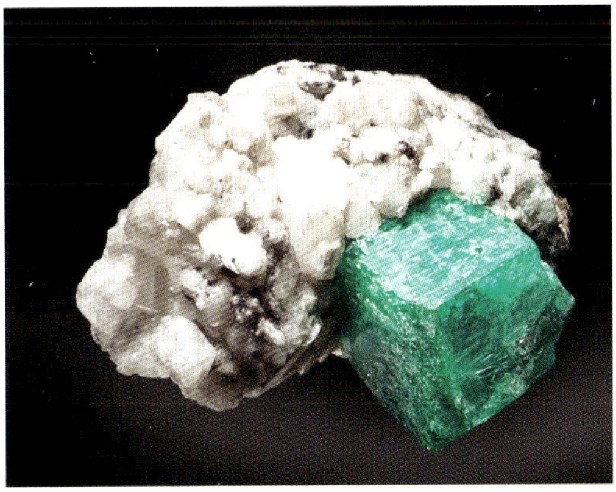

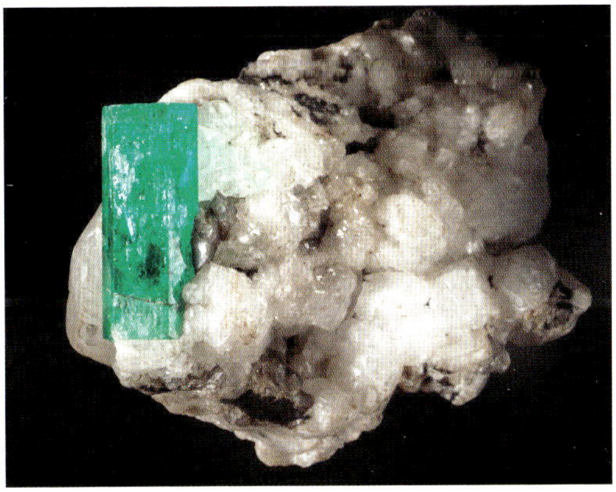

9/14 The gem-quality beryl crystals used by Sowerby for his 'Emerald' plate in *Exotic Mineralogy* and which he said 'are the pride of the collection at the British Museum'. These are now on display at the Natural History Museum, London.

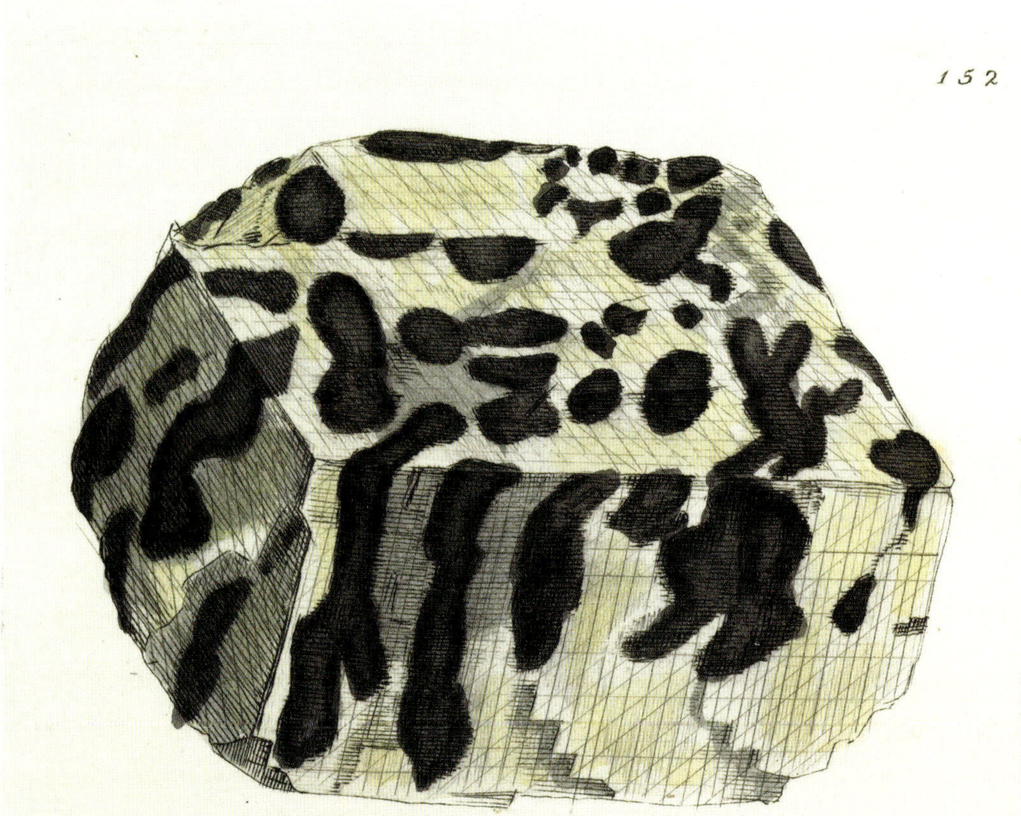

9/15 Selenium ore. Sowerby used a final issue of *Exotic Mineralogy* for a preliminary report on a new Swedish mineral involved in the discovery of the element selenium (by Berzelius and Gahn) only the year before (1817). The black ore was a copper, silver selenide, within a calcite host.

9/16 Gadolinite crystals in a matrix. A mineral silicate containing yttrium and also quantities of the rare earth elements, first discovered near Ytterby, Sweden in 1787 (so also called ytterbite.) It was named after the Finnish mineralogist , Johan Gadolin, who isolated yttrium oxide from it in 1792. *Exotic Mineralogy.*

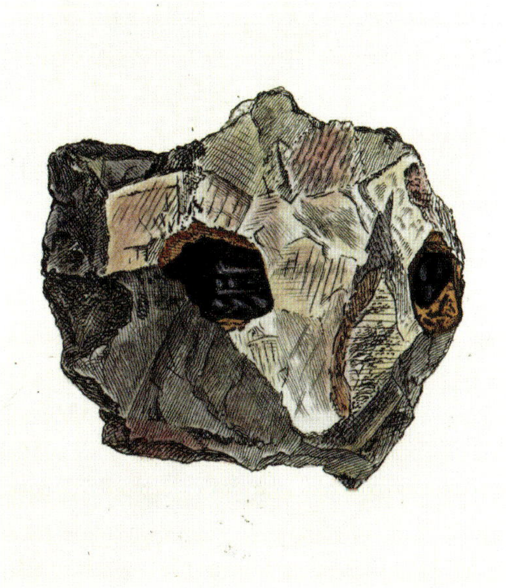

three years. With only 168 illustrations it was not on the same scale as *British Mineralogy*, but it is a fine and important production. Sowerby was not losing his touch. And this was while he was also busy with other projects including *English Fungi*, *English Botany* and the start of *Mineral Conchology*. He was also very active in learned societies, having been made a Member of the Geological Society the year after its foundation in 1807, and in aiding the production of the *Transactions of the Linnean Society*.

It was a courageous thing to undertake the publication of *British Mineralogy* and *Exotic Mineralogy*. Success was dependent on the artistic skills and scientific capacity of the author, on cooperation of the people with access to specimens and on the sale of the numbers as they appeared, and all of this over a long period of time – some sixteen years. Dedication on the part of both Sowerby and his subscribers was paramount. Now, with the luxury of the completed books before us we can look up any species, but at the time subscribers might have waited long for a specific mineral to appear. What was important was that they complemented the many other books on mineralogy which were coming off the presses around the same time. Texts such as Richard Kirwan's well-regarded *Elements of Mineralogy* or Robert Jameson's *System of Mineralogy* do not have mineral illustrations. Nor for the most part do they have easily usable identification keys. It could be hard work to identify a mineral with these works alone. Sowerby's accessible plates and text, together with the provision of synonyms of mineral names as used by recognised mineralogists, provided a very welcome tool for the collector or analyst. And this tool was attractive; a pleasure to own and to use whether for seeing the wonderful variety of different minerals or for learning of new discoveries of the time. The sequential publication of the numbers meant that the enterprise was a 'work in progress', almost like a scientific journal.

There were other important attributes. Contributors felt they were part of the whole

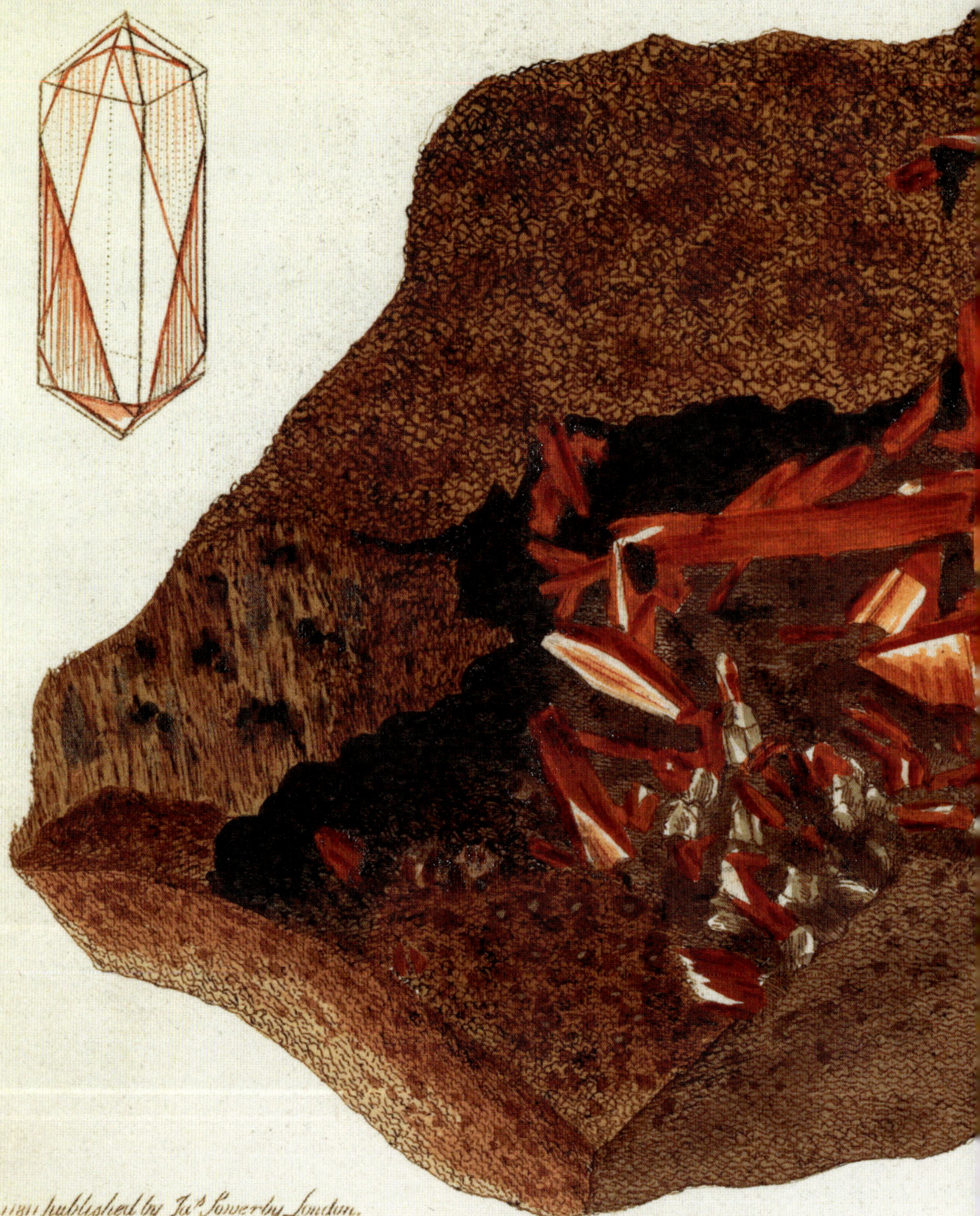

Jan 1 1811 published by Jas. Sowerby London.

4.

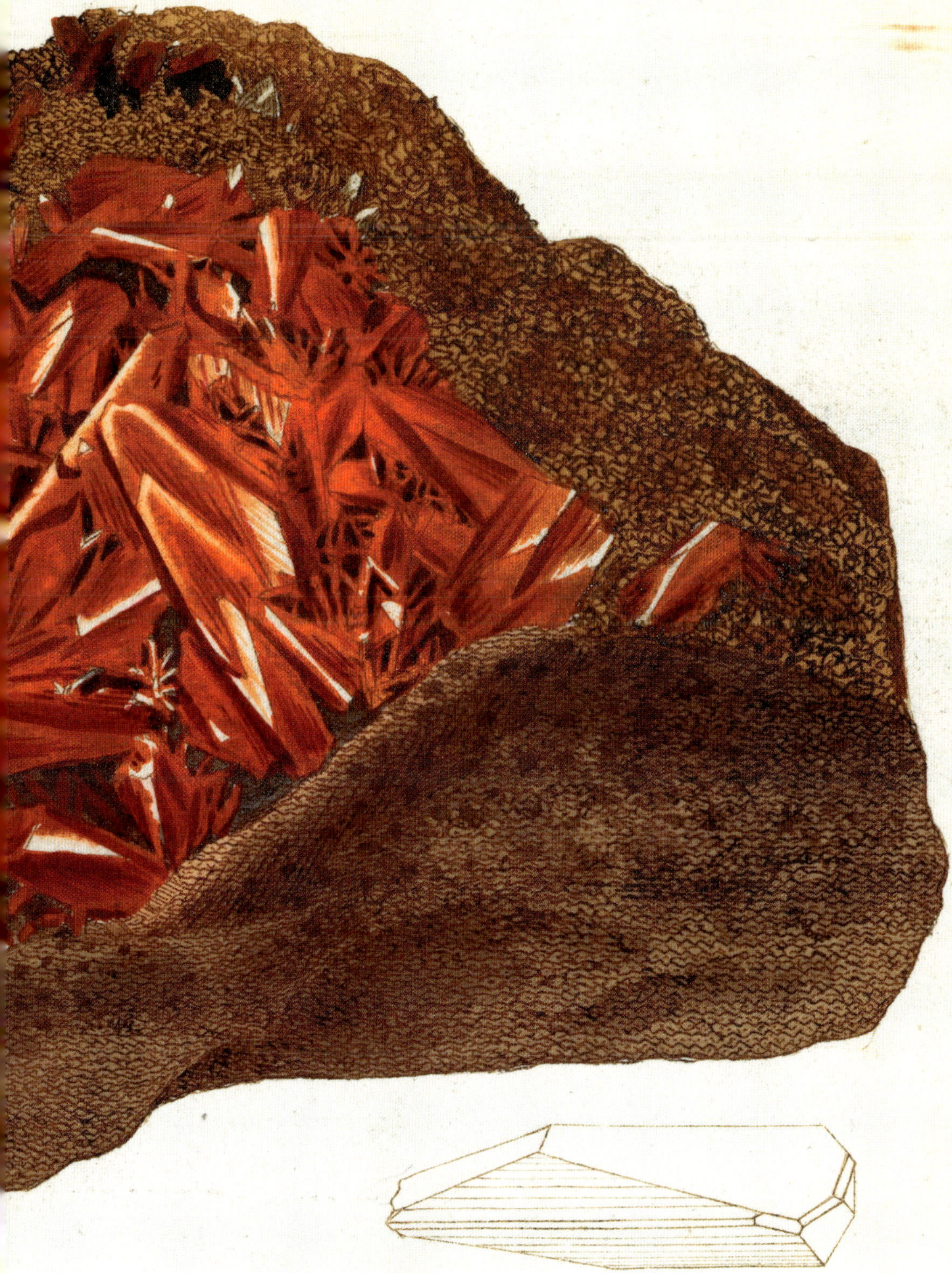

9/17 (previous spread) Crocoite crystals (a lead chromate) from Siberia. *Exotic Mineralogy.*

9/18 (below) The specimen of crocoite, from Charles Greville's collection, used by Sowerby for his illustration in *Exotic Mineralogy.*

9/19 Rubellite specimen used by Sowerby. Originally from Burma (at the time of the Ava Kingdom), it came into Greville's collection and was then purchased for a large sum by the British Museum, where, Sowerby claimed, it was also 'the pride of the mineral collection'.

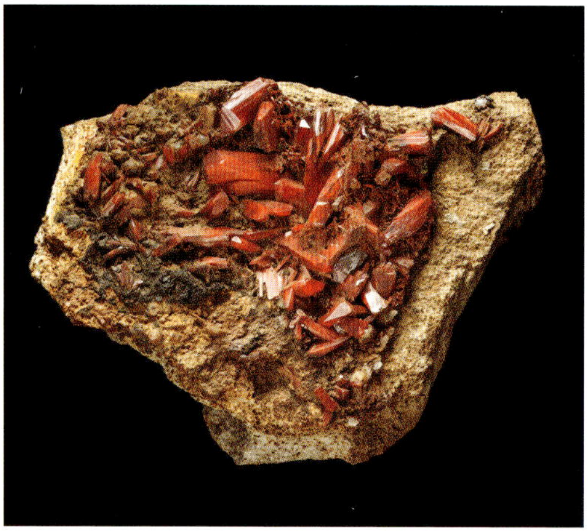

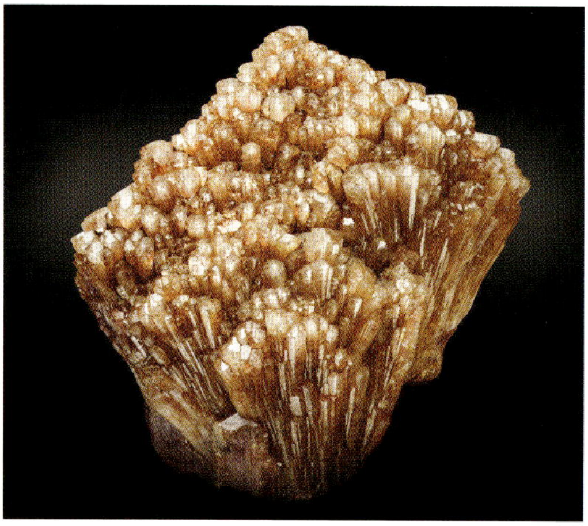

process, with their names sometimes appearing in print or by them knowing from Sowerby himself that their efforts were valuable. People had confidence in the accuracy of the representations and in the reliability of the sources. Indeed, it is possible today to observe the precision by comparing the plates with a few of Sowerby's actual specimens which are still in the Natural History Museum, London. It was also evident that the important mineral collectors of the day were, in one way or another, involved. The works had their own standing and there were no direct competitors. Certainly the volumes exemplified the nature of the mineral cabinet and were similar in style to *English Botany*, although they lacked the same longevity as they were not produced in subsequent editions. Mineralogy and geology were developing fast and becoming more technical and professional, factors which did not sit so well with the aims of the private collector. Furthermore, artistic portrayals of minerals did not hold the same attractiveness as those of many plants. But they were works of their age and aided considerably the pursuit of mineralogy, as well as encouraging a much wider general interest.

At the time of the first appearance of *British Mineralogy*, Sowerby had been married eighteen years and his wife had given birth to eight children. His business seemed secure, although it required consistent outputs to ensure an adequate income. Besides his own publications, his contributions to natural history had been primarily through the provision of high-quality illustrations for other authors' works and his contributions to the Linnean Society. His modus vivendi seemed set. But this would be a wrong interpretation, for his interests were wider and

9/20 Rubellite variety of tourmaline. It was considered, when Sowerby illustrated it, to be the 'most famous specimen in the world.' *Exotic Mineralogy*.

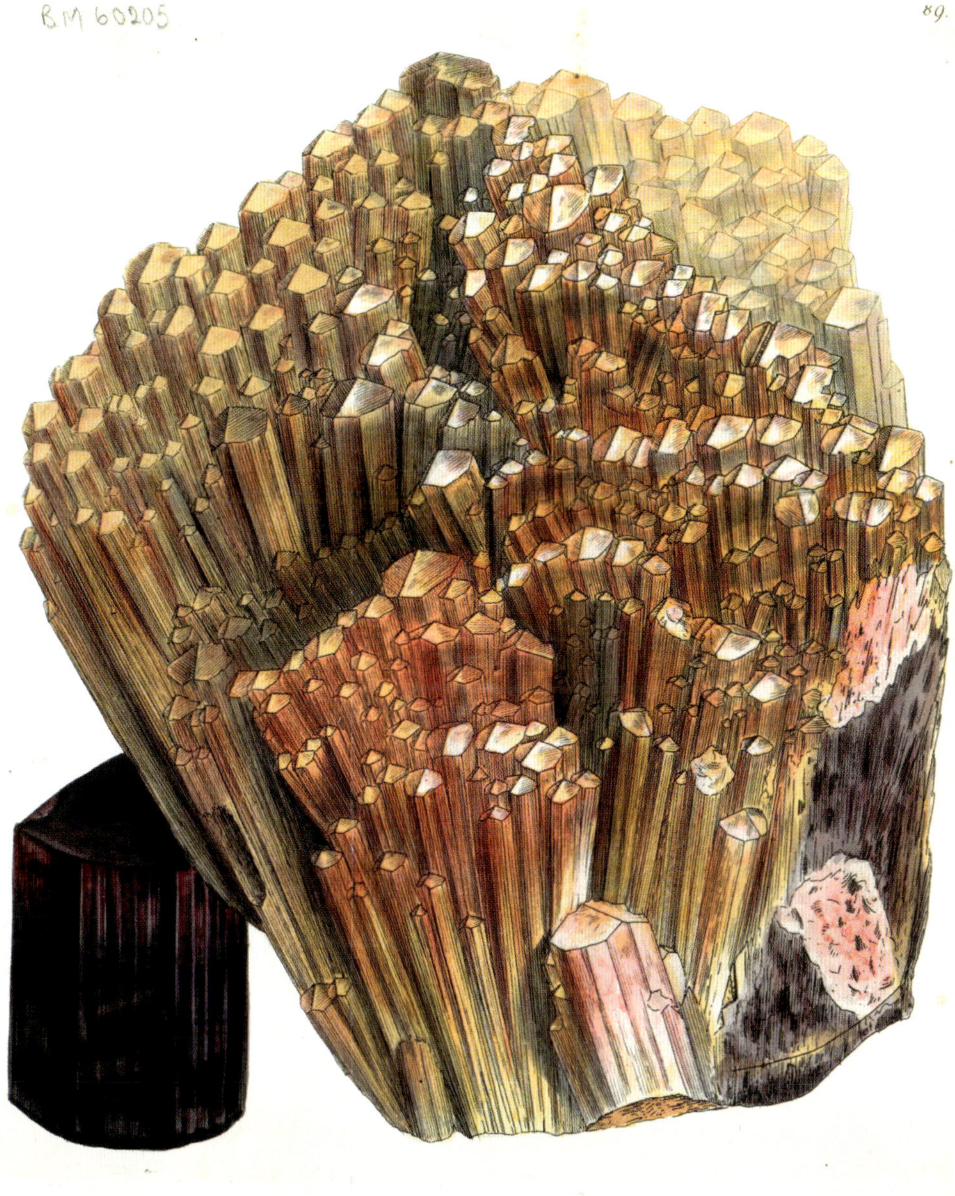

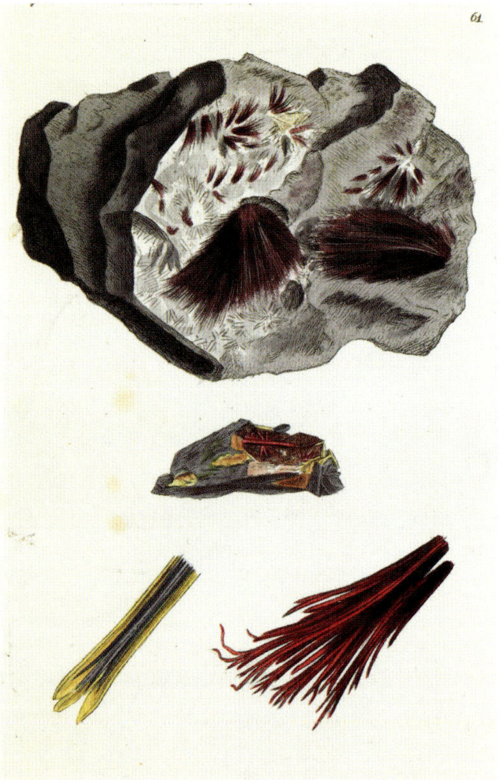

9/21 Stibnite (antimony sulphide). A fine illustration in *Exotic Mineralogy* showing the crystalline form, the feathery form and one specimen of bundled crystals with iridescence.

he was still driven by a desire to instruct others in the ways of nature, including the physical. Two separate areas were of special interest; the first was crystal form and the second was about an age-old problem, that of the production and defining of colours for accurate use by the artist.

Advances in crystallography, including those by Jean-Baptiste Romé de l'Isle and René-Just Haüy, were well known to Sowerby and he was keen to promulgate them through a novel educational device. Soon after the start of *British Mineralogy* and with his usual enthusiasm and energy, he developed a set of crystal models made from high-quality cut glass, contained in a neat ebony wooden box accompanied by an explanatory booklet. The glass models were of simple crystal forms (such as a tetrahedron) which could be combined to produce other simple or more complex forms, with the aim of showing how minerals of certain crystal shapes might arise. The booklet, with the title *Description of models to explain crystallography*, explained how to experiment with the models so as to see the results of different combinations of crystal forms.

This was an innovative and potentially useful idea but unfortunately Sowerby injudiciously chose to devote the first part of this new work to coal. A piece of coal may indeed contain some crystalline material but it does not have a coherent crystal structure of its own and so will not show true crystal forms, although it may show some regular fractures or other structures.

He does not seem to have realised this. Why he chose coal is difficult to understand other than that he was influenced by the Cambridge mineralogist Edward Clarke, who introduced Sowerby to the unhelpful idea that certain coal often had 'five sides'.[30]

He was on firmer ground when he published the second part (with its own box of models) three years later, this time on diamond. It was to have been on sulphur but Sowerby decided that diamond was an easier subject matter for the beginner. Although diamond is genuinely crystalline, this again seems a poor choice not only because of its scarcity in the lives of most people but also because diamond specimens only rarely exhibit good crystal forms. To help would-be students of the subject, Sowerby offered occasional lectures at his home for going through one or other of the parts, at a guinea an hour for one to four people and with the use of actual mineral specimens. A third part, just of the models, was produced but the initiative was not a success and it went no further. One correspondent wrote saying that he had had great difficulty in getting the glass models back into the box![31]

Sowerby could be justifiably criticised for taking insufficient pains before offering his crystallographic initiative but he could not be over another topic close to his heart, as well as to many others – that of standardising the language of colour. His aim was simple in concept but known to be frustratingly difficult in practice: to devise a scheme of reference so that when he described the petals of a flower as being a particular colour, say, 'yellowish orange', others would then know and agree what it was.

A New Elucidation of Colours, publishedin 1809, turned out to be more than a communication on the labelling of colours. It was a work which was meant to impress: dedicated to the memory of Isaac Newton (who had died eighty-two years previously), produced in quarto size, printed on fine-quality paper, carefully coloured and costing a guinea. Sowerby took, for him, an unusual approach by writing a more philosophical and lengthy text, possibly to show he had thought long and hard about the topic and that he was as much a master of it as anyone. He had been stimulated by the work of Thomas Young on the mechanism of the eye and on the theory of colours, and also by William Wollaston on refraction.[32] His aim, however, was to produce a workable system rather than an intellectual advance. It was not, moreover, presented through any learned body, such as the Royal Society, and it was really addressed to the rather limited audience of artistic practitioners, particularly those of natural history.

Sowerby had noted and recorded the interference colours produced within the cleavage planes of some specimens of mica and gypsum in his collection. He also recognised that there are three primary colours – red, yellow and blue – in

Tab. 1.

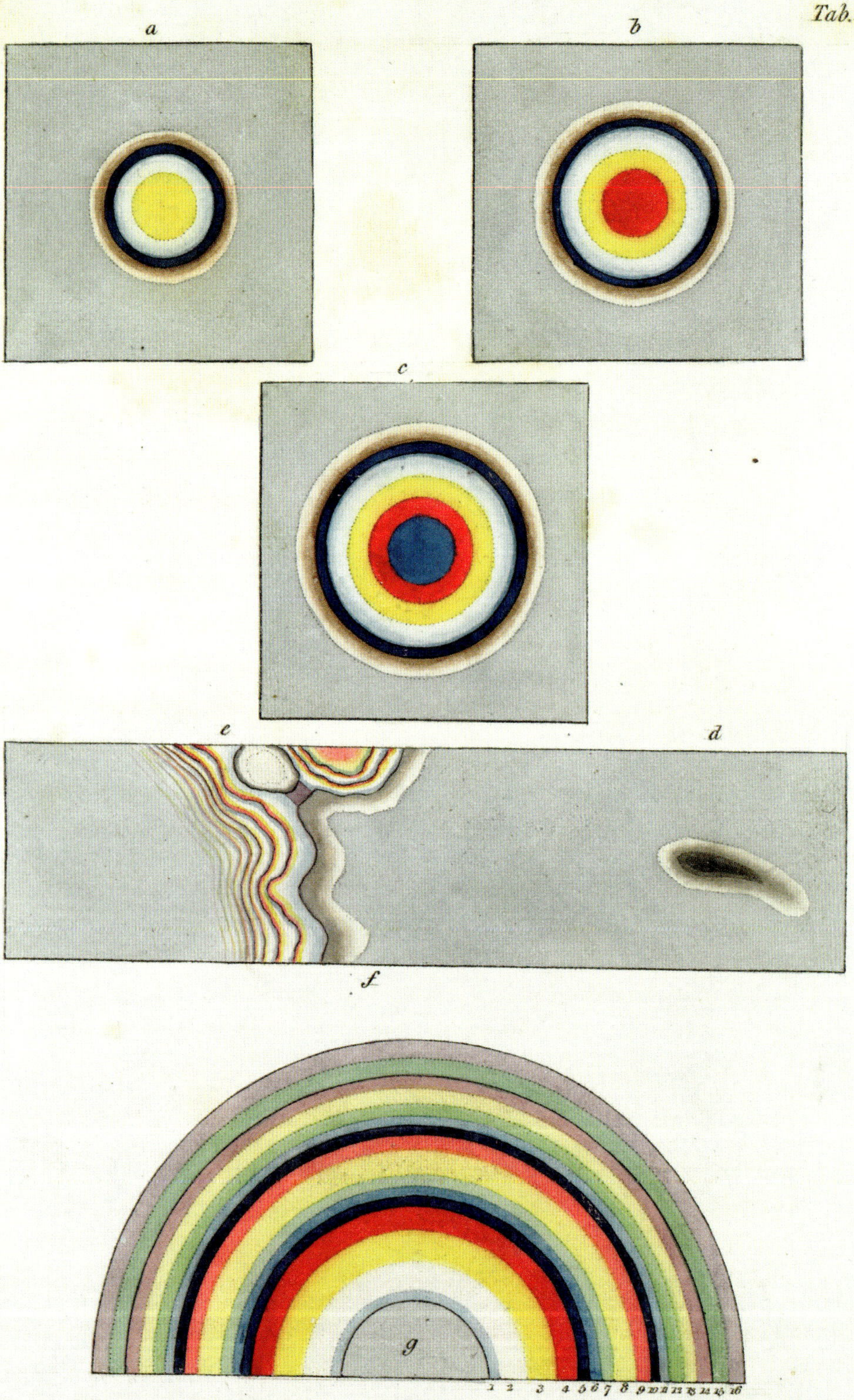

Oct. 1. 1807. Publish'd by Jas. Sowerby London.

9/22 Interference colours observed within cleavage planes of the mineral mica. *A New Elucidation of Colours.*

9/23 Sowerby's chromatometer, constructed to show the effect of different thicknesses of black bands or wedges (against a white background) on the spectrum of colours observed through an optical prism.

contrast to some of Newton's views that there are seven. Sowerby showed that the mixing of these primaries could produce a series of 'binary' colours, such as green, which in turn could be used to derive 'ternaries' such as browns and purples. With the aid of an illuminated chart of black stripes or wedges on a white background, which he called a 'chromatometer' and which was viewed through an optical prism, he demonstrated that an ordered range of colours could be observed with each in a particular and definable position. He was using the effect of strong contrast (black on white) and of different thicknesses of dark bands in an experiment that is relatively easy to perform but difficult to apply in practice.[33] He did not really show how a user could readily measure the position of a colour on the chart so that the result could be communicated to other artists who, in turn, could reproduce the colour for themselves. Difficulties such as the distance of the observer from the chromatometer and different refractive indices of optical prisms were just some of the obstacles in the way of a viable technique.

His device, however, did show the different sequences of colours, which he then used to 'form a list of tints in as simple an arrangement as possible, selecting the usual terms or commonest nomenclature, that it might be easily understood'. This list, with an accompanying coloured and rather fine-looking chart, contained sixty-three colours (his 'chromatic scale'). These were carefully painted in the production of *Elucidation of Colours* but were, of course, liable to the usual variations during colouring and to fading or change over time. Sowerby well recognised these problems and so, rather typically, he included a helpful section on 'Material colours useful to artists' giving advice and hints on the use of 'colours' such as carmine, chromate of lead, burnt ochre, gamboge and Naples yellow, whether as primaries, binaries or ternaries.

9/24 Chromatic scale of sixty-three colours. The upper main triangle contains the three primaries (yellow, red and blue) and their binary mixtures (made from a mixture of two primaries) to give thirty-six colours. The ternaries (37–63) are in the sectors below the main triangle.

This made the book of real use and interest to other artists or natural historians. For example, William Kirby (the so-called founder of entomology and author of an illustrated work on bees) questioned Sowerby over the use of the colours and suggested that there should be additional ones, to which Sowerby indicated how these could be produced by further combinations using his chart.[34]

Several other authors had attempted to produce systems for defining colours. One notable individual was the entomologist and engraver Moses Harris who produced *The Natural System of Colours* in 1766. He used the three primaries red, yellow and blue in a coloured and graduated circular scale to give six hundred and sixty 'teints or colours' – each one definable in its space. This was a simple, elegant and practical reference work but with the usual risk of colour fading. Harris made clear his intended readership by dedicating it to the painter Sir Joshua Reynolds. By contrast Sowerby's was based, in part, on the fundamentals of the spectrum and was more readily available. Here was Sowerby revealing his keen interest in experiment to achieve greater accuracy. He was proud of this – hence the fine quality of the book – and, as usual, had a strong urge to communicate his findings, even giving public demonstrations which, as indicated by their advertisement, must have been both enjoyable and dramatic. For five shillings a person could witness, on a large scale, the primitive tints from the sun, the same from the moon, and also from candles, and even the seven tints of the rainbow shown six feet or more in depth. And if that wasn't enough then they would see 'various Proportions to show the Mixtures, equal to many Rainbows, ten or twelve feet deep'. These one-hour demonstrations which 'might be gratifying to the intelligent to see' were held in his room or workshop at the back of his house, some fifty feet long and capable of taking an audience of up to twenty people.[35]

It is said that *Elucidation of Colours* was conceived and written during a period when Sowerby suffered from rheumatism and was unable to use his gravure, but if that is so it did not affect the production of new plates for long.[36] His botanical and mineralogical works were continuing all the while and *Exotic Mineralogy* came on stream just two years later. Many of the business activities were buoyant and supported by his wife Anne, who dealt with matters such as the orders for publications, but she was not always in the best of health and, although they had some staff, would have shouldered many household burdens, cared for the children, and then suffered a grievous, unexpected loss. In April of the year of the publication of *Elucidation of Colours*, their second daughter died. Felicia Anabella was only nine and although she seems to have been a somewhat sickly child, her loss was particularly poignant for Anne. Felicia Elizabeth Ann arrived two months later and was

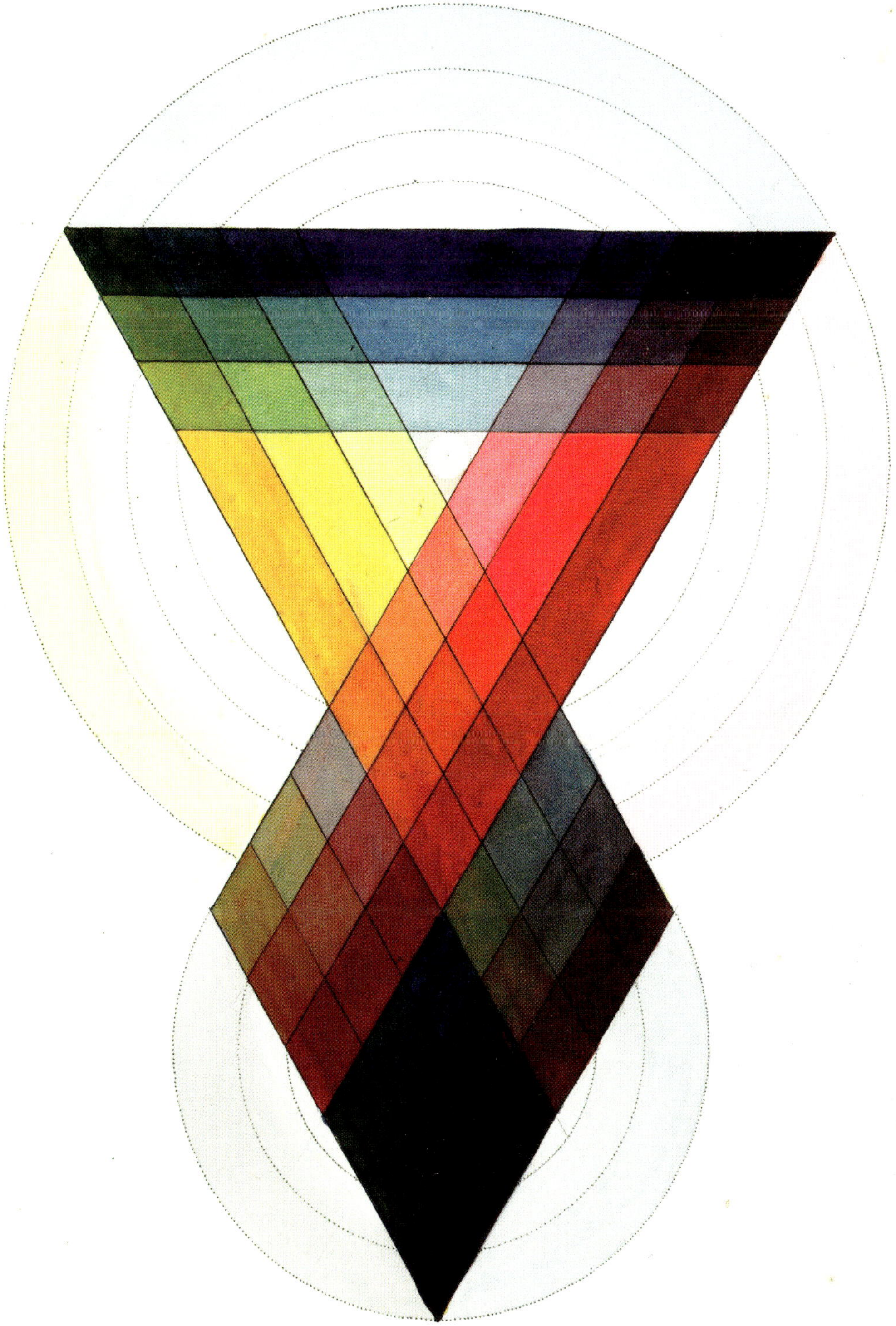

Oct 1 1807. Publish'd by Ja^s Sowerby London.

9/25 First part of the letter from George Sowerby to his parents on the subject of his marriage. 26 July 1810.

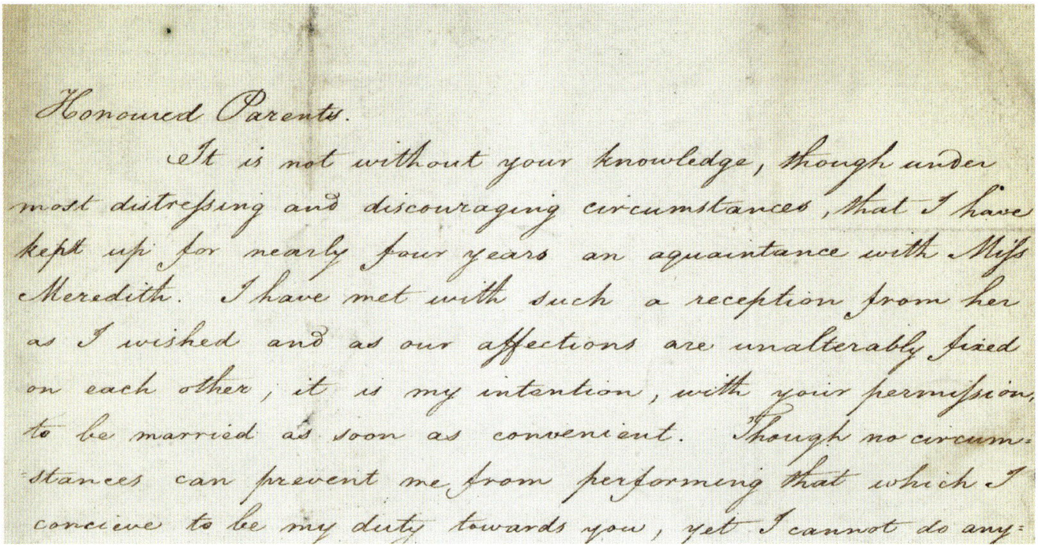

Honoured Parents.

It is not without your knowledge, though under most distreſsing and discouraging circumstances, that I have kept up for nearly four years an aquaintance with Miſs Meredith. I have met with such a reception from her as I wished and as our affections are unalterably fixed on each other, it is my intention, with your permiſsion, to be married as soon as convenient. Though no circum-stances can prevent me from performing that which I conceive to be my duty towards you, yet I cannot do any-

no doubt named in her memory. She was their last and lived to marry and have children. Now there were six of the nine Sowerby children alive, four older boys and two very young girls, and in just a few years, grandchildren would begin to add to the number.

Family stresses of a different kind also arose. George Sowerby, aged twenty, now indicated that his thoughts were turning to marriage. Such an event could have undesired repercussions on the Sowerby system for George was an excellent amanuensis as well as a contributor to many of the scientific activities, including chemical and entomological ones. If he were to leave the family home and business then a considerable gap in resources would arise. He was not quite at the age of majority and it appears that the idea of his marriage was badly received by both parents. It may have been that the intended wife was not considered suitable but there is little evidence for that, and one can only surmise about the root causes.

His first letter on the subject was sent just to his father and started 'Christmas being now come, you may very naturally suppose that I shall begin to think of marrying' and continued that he hoped there would be no more delays over discussing it and that his marriage 'would not in any way be an hindrance to business'. But if any discussions did take place they did not resolve the possible difficulties, for in the summer of 1810 when George was twenty-two, he wrote a longish letter to his 'Honoured Parents' which showed that the topic was not only difficult but required an appeal to their better sentiments:

> It is not without your knowledge, although under most distressing and discouraging circumstances, that I have kept up for nearly four years an acquaintance with Miss Meredith. I have met with such a reception from her as I wished and as our affections are unalterably fixed on each other, it is my intention, with your permission to be married as soon as convenient. Though no circumstances can prevent me from performing that which I conceive to be my duty towards you, yet I cannot do anything with pleasure to myself at present owing to my decreasing state of health caused by the protraction of this which I take upon me as a duty I own to myself & for those purposes for which marriage was ordained. As I conceive it to be my duty never to marry without your consent, I hope you will not throw unnecessary obstacles in the way, by disapproving of my choice, nor let your consent be wanting to make me, your son, happy; but afford me such advice and assistance as you think proper. It is impossible I should experience more trouble and anxiety than I have already & it is in your power to draw me out of all or to keep me still very unhappy. I shall not again notice this to you until you mention it to me, I hope you will not keep me long in suspense but let me have as soon as possible your determination, and whatever that may be not to let it be made public. I am Honoured Parents Your affectionate (I hope I do not flatter myself to say) and dutiful Son G. B. Sowerby.[37]

This letter was sealed, not simply placed into the hands of his parents, but its effect seems to have been limited. The issue smouldered on for several more months until marriage was seen as inevitable. George now needed help over his future circumstances, such as equipping his married quarters in Mead Place. Again his parents, or at least his father, showed few signs of cooperation for when George requested support over the purchase of furniture he felt it necessary to add the resentful sentence: 'surely as a parent you cannot refuse to provide in such a manner as you can afford for the lawful exigencies of your own family'.[38] Was George sometimes a 'difficult' and demanding son or was his father wishing to maintain full control over him in a selfish manner? Whatever the underlying factors they were not revealed and the family business continued much as it had done before George's marriage to Elizabeth Meredith in April 1811, some five years after the start of their courtship and less than a year before the birth of their first son. It would not, however, be an end to the writing of difficult letters within the walls of Mead Place.

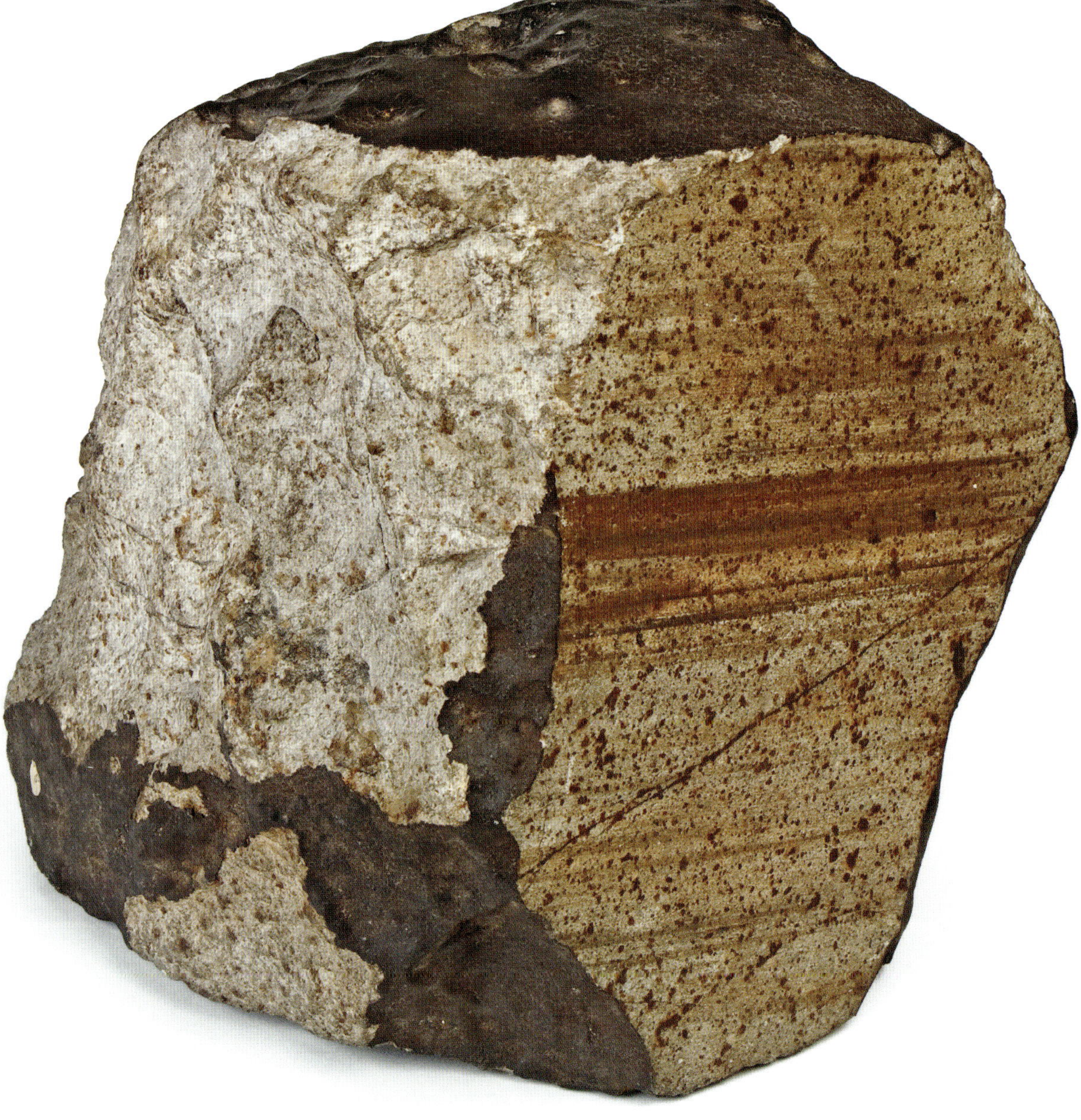

CHAPTER TEN

Meteorites and a 'sword from heaven'

10/1 Wold Cottage meteorite, showing the cut and exterior surfaces. Credit: The Natural History Museum, London.

A meteorite fell from the sky onto the land near Wold Cottage in Yorkshire, England, on 13 December 1795, burying itself nineteen inches into the earth and chalk having just missed hitting a farm worker in the field. The drama of its arrival and of its unusual, somewhat curious nature immediately caused a flurry of interest in the neighbourhood and beyond. The landowner, Major Edward Topham, was not at home at the time but some of his staff and other locals witnessed the fall or heard the noise of the stone's flight. They dug it from its resting hole, took it into the house, looked it over and weighed it at 56 pounds. This was a remarkable event yielding a magnificent specimen; and the colourful Major Topham was just the sort of person to capitalise from it.[1]

Stones which fell apparently from nowhere received a mixed reception. Accounts of such falls were dismissed by many as sheer fantasy while others attempted various explanations with origins involving volcanism, lightning, the moon, or even aggregation of matter in the atmosphere. A German scientist, Ernst Chladni, published historical and other evidence just a year before the fall of 'Wold Cottage' (as the meteorite was to be called), and proposed an extra-terrestrial origin, only to find this being initially ridiculed. A significant shower of stones falling on Siena in Italy at about the same time was also easily explained away by the eruption of Vesuvius some eighteen hours earlier, despite its distance of two

10/2 Major Edward Topham, portrait by John Russell (c.1795).

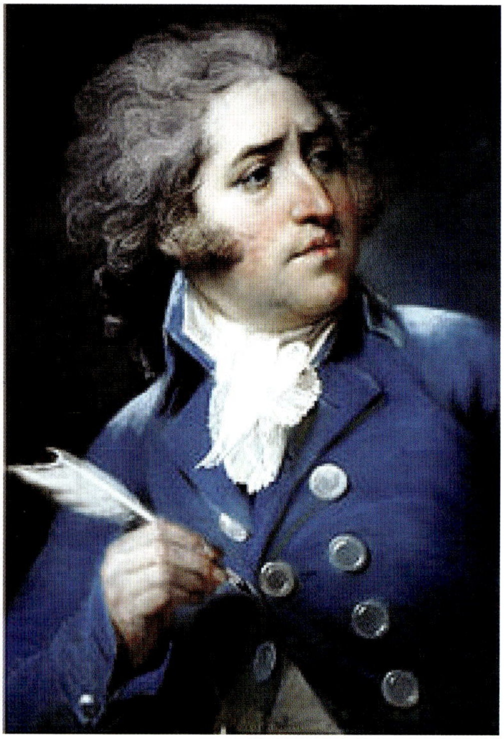

hundred and thirty miles. But such falls were a hot topic of conversation and Wold Cottage inevitably added to the debate, especially as it was nowhere near a volcano and the stone had arrived on a clear day. Would the words of the peasant witnesses be carefully considered or dismissed as coming from those with wild imaginations?

Major Topham was prepared to believe what he had heard. He was a celebrated military man, newspaper proprietor, writer of plays, landowner and a local magistrate, proud of his status as well as of his integrity but with a certain propensity for the limelight and for the ladies, and already much in the public eye. He carefully recorded accounts of the fall from various witnesses and had soon agreed for the meteorite to be displayed in London in July 1796, where visitors could pay just one shilling for admission plus an engraved illustration and handbill containing the sworn testimonies from various people. James Sowerby might well have visited, for he had been collecting 'meteor stones' for some time and was always intrigued by unusual or new specimens, especially any which must surely feature in his *British Mineralogy*. Joseph Banks certainly did see it and it was not long before a portion of it was in his hands, allowing him to arrange for it, together with other meteorites, to be expertly analysed for a report to the Royal Society. The crucial results were the similarity in the chemical composition and constituents of all the different specimens and of their dissimilarity with normal rocks. Most importantly they contained iron which was rich in nickel – a type of iron absent from terrestrial rocks. The tide was turning for accepting their origin from outer space.

Within a few years Topham erected a commemorative pillar, which still stands today, on the spot where the meteorite fell. While the limelight had been enjoyable and fired his extrovert personality it had become tedious

10/3 Monument at Wold Newton, Yorkshire, erected by Major Topham to mark the spot where the Wold Cottage meteorite fell in December 1795. Watercolour (1812) by George Nicholson.

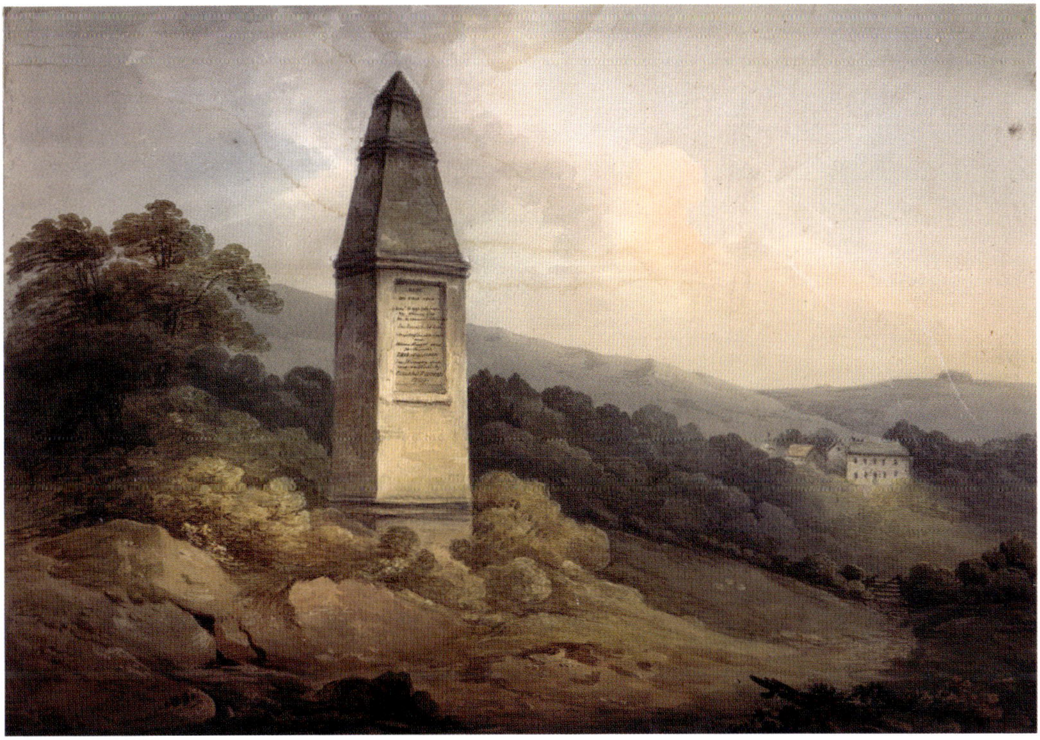

through the workload it caused. Sowerby provided a good exit and there is little doubt that he had been desirous of acquiring it for a long time. He wanted to have the fine specimen in his museum for the public to see and to have an account of it in his *British Mineralogy*, even if some people might look askance at its inclusion, for he later wrote:

> to introduce a substance, however curious, as having fallen like a meteor from the skies, or as Phaeton from the heavens, might seem absurd in a work on British Mineralogy. But, whatever may be the extent of this term mineralogy, it is pretty universally understood to include a knowledge of stones and metals; among the latter of which we place this production, and feel much gratified in adding so great a rarity to the British catalogue. [2]

In answer to a request, Topham sent Sowerby a piece in April 1804 but this was never going

10/4 Wold Cottage meteorite. Engraving of pieces including some magnified ones. British Mineralogy.

to satisfy the author of *British Mineralogy* and museum owner – he wanted the whole thing. Topham, however, was being a reluctant donor. By September, Topham sent drawings of it by his daughter in an effort to assuage Sowerby and in the expectation that this would be sufficient for him to do a suitable engraving to the right scale. He was naturally reluctant to risk sending the whole meteorite by any of the postal systems of the time. But Sowerby was not content with someone else's drawing and he was already telling his subscribers that a description and illustration of a 'Lapis meteoricus, or Sky-fallen Stone' was intended to appear as the first plate of the second volume. He wrote imploringly, almost bullyingly, with a sense of urgency and of shared moral obligation to the public:

> My Dear Sir,
> I am in a very difficult situation between your generosity and my duty to you and the publick, and it is not a little enhanced when I feel the gratitude due to your kindness in getting the pretty sketches your daughter has been so good as to make for me. I feel it gratefully and but for the credit I hold with the publick to do my works myself I am restrained under the severe penalty of sarcastic criticism to keep my word which I have never forfeited and I hope ever to treat them handsomely. I have perhaps unfortunately promised them from your generous manner the most authentic account and figure, from my own critical inspection and best abilities. The time of promise is elapsing very fast as it was to be in the first number of my second volume of British Mineralogy to be published in Dec 1 this year.[3]

Whether or not Topham needed to be so reminded of his own generous manner is uncertain but the letter had the desired effect and the whole meteorite came into Sowerby's hands in time for the publication. Topham asked for only ten guineas and an elegantly bound and lettered *British Mineralogy* in return – a generous settlement. He also provided his account of the fall and of his desires for the change of ownership, recognising that Wold Cottage was a 'very troublesome companion' through the demands of enquirers often with associated degrees of scepticism.

Despite the hullabaloo, the ensuing illustration in *British Mineralogy* was a weak effort. Only two sides of a fragment of the meteorite were shown, together with four small pictures of the components, some enlarged to show a little detail. The artwork lacked Sowerby's usual flair and attention to detail, the whole being rather dreary and showing signs of haste. The text gave Sowerby's observations on the texture and components, including the presence of nickel-bearing iron, as well as Topham's account. The important fact was the

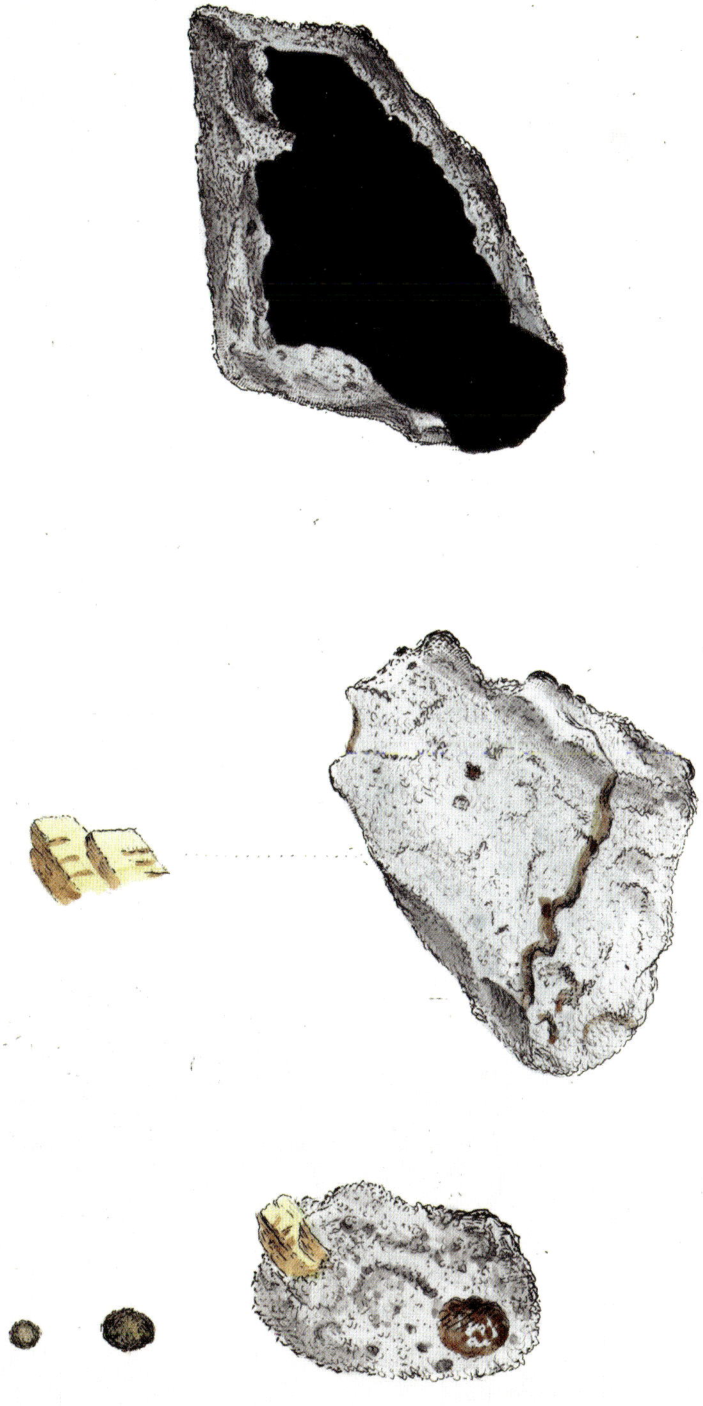

similarity of this meteorite to others that had fallen elsewhere, and, furthermore, all were unlike the rocks on which they fell. If Sowerby thought they came from outer space he chose to mention it only in passing, preferring instead to discuss an atmospheric origin. In this, he was influenced by papers and pamphlets written by several others at the time, many of which lacked critical analysis.[4] So it was no surprise to find that his account was rather incoherent and 'full of conjecture', although he thought that 'time may bring the truth', as indeed it did and well within his own lifetime. But his publication widened interest in meteorites and increased knowledge of them, although it was probably criticised by some over the meanness of the illustration, for Sowerby soon told his readers: 'a full-sized outline of the Yorkshire Atmospheric Stone will be published in the course of the Work. The small piece was figured, because Mr. Sowerby was not sure of getting any other British Specimen in the time promised.'[5] Only much later did he make a fine, large and coloured illustration of Wold Cottage, placed at the centre with specimens from Scotland and Ireland either side. Apparently a copy adorned the walls of Wold Cottage until not long ago.[6] The Wold Cottage meteorite is now one of the Treasures of the Natural History Museum, London, and is on public display. It was also one of Sowerby's 'treasures', for it features prominently in his portrait of 1816. Sowerby suffered the same consequence as had Topham; too many visitors to his museum led him to restricting visits 'to the first and third Tuesday of every month, from eleven in the morning to three o'clock in the afternoon'.[7]

Sowerby was always actively collecting for his museum and helping others through exchanges and purchases. When a meteorite fell in County Tipperary, Ireland, in 1810, he used the good offices of the Dublin Society to acquire a specimen: the Society was only too pleased to help, knowing of Sowerby's standing, while he also ensured they had a piece of Wold Cottage.[8] Others would get him specimens or assist his purchases. Even much later (1820) an apothecary in Hanover told Sowerby he expected to obtain some of a meteorite which had fallen in a village near Göttingen and said 'if I do, no other person shall have it but you'.[9] Interest in the subject was high. Humphry Davy was provided with analytical results of meteorites for his chemical work and lectures. Yet there was a clash of interest for while Sowerby was promulgating natural history he was unwilling to loan his Yorkshire meteorite to the Professor of Mineralogy at Cambridge, the Reverend E. D. Clarke, who wanted it for a lecture saying that he would not 'hurt a hair of its precious head'. Sowerby was unwilling for it would 'baulk many' who were already coming to see it – but he was also worried about its possible loss. Clarke tried a year later, writing 'I am upon my knees entreating the Loan' and, noting that Sowerby had lent it to Humphry Davy

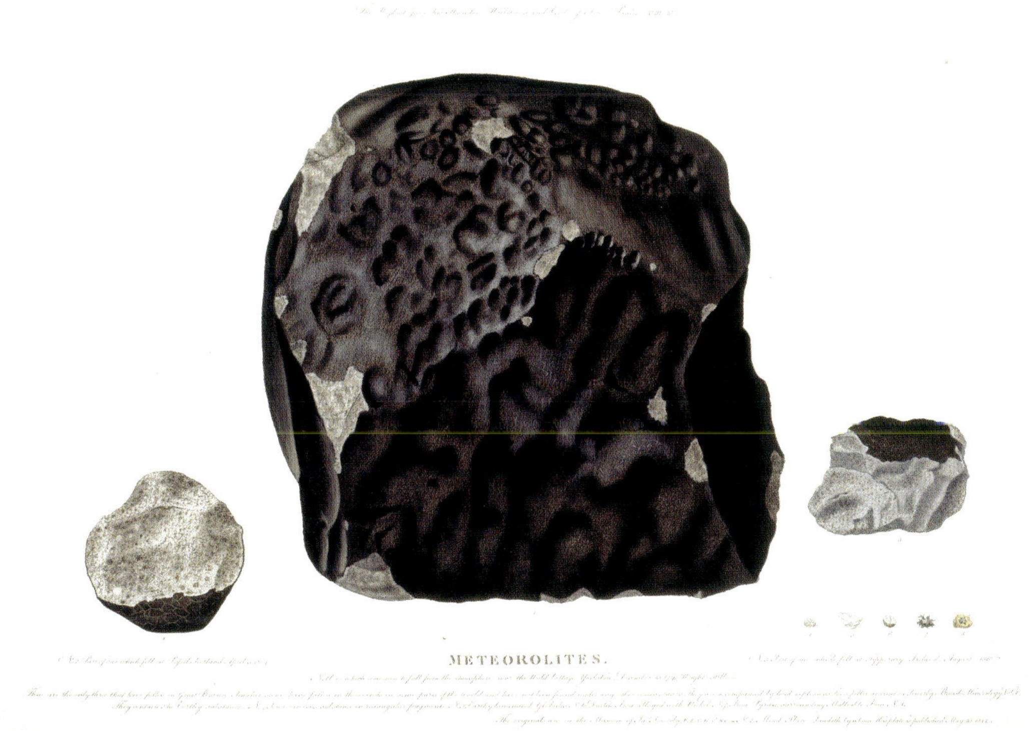

10/5 'Meteorolites'. A large (370x490 mm) hand-coloured engraving of meteorites and their constituents. In the centre, the large Wold Cottage meteorite, on the left a piece of the Possil meteorite which fell in Scotland in 1804, and on the right the Tipperary (Ireland) meteorite which fell in August 1810. At the bottom right-hand side are enlarged examples of the constituent minerals (1812).

for a lecture at the Royal Institution, suggested that surely showing it to others would increase the 'celebrity of your acquisition'. Sowerby again demurred and after three more years Clarke announced he had given up and had borrowed specimens from others who had the 'most disinterested liberality' (unlike Sowerby!) and that he did not want the Wold Cottage meteorite anyway.[10] Any antipathy between the two men would, however, be removed a few years later when Sowerby loaned Clarke his important Cape of Good Hope iron meteorite.

Work on meteorites would occupy only a small proportion of Sowerby's time. He probably slept for a relatively short time, and his waking hours were full of numerous demands, many continuous and others more spasmodic and unusual. Continuity was maintained by living in the same house since

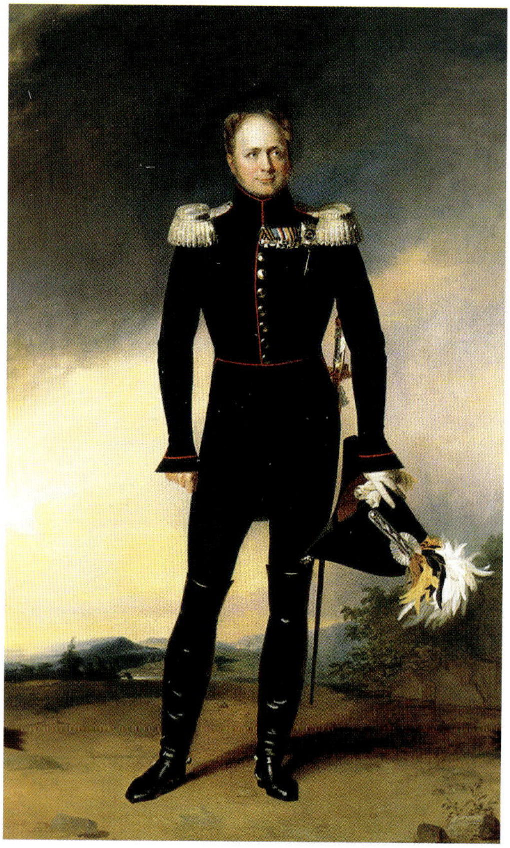

10/6 Alexander I, Emperor of Russia 1801–1825. Portrait by George Dawe (1824).

marriage with an established working system and the regular production of publications, some lasting for many years. Some events undoubtedly disturbed this pattern, like the tour with Dawson Turner to the south-west of England, or the sudden death of a child, or the debacle over the production and sale of *Flora Graeca*. A few were initiated by Sowerby himself, such as the experiments with light and colour and his production of crystal models. Two separate events linked to the war with France were of different interest: the first in 1812 was an urgent demand on Sowerby to help solve the problem of rot in the timbers of important ships of the British Fleet. The second was Sowerby's idea to celebrate the victory over France by the Allied forces in 1814 by making and presenting to the Tsar of Russia a sword made from the iron of a meteorite in his own collection. This was the diverse nature of Sowerby's world of natural history.

It is of course true that the arrival of peace in Europe was a huge relief to the citizens of Britain. The long years of conflict had taken their toll on the Government's finances, taxes had increased and diversified, the threat of a summons to serve in the militia had often been present and the real fear of an invasion was almost tangible. As early as 1803, Dawson Turner had written from his home town of Yarmouth to Sowerby that people were 'frightened out of their wits about the French & most families who can afford it, are removing their wives and children out of the

Town'.[11] Sowerby had received summons to serve in the militia.

The event that spurred Sowerby into action was the visit to London in June 1814 of the Allied Sovereigns, the King of Prussia and the Russian Emperor, Alexander I, and some distinguished generals. The celebrations marked the end of the war with France and the signing of the Treaty of Paris. Sowerby, who normally expressed little about current affairs or politics in his letters, was delighted at the onset of peace and of the role that Alexander I had played in bringing it about. He wished to demonstrate his appreciation and to do so in a way that showed something of the wonders of natural history. What could be better than to use one of the extraordinary items of his own collection transformed into a symbol of war and peace? Thus his idea of a ceremonial sword made of meteoritic iron was born. His problems were twofold – one was the limited time to arrange for the making of the sword and the second was to secure a means of presenting it to the Emperor.[12]

The distinguished entourage arrived in London on 7 June, passing along Westminster Bridge Road, near to Sowerby's house. Sowerby planned to display a large Union Flag. Several grand receptions and banquets were held, the Emperor was honoured by being admitted a Knight of the Order of the Garter; the King of Prussia and Alexander attended the races at Ascot; and Alexander received an Honorary Degree at Oxford. Alexander particularly pleased the crowds with his easy manner and the style of his mounted guard in parades in the parks: 'As the white plumes of the Emperor's guard danced among the trees, the people all ran first to one side and then to the other', wrote one enthusiastic observer.[13]

But Sowerby had been busy before the arrival of the grand company. He chose one of his fine meteorites – the Cape of Good Hope iron – from which to make the sword. A piece was cut off and then shaped into a blade at the 'Patent Press Steam Engine and Machine Manufactory of Henry Maudslay and Co., Lambeth', not far from his home.[14] The final shaping, sharpening and mounting onto a steel handle and the making of the scabbard was by Jon Prosser, 'Manufacturer of Arms and Accoutrements to the King and their Royal Highnesses the Prince of Wales, Dukes of York, Kent and Cumberland, No 9 Charing Cross.' A special mahogany sword case was also provided. The blade was decorated with engravings of the Imperial Crown, a wreath of laurel and palm and the two-headed eagle of the Russian Emperor on one side and with the inscription 'This Iron having fallen from the heavens was, upon his visit to England, presented to His Majesty ALEXANDER EMPEROR of all the RUSSIAS, who has successfully joined in battle, to spread the Blessings of PEACE throughout EUROPE.' The result was a handsome sword, slightly curved, with a decorated grip and a

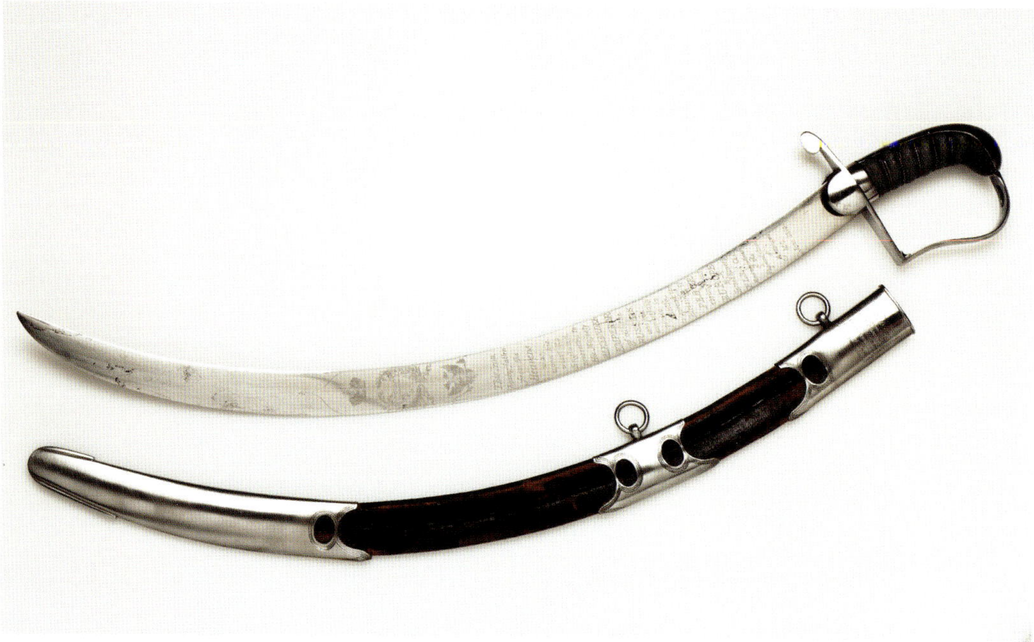

cleanly shaped rain guard, the whole being about three quarters of a metre in length. The sword, with its meteoritic iron and its well-made scabbard, was a truly worthy item for an Emperor.

Sowerby's own account of the manufacture proudly declared how quickly the whole job was done:

> Blade hammered at red heat out of a single piece of an inch thick, ground and polished; its spring was given it by hammering it when cold; the haft was lengthened by welding on a small piece of steel; it was found to work very pleasantly, the whole operation taking about 10 hours; the mounting & engraving occupied the following two days; thus no sword was ever completed from the crude material in so short a space of time.[15]

Sowerby did not stint in his efforts or costs to achieve a good result but if the sword's preparation had been achieved successfully, its presentation was to be a much more difficult matter. He sought permission for the presentation four days after the arrival and only three days before the possible departure of the Emperor. Attempts to get to the Emperor or his officials were either thwarted or unsuccessful and the sword was finally taken to Russia some days after his departure, with the intended aim of a later presentation. But before the sword left British shores it was displayed at a reception for many of the distinguished visitors and others at Sir Joseph Banks's house. One guest wrote to a relative:

> the most interesting object of the evening was a sword come down from heaven on purpose for the Emperor! Let the Prince Regent and his garters and his orders, and the merchants and the aldermen and everybody hide their diminished heads! What are they and their gifts to the Philosophers?
> This is literally a sword made by Sowerby

10/7 Meteoritic sword, with scabbard, for Alexander I, showing the dedicatory engraving. Copyright © The State Hermitage Museum.

> from the iron from some meteoric stones lately fallen of course in honour of the Emperor. There is an inscription on it something to this effect, but not so neat as the subject demanded, and it is to be presented to Alexander – who does not deserve it, by the by, for having entirely neglected Sir Joseph amongst all the great sights and great men, which has rather mortified the poor old man.[16]

But Alexander had not seen it and it was to be a considerable time before he would, and even then it was only because of Sowerby's persistence. When the sword left for Russia, it went with its scabbard and case, a coloured print of British meteorites including the renowned Wold Cottage stone, and a letter to the Emperor referring to another iron meteorite that had fallen in Siberia and an analysis of the Cape of Good Hope iron by the chemist Smithson Tennant:

> To His Imperial Majesty, Alexander, Emperor of all the Russias
> May it please your Imperial Majesty
> Part of a mass of Iron, of the same nature and celestial origin as that so celebrated which Proffessor Pallas discovered some years ago in Siberia and which is now placed in your Majestys Museum, being in my possession, I have presumed it would be agreeable to your Majesty to accept a Sword made from it, as a mark from an individual of that gratitude every Englishman is so anxious to express and respect for the familiar way in which your Majesty has been pleased to visit my Country.
> May it please your Imperial Majesty
> The Meteoric Iron of which the Blade has been hammered was found about 200 miles within the Cape of Good Hope by Captain Barrow. It has been examined by my Countryman Smithson Tennant who established its nature by discovering about 10 per Cent of Nickel in it. It is the only Sword ever made of that rare extraordinary material. That your Majesty may be graciously pleased to honour an humble individual by receiving it is the ambitious hope of Your Majesty's most obedient and ever grateful Servant.
> Jas. Sowerby
> 3 July 1814[17]

But then everything went quiet. If Sowerby was expecting to receive at least an acknowledgement he was to be disappointed. He kept in touch with some Russian contacts as much as he could but the importance of his sword was clearly not shared by all. However, one particular contact, Dr Alexander Crichton, was not only a mineral collector and someone interested in Sowerby's publications but, most conveniently, also 'Physician Extraordinary to the Emperor of all the Russias'. So in a letter about possible mineral exchanges he also explained

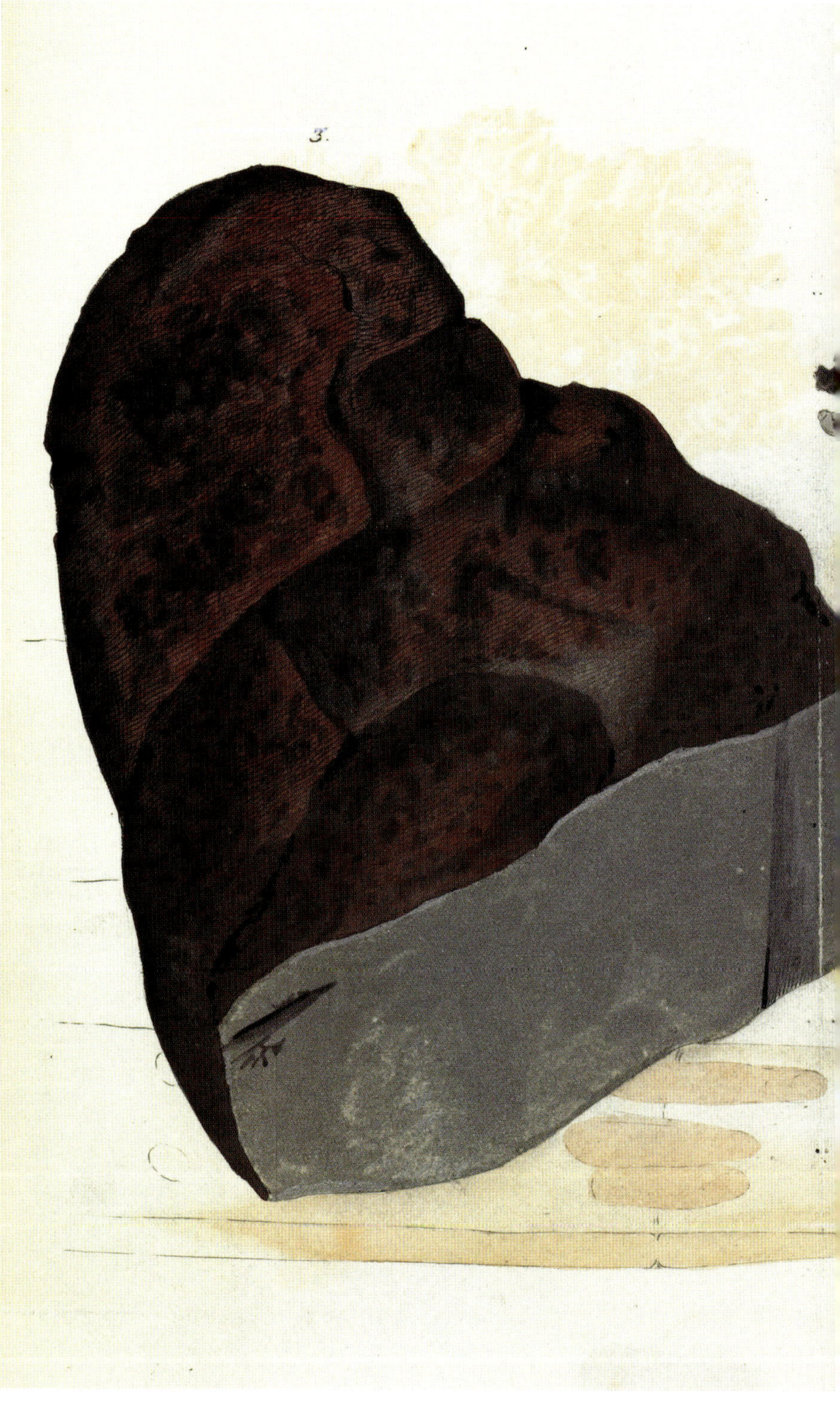

10/8 'Meteoric Iron'. Engraving in *Exotic Mineralogy* of the cut Cape of Good Hope meteorite (on the left) together with iron meteorites from various countries, and two knives, made by Eskimos, of bone inset with pieces of iron meteorite. These last were discovered by the Arctic explorer John Ross, in 1818.

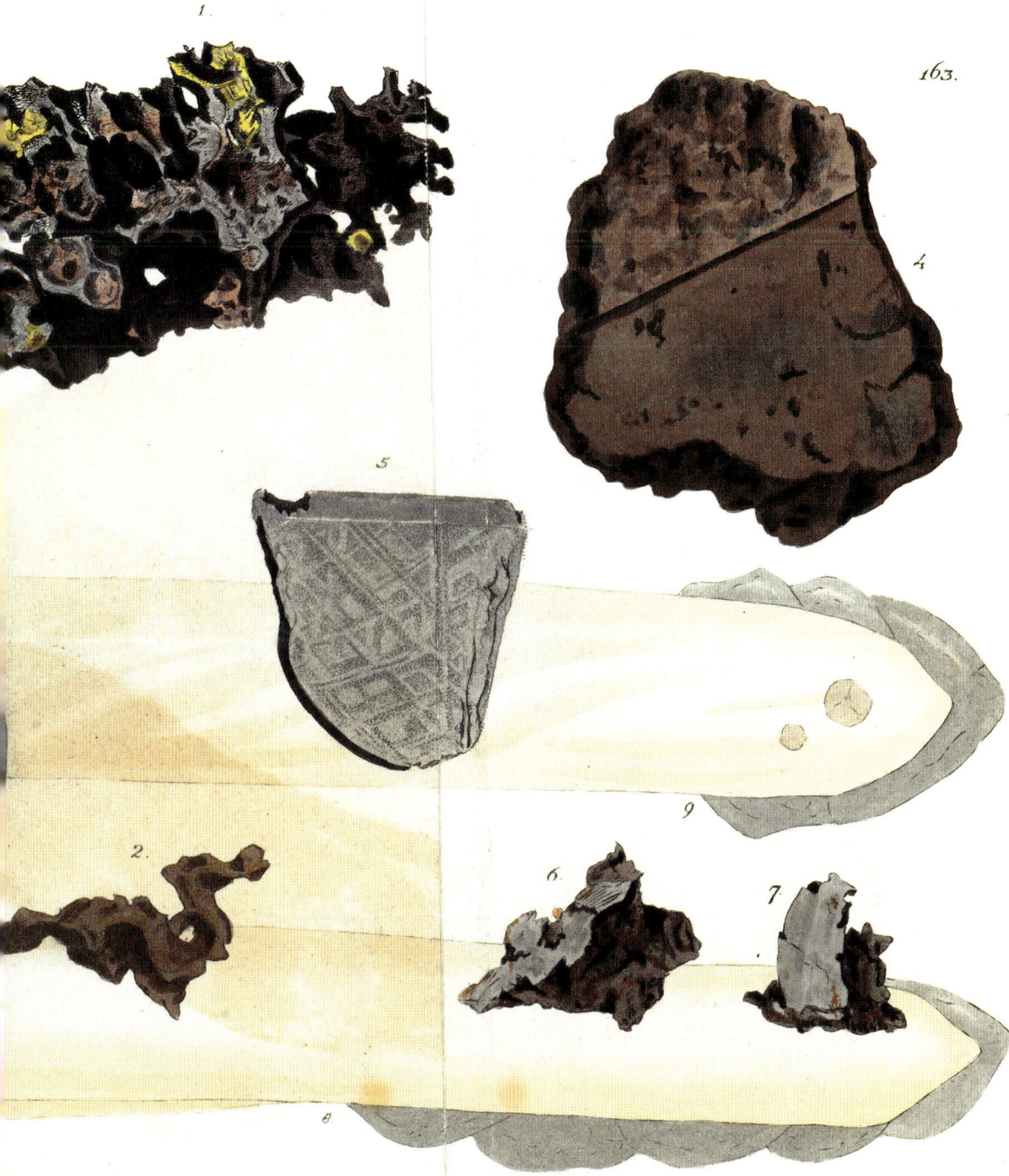
163.
1.
4
5
9
2.
6.
7
8

about the sword and discreetly sought Crichton's help: 'I hope it will soon be in his hands and when I find it is I shall be glad to know from you what is said about it.' With the passing of time the meteoritic sword must have faded from most people's memories – except Sowerby's! Over four years after the Emperor's visit to London, Crichton asked the Emperor himself about it but was told he had no recollection of it. He presumably had other much more important matters on his mind. But a possible route to success was offered. It was suggested that Sowerby write to the Minister of the Interior, a certain Mr Kosodawlew, who was supposed to have originally given the sword to the Emperor, so that he, in turn, could have an excuse to remind the Emperor of the event (and perhaps even show him the sword again). This suggestion was, of course, included in a letter about minerals and their prices – the sort of thing that collectors really wanted to write about.

The suggestion worked. Nearly five years after the Emperor's visit, Sowerby finally received a most welcome letter from the Minister of the Interior, Mr Kosodawlew, with the good news that not only had 'his Imperial Majesty been pleased to read your letter and to accept your sword with particular benevolence' but that he wished also 'as a mark of His satisfaction, to make you a present of a ring adorned with diamonds and a large green precious stone set in the middle of them'.[18] The ring was to be delivered by Dr Crichton, who was coming to England, but getting the ring onto Sowerby's finger was also subject to delays, since Crichton first visited relatives in Scotland and had other duties to fulfil. But by November, Sowerby had received his gift and the saga prompted him to write a short report for the *Philosophical Magazine* and to include a coloured plate and description of several iron meteorites in his *Exotic Mineralogy*. People soon learned of his success. Edward Clarke, the Professor of Mineralogy at Cambridge University, asked to borrow the Cape iron for a lecture, to which Sowerby immediately agreed and sent him other specimens as well, in contrast to his earlier refusal over the Wold Cottage stone meteorite. The science of meteorites had come a long way. The entry in *Exotic Mineralogy* made a clear distinction between stone and iron meteorites and was no longer ambivalent about their extra-terrestrial origins. Sowerby's collections were well regarded and he was often the first person to whom people such as Edward Clarke or other collectors would turn for advice or specimen exchange. Meteorites featured significantly in the auction sale of his museum several years after his death.

There is an interesting epilogue to the sword and its saga. The sword faded into obscurity after the deaths of Sowerby and the Emperor. A search was made for it in Russia in 1947 but no trace of it could be found. It was assumed to have been destroyed, lost, or forgotten for a second time, but enquiries in 2012 revealed its presence in the State Hermitage Museum, St Petersburg, still in good condition and a fitting reminder of the events of 1814.[19]

None of Sowerby's contemporaries moved on as many fronts as he did. Some would have probably seen this as a weakness, for he could never be the true master of a subject. But his vision was different and he achieved real influence in the study of meteorites as he had done in plants, fungi and would do later in fossil shells. Between the fall of the Yorkshire meteorite and the fall of Napoleon, he was working on his *Elucidation of Colours*, *English Fungi*, *British Mineralogies* and *Exotic Mineralogies* as well as illustrating some authors' works, including James Smith's introduction to physiological and systematic botany. *English Botany* only came to an end in 1814, and was being 'replaced' by *Mineral Conchology*. He was a contributing member of the Geological Society and the Linnean Society as well of a newly formed Lambeth Chemical Society. He was a driven man; life continued to centre on his work at Mead Place.

The scientific world that he had entered many years before had certainly changed, yet the methods of botany and of zoology were much the same. Between the years of the fall of the Wold Cottage meteorite and of his receiving the ring from Alexander I, there were major advances in the fields of chemistry, geology and astronomy. The *Transactions of the Royal Society* show that

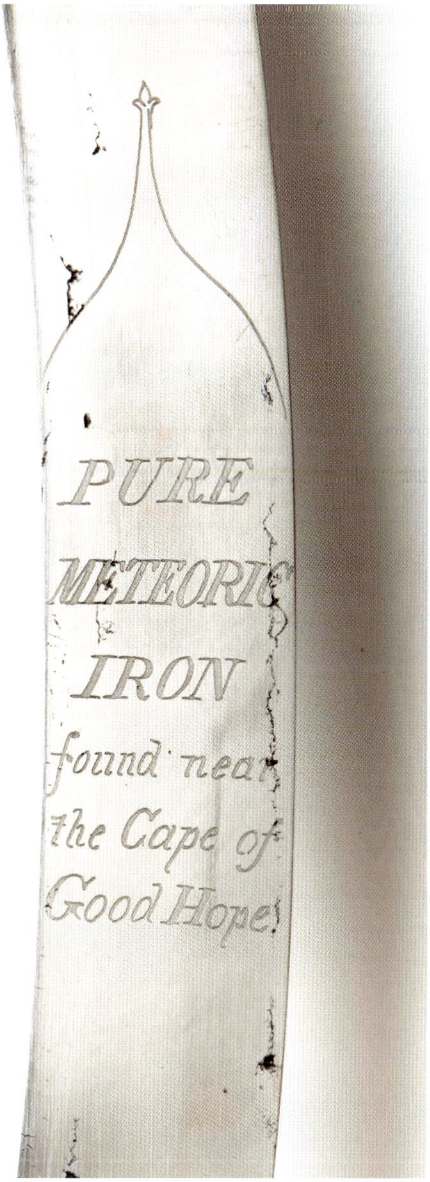

10/9 Engraving detail on the blade of the meteoritic sword. Copyright © The State Hermitage Museum.

chemical elements were being described and discovered by people such as Humphry Davy, William Wollaston and Smithson Tennant. William Herschel was reporting on celestial bodies, on the movement of the solar system and on double stars. Count de Bournon, William Gregor, Humphry Davy and James Smithson published their investigations of minerals, while David Brewster revealed how light was affected when it passed through certain crystal substances. These *Transactions* contained few papers on plants or even on animals. These were the predominant subjects of the Linnean Society with James Smith, as its President, choosing to publish his work there, although he was also a Fellow of the Royal Society. Sowerby occasionally contributed a note or paper. Most Linnean Society papers were descriptions and observations on fossil shells, fungi, mosses, lichens, birds, bats and badgers, among other subjects. The biodiversity of other countries, especially New Holland (Australia), also featured. The Reverend William Kirby in his paper 'A description of several new species of insects collected in New Holland by Robert Brown' wrote: 'No country has produced more novelties in every department of Natural History than New Holland; and the plants and animals of that insular continent, if I may so speak, are most of them of a peculiar character.' The pursuit of natural history still suited James Sowerby, even when it involved minerals and meteorites.

Outside the learned societies, science and technology were developing in their particular ways. The Industrial Revolution, with its advances in the harnessing and use of water power and steam power and in the development of machine tools, would have been known to Sowerby but did not really impact on his own activities. It was, however, William Smith, often called the 'Father of English Geology', who helped to bring Sowerby into a new field of enquiry. Smith had published, in 1815, his now famous map of the geological strata of England, Wales and part of Scotland based on the fossils they contained.[20] The following year his important work on the identification of strata by their most characteristic fossils appeared, with coloured plates by Sowerby, to which we will return. Geology and fossil shells were increasingly part of Sowerby's interests, as they had been for several years. In 1808, close to the time of the founding of the Geological Society, he assisted in a demonstration at Mead Place of a new mine-boring apparatus invented by the Irish mining engineer James Ryan, attended by several members of the Society including the President, George Greenough, Arthur Aikin and Sir Joseph Banks. Sowerby helped introduce Ryan to Willaim Smith's methods and Ryan contributed specimens for Sowerby's work on *Mineral Conchology*, which was now in preparation. Sowerby was a stronger supporter of William Smith's methods than were many of his colleagues in the Geological Society.[21]

10/10 Cedar of Lebanon (*Pinus cedrus*). Part of Sowerby's folio plate for Aylmer Lambert's *Pinus* – a magnificently illustrated work produced over more than twenty years. He wrote: 'This is too remarkable a tree to be mistaken or confounded with other species... the cones are very large.'

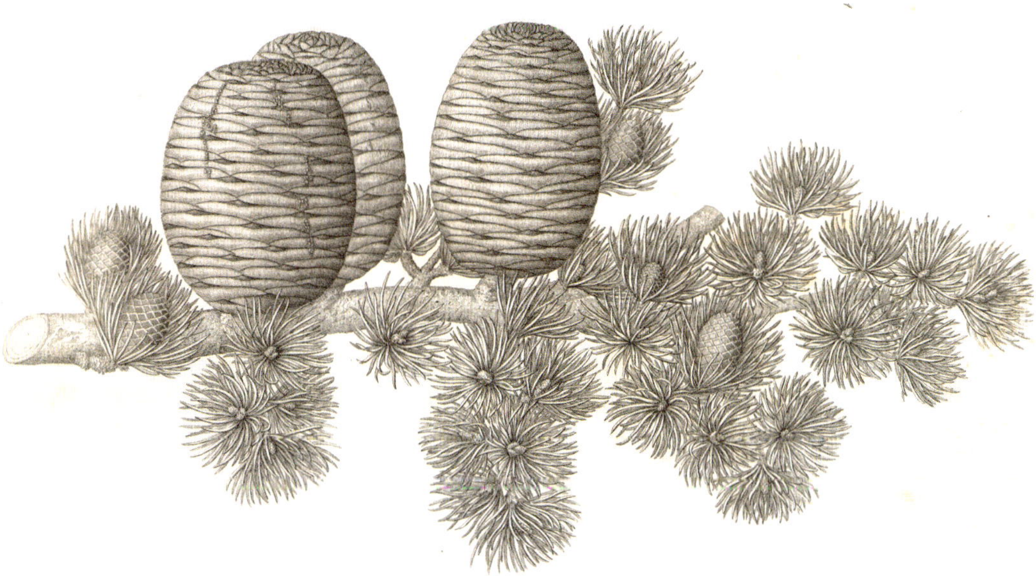

The ending of the Napoleonic wars and the arrival of peace brought increased sales of his works, especially *English Botany*, in other parts of Europe. Money to support the family was also gained by some small commissions for illustrations such as for pamphlets and books by the traveller and London mineral dealer John Mawe, for Aylmer Lambert's fine folio work on conifers, for William Leach's on crabs and lobsters and so forth, and for T. Purton's book on plants in the Midland counties of England.[22] He was also busy with production work for the *Transactions of the Linnean Society*, as well as being a reasonably active member of other new philosophical and chemical societies. No wonder that sometimes he had to apologise to his friends and correspondents about errors or haste. To the surgeon and natural historian Gideon Mantell he wrote thanking him for a large fragment of an ammonite and also begged 'excuse for my hurry as sometimes I cannot get all I want done although I begin at four o'clock in the morning'. Yet he was fully prepared, when the need arose, to stop his planned work and to give immediate attention to an urgent request for knowledgeable assistance. Such was the case of serious rot in the fighting ships of the British Fleet during the final years of the war with France.

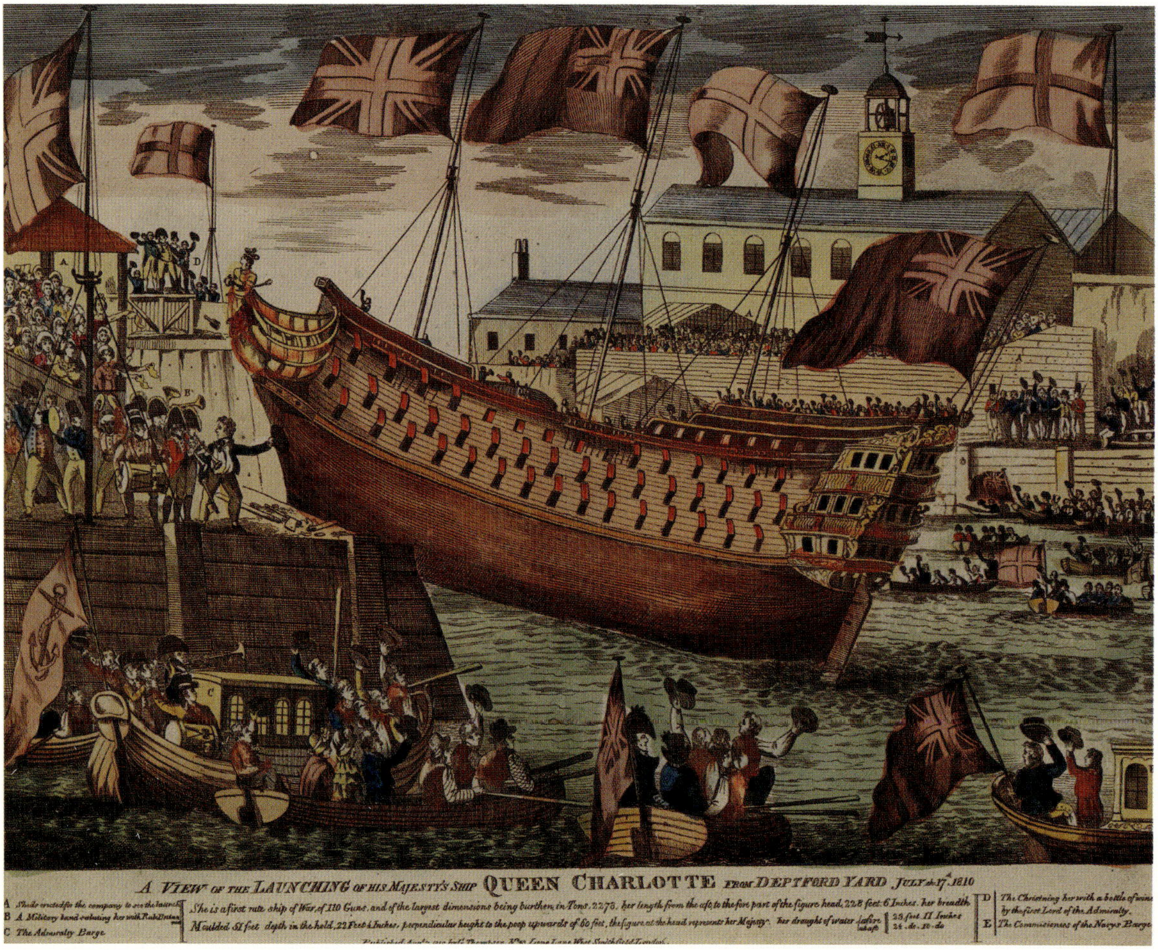
A VIEW OF THE LAUNCHING OF HIS MAJESTY'S SHIP QUEEN CHARLOTTE FROM DEPTFORD YARD JULY the 17th 1810
A Sheds erected for the company to see the launch
B A Military band saluting her with Rule Britannia
C The Admiralty Barge
She is a first rate ship of War, of 110 Guns, and of the largest dimensions being burthen in Tons, 2278, her length from the aft, to the fore part of the figure head, 228 feet 6 Inches, her breadth
Moulded 51 feet depth in the hold, 22 Feet 6 Inches, perpendicular height to the poop upwards of 60 feet, the figure at the head represents her Majesty. her draught of water
23 feet 11 Inches
D The Christning her with a bottle of wine by the first Lord of the Admiralty.
E The Commissioners of the Navy's Barge

CHAPTER ELEVEN

A lot of rot

11/1 A view of the launching of HMS *Queen Charlotte* from Deptford Yard. 17 July 1810, by George Thomson. Credit: National Maritime Museum.

In 1812, during the Napoleonic Wars, His Majesty's Flagship the *Queen Charlotte* was causing significant concern. This was not the *Queen Charlotte* which engendered consternation in 1797 when, as it was preparing to weigh anchor to pursue the French, its crew mutinied over pay arrears, bad food and, no doubt, much besides. That *Queen Charlotte* caught fire and sank in 1800. Now the consternation was of a different sort: the new ship itself was sick, and the sickness was dry rot. It had been launched as a first-rate ship of the Line with 110 guns in 1810, so was only a couple of years old when James Sowerby was called on to inspect the patient and administer a cure. Ships such as these were mighty wooden structures, impressive in their scale and design, yet fraught with difficulties in construction and use. They played a critical role in the war with France.

Wood rot was a common and well-known issue for His Majesty's fleet and the *Queen Charlotte* had more than its share of problems during its early life – the costs of repair soon exceeded those of the initial construction. Construction had started in October 1805 with the laying down of the keel. By the beginning of 1810 it was nearly ready and heating stoves were placed in the main hold and other areas to dry the ship out before its launch in May. The rest of the fitting-out was done by the autumn and in May the following year the *Queen Charlotte* was taken to Plymouth where she was found to have become so rotten that 'all the planking within and without board together with many of the timbers and beams had to be replaced'.[1] The Navy Board had attempted over time to deal with this important issue, so at first it is a little surprising that it suddenly and even urgently summoned Sowerby to serve his country in this way. But Sowerby moved quickly, as was his nature, and seemed initially content, perhaps even proud, to have been asked to fulfil such a task, the like of which he had never done before. It was, for Sowerby, of relatively short duration but its passing revealed not only hidden depths to his competence and character but also something of the Navy Board's special ways of working: both officially and behind the scenes.

On Wednesday 15 July 1812, Sowerby received a letter summoning him to attend the Navy Office at the Admiralty the next day at 3 o'clock. There he learned of the significant problem of dry rot and saw examples of its effect. Two days later he had written instruction to

> proceed with Mr Lionel Lukin to Plymouth as early as possible and inspect the state of His Majesty's ship Queen Charlotte, which has received very considerable injury by Dry rot, with a view to ascertain what measures it may be advisable to adopt to stop its further progress if possible, by preventing the vegetation & spreading of the Fungi. We further desire that your report on this subject may be extended to the measures necessary to prevent this evil in any of His Majesty's Ships; and that you will make it either in conjunction with Mr Lukin or separately as you may find most convenient.[2]

Given the long-term nature of the problem, this was a tall order.

As was customary, there was no written mention of any payment or other sort of recompense, nor any indication of assistance besides the presence of Mr Lukin (who had already made some investigations but was deemed, by some at least, to not be a suitable expert), nor of when they expected the report. However, Sowerby also received a separate confidential letter from Mr John Knowles, Secretary to the Committee of Surveyors of the Navy Office, advising him that his report will be, as he put it, 'of importance to your own interests as well as the public service' and to 'make the report singly without the interference of Mr Lukin'.[3] No indication of how the report would be of importance to Sowerby was given, although it seemed as though Knowles might look after him more than other officials would. Indeed, the following day Knowles sent him confidentially much information on observations held by the Navy Office on dry rot and the heating of timbers, including the fact that many ships had been destroyed by the fungus *Boletus lacrymans*, some after only a few years in service.[4]

Mr Lukin was a maverick. He had ingratiated himself with the Navy Board from whom he received various commissions, some suggested by himself. The evidence also suggests that he used His Majesty's ships and dockyards as a laboratory for his crude experiments rather than providing advice based on sound and tested knowledge. And some of these experiments had been serious failures. One concerned Lukin's proposal to impregnate timber with oily substances as a protection against rot. For this, a special kiln was built at the Woolwich Dockyard in which timbers were exposed to oily vapours from the heating of pitch pine sawdust. The plant exploded, killing six men, injuring many others and doing serious structural damage. The treated timbers were also a failure. A certain Mr Benjamin Tucker, who was later to become a Surveyor of the Navy was, as early as 1811, expressing his amazement at the faith placed in Lukin, who he described as a 'sort of Dr Brodum', an infamous quack and medicine dealer of the time.[5] Lukin, unlike Sowerby, had no proper mycological experience.

By Tuesday evening, Sowerby and Lukin were on their way to Plymouth in south-west England, passing through Staines, Bagshot and Basingstoke, dining at Andover, and sleeping and breakfasting at Amesbury, near Stonehenge in Wiltshire. Then on Wednesday and Thursday they passed through Wincanton, Ilminster, Honiton and Exeter before arriving on Friday by 11am at the dock and the *Queen Charlotte*.[6] Travel, of course, was a time-consuming business in 1812. Sowerby must have had many opportunities to discuss dry rot with Mr Lukin but he was to write his report 'singly', as Mr Knowles had advised. He examined the *Queen Charlotte* the rest of that day and some of the next. He found that great quantities of timber had already been removed and that it was intended to take out more. Several fungi species were indeed present, especially from the middle deck downwards, including *Boletus hybridus*, *Boletus medulla-panis*, *Xylostroma giganteum* (a leather-like fungus peculiar to rotten oak) and *Auricularia pulverulenta*. *Boletus lacrymans* (as mentioned by Mr Knowles) was present but not common.[7] He used a hygrometer, noting that the air within the ship was very damp; just right for fungal growth. As soon as his examination was finished he sent his draft report to his son George at Mead Place asking him, with the help of his brother James, to copy it as well as he could. By the afternoon he was looking at some quarries in the neighbourhood. Sowerby certainly worked fast and hard.[8]

His return journey was a very different affair from the outward one and with more than twice the number of days travel. Now, on his own, he had the time to pursue the sort of natural history he loved. He first headed north to Tavistock and the village of Lydford on the edge of Dartmoor, probably to make observations of the local copper, lead and silver mineralisation, and also to view the fine waterfall at Lydford Bridge, of

11/2 *Boletus lacrymans* (now *Serpula lacrymans*) fungus – the cause of dry rot in HMS *Queen Charlotte*. The word *lacrymans* refers to the fact that the fungus often exudes tear-like drops of water. *English Fungi*.

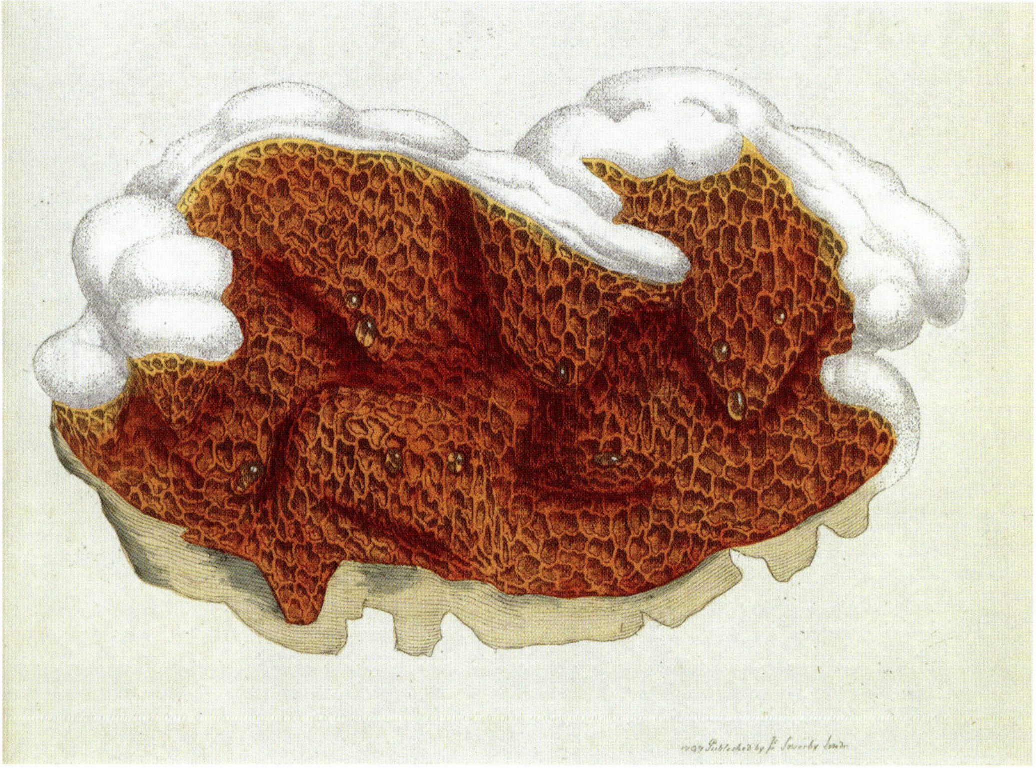

which he made a rough sketch. His homeward route included examinations of flints, gravels, sandstones and iron stones among other rocks and deposits. He arrived home on Sunday 2 August, some twelve days after his departure and with less than two actually at the dockside. He delivered his report just two days later, stressing the urgency of the problem.[9] Staying away from home for such duration was a very unusual thing for Sowerby and is especially significant since he was busy with so many initiatives. He was at this time preparing material for, and publishing, his *English Botany*, *English Fungi*, *British Mineralogy* and *Exotic Mineralogy*. He was also just starting his work on fossil shells – *Mineral Conchology* – and to further confound things his first grandson was born to George who was still living at Mead Place but having had a difficult period in the relationship with his father. Even his first and very dutiful son, James De Carle, was away from home and staying with Dawson Turner in Yarmouth. So the workload must have fallen on his wife Anne and one or two of his other children such as Charles, who would have been seventeen years of age. His return must have been greatly welcomed.

He opened his report with his shock at learning of the state of many of the Country's 'National bulwarks' and somewhat fawningly expressed his resolve to fulfil his duty especially when: 'I perceived the penetration and anxiety of your Honourable Board by the questions you asked me; my answers to which, when in any degree to the purpose were treasured up in a manner that convinced me how very desirous you were to execute with fidelity your most arduous trust.' He was, he said, somewhat diffident over his commission but could think of no one else who had the right experience so 'while I might feel proud in a concern of such magnitude of the hope of being able to do good, I could only consider myself as an instrument in the hands of Providence selected for a particular department at a crisis so momentous as the present', and he apologised for any mistakes occasioned by the haste with which he had drawn up his observations and to the fact that this was his first ever official report. He listed the offending fungi and gave reference to their illustrations and descriptions in his *English Fungi*.[10]

Not surprisingly his recommendations were straightforward and sensible, yet were ones not already practised: always use seasoned, dry timber in the repairs; do not paint wet wood; have a good circulation of air throughout the ship, and use awnings over the ship as protection from rain during the repair. He rejected the use of heating for drying timber as it could encourage the very same undesired fungi, and also rejected the use of 'acid or saline impregnations' as 'being detrimental to the wood, bolts and workmen'. The regular use of hygrometers to assess the humidity in the various levels of the ship would also help to identify the need for remedial actions. His report was different from recommendations made several months earlier by Mr Lukin who, while also stressing the importance of good air circulation, advised on the use of 'Mundic' in a preparation he had concocted.[11] Mundic is a naturally-occurring mineral of iron sulphide (pyrite) but which, in the south-west of England, can contain very high quantities of arsenic. Lukin proposed treating the timbers with his mixture, in the 'state of vapour', and also with holes drilled to allow greater penetration, but provided little detail of the method of application or of any protection for the workmen. He further proposed a greater removal of faulty timbers than others thought reasonable. Plymouth Yard also reported a few days later that after following the recommendation by Sowerby and Lukin to remove one plank from head to stern at the bottom of HMS *Queen Charlotte*, several timbers bearing fungi were found, with some so decayed that they were no longer fit for purpose.[12]

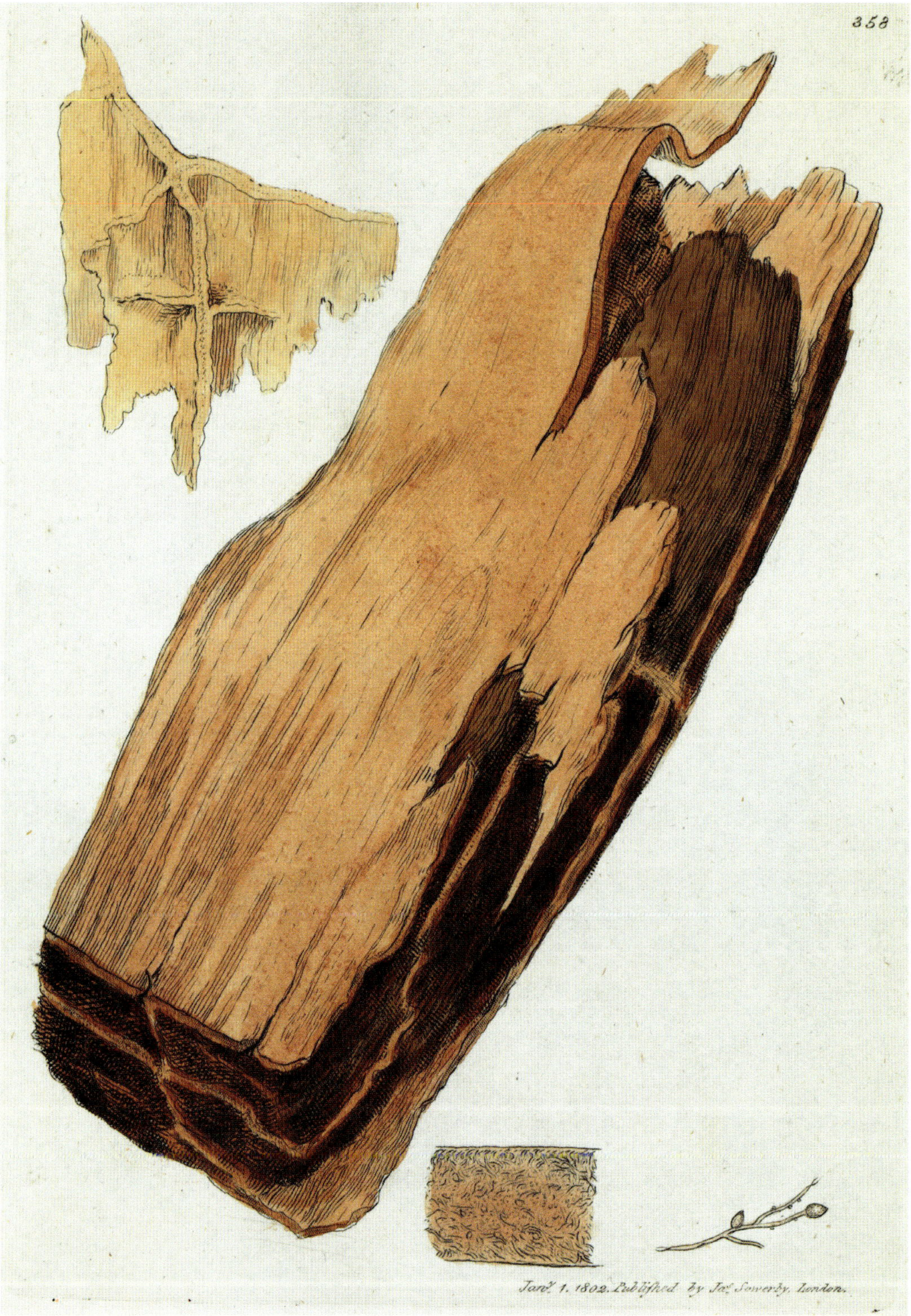
358
Jan.y 1. 1802. Published by Jas. Sowerby. London.

11/3. The wood-rot fungus *Xylostroma giganteum*. Sowerby wrote: 'This may truly be named giganteum, from the gigantic strides it takes through some of our largest oaks.' *English Fungi*.

His report was quickly accepted and he was informed, in a private letter from Mr Knowles, that all his recommendations would be put into effect, without any further reference to Mr Lukin's earlier reports, and also that 'their Lordships' had directed the Navy Board to 'acquaint you of their Lordship's approbation'.[13] There was no mention of any financial remuneration; their Lordships thought they could summon anyone to serve their purposes without any concern of the impact. The official letter had an even more condescending tone in that the administrators were pleased just 'to mention' that the Lord Commissioners '*consider you entitled* to their thanks and approbation for your attention and Report' (added italics).[14] The same letter asked him to recommend a make of hygrometer! Mr Lukin had written separately to say that he had nothing more to say, other than that he agreed about the circulation of air and still advocated the use of his chemical preparation.[15] Indeed, the use of mundic in a paste form was applied to some timbers but had to be stopped because of the damage done to the workmen's health.[16]

So did Sowerby complain about his treatment? It was clear that he was very unhappy about the lack of proper recognition and had made some enquiries of people associated with the Navy Office. He also knew too his report had been well received and that Mr Knowles and others had visited the dockyard to ensure implementation. Within a couple of weeks he had drafted a letter, intended for the Navy Board, which he sent to his son James in Yarmouth (staying and working with Dawson) for comment. But he himself was not happy with it and made changes such as removing the sentence 'neither shall I consider my time as lost, and hope that the interruption my avocations have suffered in this business will not make me repent of what I have done'. It seems the letter was never sent but the content revealed his frustration – and this was not going to disappear without some resolution.[17] Yet, early the following year he wrote with further advice and sensible recommendations to the Commissioners on matters such as the seasoning and storage of timber for use in construction, on the use of hygrometers, and even a little on the best types and sources of timber. For this he did not seek any financial redress. He did, however, remind the Board of his achievement with his customary style: 'Should the operations I have recommended for the Queen Charlotte continue to be attended with the success I anticipated, they will give confidence, exciting more particular attention and becoming a powerful stimulus to further exertion.'[18] The Navy Office, for its part, felt content with the proceedings for it asked Sowerby in the autumn to attend the Deptford and Woolwich shipyards to discuss and make recommendations on the mode of stowing timber. Sowerby, as was his wont, duly obliged by soon visiting the yards, identifying the serious failures and making sound proposals for tackling the problems. In his letter of recommendations he even went so far as to say that he felt the highest esteem for the judgement of the Commissioners such that 'as an individual who, partaking of the blessings of being born under a British Government, must be happy to subscribe himself... Your very faithful & humble Servant'.[19] He still did not request any recompense but that situation was soon to change.

Even if he had been happy so 'to subscribe himself' he had also discussed his plight with Mr Knowles, whom advised him to compile a statement for the Navy Board, listing all the work he had done and asking for payment; this would then bring the subject to their attention as if they would never have thought about it before. So in December 1813, some fifteen months after his visit to HMS *Queen Charlotte*, he detailed his many items of work, advice and correspondence in a statement of 'attention to a subject of such National consequence that I cannot for a moment forfeit it, and the urgency of which has even interrupted some of my periodical works to the disappointment of my Friends & consequently my own loss'.[20]

This had the desired effect. A very short letter arrived from the officials of the Navy Office telling him of the award of two hundred guineas for his efforts but also making sure that Sowerby was not given too much credit through the added sentence:

> At the same time we deem it necessary to apprize you that in awarding this sum we do not attach to you the merit of having saved the Queen Charlotte by your advice, all the defective materials having been taken out of that ship at Plymouth when she was repaired there.[21]

Such an untrue statement clearly rankled Sowerby.[22] Mr Knowles wrote to congratulate him but wished that the recompense had been more, and he even advised that Sowerby thank the Board! This he did but surprisingly with the somewhat obsequious finish: 'May you enjoy a continuance of happy years and the British Navy flourish under your arduous endeavours.'[23]

The financial reward was certainly very low. The Navy Office on just one occasion had recommended Mr Lukin be paid £500, in addition to sums 'already allowed' for his services before Sowerby arrived on the scene; but then Lukin was active in seeking payment, had become well known to the officials, and would not suggest that he was ever in a servile position. But more significantly Sowerby had overplayed his hand. In his statement to the Navy Board on the services provided he also gave strong hints of the sort of recompense he was expecting and in a form that would undoubtedly be troublesome to the Board:

> It was generally supposed that the Dry Rot had rendered the Queen Charlotte incurable and

> she is now on service, providentially saved by my directions; I need not mention the value of such saving to those conversant with the subject, it would certainly be enough to afford ample indemnification for my past services; and, also upon future ones, reckoning upon other first rates etc., assuredly saved and to be saved by such an example properly applied the amount would be immense.

These were strong words implying that Sowerby could be demanding rates of salvage and the Navy Board inevitably took a defensive position.

As soon as the Board received his statement, it simply asked the Surveyors office for 'their opinion of the extent of the benefits which have been or may be expected to be derived from the modes recommended by Mr Sowerby for the prevention of the Dry Rot'.[24] The Surveyors' short reply was that Sowerby's recommendations on keeping timbers dry had proved useful as had his recommendation to discontinue the use of stoves for heating, but that clearly the current state of the *Queen Charlotte* was nothing to do with Sowerby as all defective timbers had previously been replaced.[25] On the same day the Board met and agreed the sum of two hundred guineas and that Sowerby should be told that the final state of the *Queen Charlotte* was not owing to his advice. The minute of the meeting also changed the wording on the heating stoves from 'discontinuance' of use to 'continue the moderate use', something which Sowerby was against.[26] The Board was clearly in a defensive mode over its own actions.

So that was that; but it wasn't quite. Sowerby continued to give some advice or help and still suffered significant delays over payment even when he had sent a bill,[27] and the Navy continued to have major problems over rot in its fleet by not maintaining the good working practices Sowerby had advised. He had been well chosen to do the task but the Navy Board, despite accepting all the recommendations, failed through its various inactions to give due credit and to gain the full and lasting benefits. Sowerby had been treated shabbily; he had stood up for himself but he did it too late and in an ill-advised way.

Paying for good advice, even when it had been sought, was not a guaranteed part of the culture of the time. Sowerby's mycological knowledge was recognised by officials and friends alike but that might have led him to supply special favours. On one occasion, Dawson Turner wrote, regarding dry rot at the grand Palladian-style Holkham Hall in Norfolk, owned by Thomas Coke, the Earl of Leicester that 'Mr Coke, begs me to ask if you can make it convenient to come down to his Palace at Holkham in Norfolk in the course of July', but with no mention of possible remuneration, so Dawson felt obliged to add the suggestive phrase 'well knowing that with his princely fortune... '[28] We do not know whether Sowerby ever went, nor whether, if he did, he was paid.

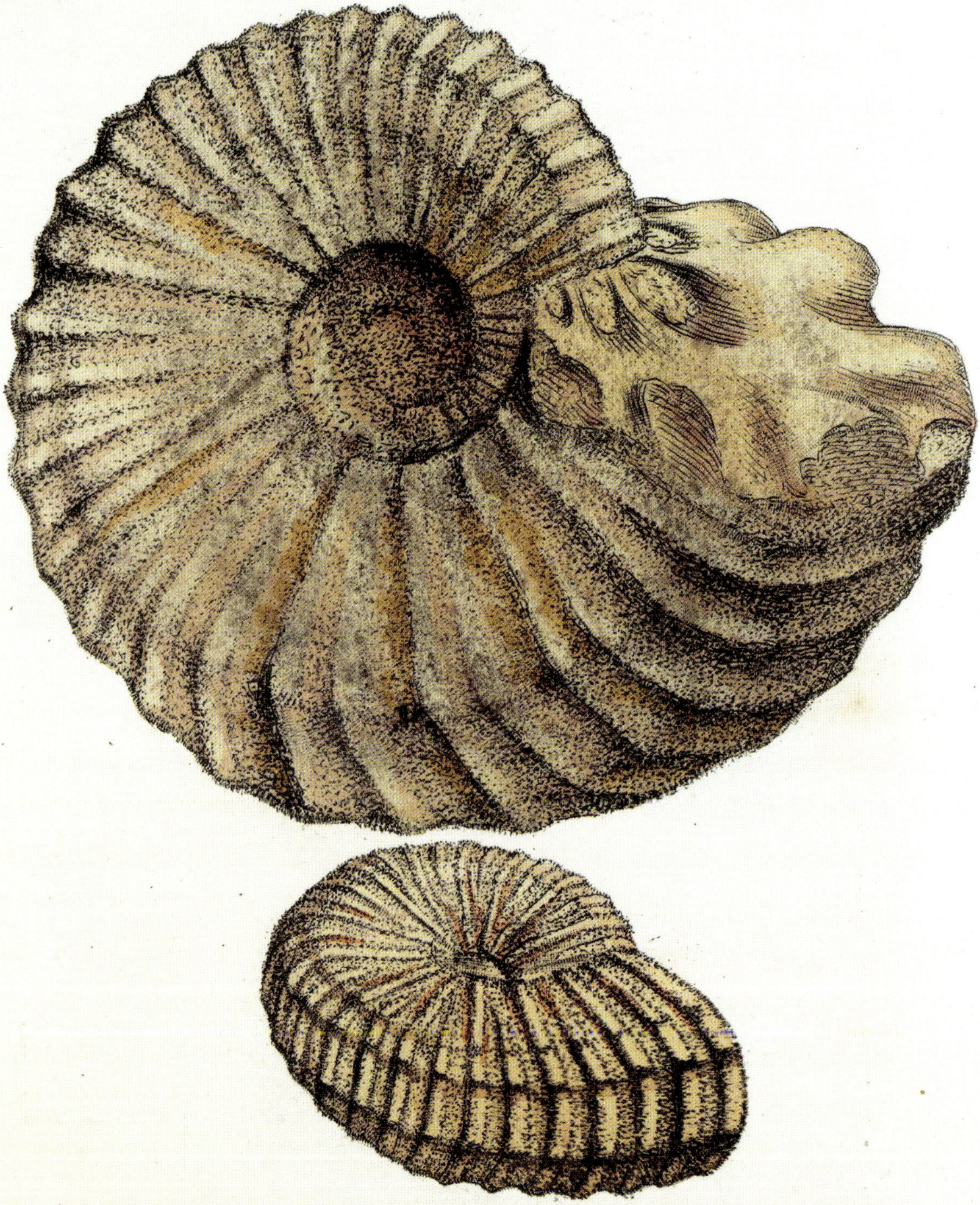

Ap. 1 1814 published by Jas. Sowerby London.

CHAPTER TWELVE

Mineral conchology

12/1 *Ammonites mantelli*. Engraving, in *Mineral Conchology*, of the ammonite Sowerby named after his friend Gideon Mantell who supplied him with many fossil specimens, especially from the area of Lewes in Sussex. *Ammonites mantelli* is now named *Mantelliceras mantelli*.

In August 1797 James De Carle, aged just ten years, received a box from his father who was away with his wife Anne visiting relatives and friends in Norwich. A short accompanying note told him that it contained 'some 'curiosities'... which I give you in charge to take great care of for me'. Among these were some fossil belemnites, which his father said must be very carefully preserved, and he provided him with a simple sketch so there could be no doubt in the young boy's mind as to which they were.[1] A few days later, before Sowerby took the coach from Norwich to visit Dawson Turner in Yarmouth, he sent another package containing seaweeds and instructions on how to dry them. James De Carle's training in matters of natural history had started early.

Sowerby knew that the bullet-shaped belemnites were organic remains and not the thunderbolts of folklore. The visits to Norwich always included his scientific interests; the geological strata of the area naturally provided interesting opportunities for investigation. His collection of fossils was already underway and his museum at the back of his house had been built the previous year to house them and other specimens of British natural history. The fact that he had just started his own publication on *English Fungi* and that *English Botany* was continuing as a very successful one did not deter him from examining other areas of natural history that might, at some future date, be of use and significance. All natural history was open to Sowerby; his main difficulty was insufficient time to pursue all that he wished. Indeed, mineralised fossils were to feature just a few years later in his *British Mineralogy*.

Now, some fifteen years later, the very publications that had involved so much of his time and energy were showing signs of coming to an end and there were no significant commissions in sight from other authors even if he wanted them. The mineralogical works were in full flow and with several years to run, but despite their quality they were not selling as well as he would have wished; money had to be earned, especially in the difficult economic climate of the time, and the creative urge had to be satisfied. Starting a work on fossils, especially fossil shells of which he had a good collection as well as drawings, was a natural consequence.[2] It was also a topic of interest not only to his two sons James De Carle and George, but to an increasing number of others.

Fossils were more than just natural curiosities in the cabinets of natural historians and collectors; they were increasingly recognised as indicators of organisms living in past eras and, furthermore, useful in the ways that they were associated with particular geological strata of interest to the likes of miners and canal builders, as well as to the members of the still young Geological Society. This was a stimulus for Sowerby; another was that his correspondents were active and enthusiastic. Possibly the first woman geologist, Etheldred Benett, an unmarried woman living in Wiltshire, had the financial means to follow her enthusiasm for fossil collecting, the results of which she wished to share with friends.[3] She was one of the few women in a usually gentlemanly pursuit to amass a fine collection through purchases and from her fieldwork, especially in her own county and Dorset. She lent or gave specimens to Sowerby, with whom she often corresponded and occasionally visited to exchange opinions on particular species. She was, much later, to write her own *Catalogue of the Organic Remains of the County of Wiltshire*. Gideon Mantell, a surgeon and geologist who was to discover the dinosaur *Iguanodon*, explored the Sussex strata and collected specimens for his own publications and for sharing with both Sowerby and Benett, among others. He was more than happy to devote much time to satisfy Sowerby's requests for fossil and mineral specimens. He and Etheldred were just two of many others desirous of a reliable and useful work on fossil shells. Sowerby was on the threshold of a significant and durable project. It was to result in a work of inestimable value to science.

The beginning of the nineteenth century could rightly be termed an exciting and transforming period in the history of geology. William Smith had for several years recognised the characteristic fossils of the strata within the sequence of rocks in different parts of the country, and understood much of the strata's structural dispositions. He had brought a predictive element into geological exploration. James Parkinson, a surgeon who identified the shaking palsy disease (to later be named after him), was also an inspirational palaeontologist. In his books, started in 1804, intended for a general audience and appropriately titled *Organic Remains of a Former World*, he illustrated fossil plants and animals of different strata and wrote fully on their origins. He was therefore able to write in his preface:

> The phenomena particularised in the latter part of this volume, yield some important knowledge respecting the structure of the planet which we inhabit. These facts would also supply, if it were needed, the strongest proof of the error of those who believe, that there has always been a succession something similar to what is continually observed; and that the human species have had, and will have, a uniform and infinite existence. With almost equal force will these phenomena oppose that system also, which considers the form and structure of the surface of this planet, as resulting from a regularly recurring series of similar mutations. The loss of whole species or genera, and the late creation of others, as is assumed in this work, are circumstances which strongly militate against both these hypotheses.

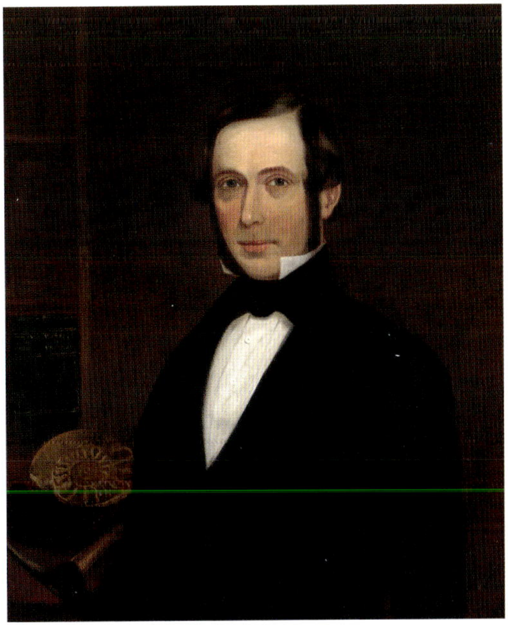

12/2 Gideon Algernon Mantell. Portrait (oil) by unknown artist, about 1835. Credit: reproduced courtesy of the Geological Society London.

Others might have strongly disagreed that extinction had occurred but he was justifiably able to brush their arguments aside. Likewise, the Parisian palaeontologist Georges Cuvier showed that the chalk of the Paris basin was essentially characterised by its fossils and that these in turn were entirely different, not only in species but often in genera, from those enclosed by the region's coarse limestone.[4] He also saw strata containing freshwater fossils and concluded they must have been laid down under such conditions. Sowerby knew these men, had their publications and letters by his side and supported their views. He was to interact, in various ways, with all three.

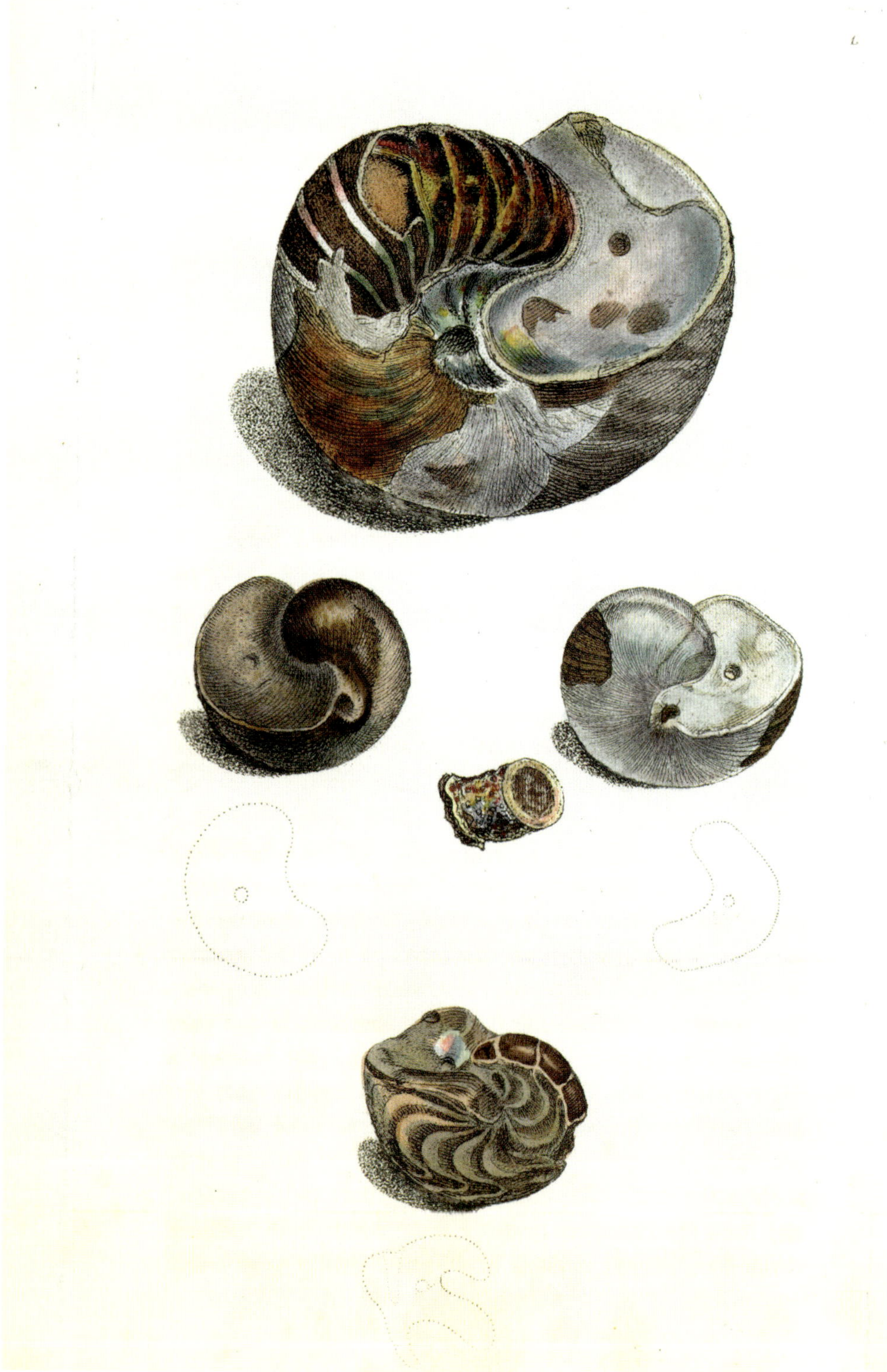

12/3 Fossil *Nautilus* specimens. The first plate in *Mineral Conchology*.

Mineral Conchology first appeared in the summer of 1812 in the familiar Sowerby publication style – that of a periodical with each number containing a few plates and letterpress, stitched together and covered with the usual blue paper wrapper – at 5s. The very first plate was of fossil *Nautilus* species, a fine choice for a beginning and made more immediate by the specimens coming from the London Clay at Highgate in North London, where major roadworks were exposing the strata and causing 'many discoveries', as Sowerby himself said.[5] The letterpress for the plate ran to three pages with details of the locations, the geology, the characteristics (both general and specific) of the fossil shells and, of course, reference to the good people who had given specimens. That first issue contained just three plates (subsequent ones usually six) and it was followed two months later with the second number. But other work, especially that over the problem of rot in the Navy's ships, intervened and it was not until the following February that the third issue appeared and a steady publication pattern began. It was to continue for another six years to Sowerby's death, and then, under the hand of his son James De Carle, for a further twenty-four years up to 1846.

The work is remarkable in many ways: for the excellence and beauty of the illustrations, for its taxonomic importance (including the description of numerous new species) for the recognition given to the significance of the geological strata which hosted the fossils, and for its considerable influence both at the time of its appearance and subsequently. Five hundred copies of each number were printed and hand coloured – a not inconsiderable number for such a publication at the time – and two pirated editions (in French and German) were later produced. It was from the start focused only on those fossils with hard shells – the brachiopods and the large phylum of mollusca, including extinct groups such as the ammonites – but this restriction was understandable with so much being discovered and described. Here was a major scientific work, appearing with little fanfare, being immediately respected for its quality and use both then and subsequently (allowing for changes in classification).

Mineral Conchology was not a review of knowledge, it was a serial that showed work in progress and so gives fascinating insight into the development of palaeontology. The specimen donors and correspondents acted almost as a research team, sharing specimens, ideas and information. Gideon Mantell was an exemplar, attempting to balance his professional life with his passion for geological investigation. A young man of twenty-two years at the start, he remained a loyal and enthusiastic contributor, sending numerous specimens from the Sussex rocks, discussing their characteristics and detailing their geological locations. He also reported on how the strata revealed the

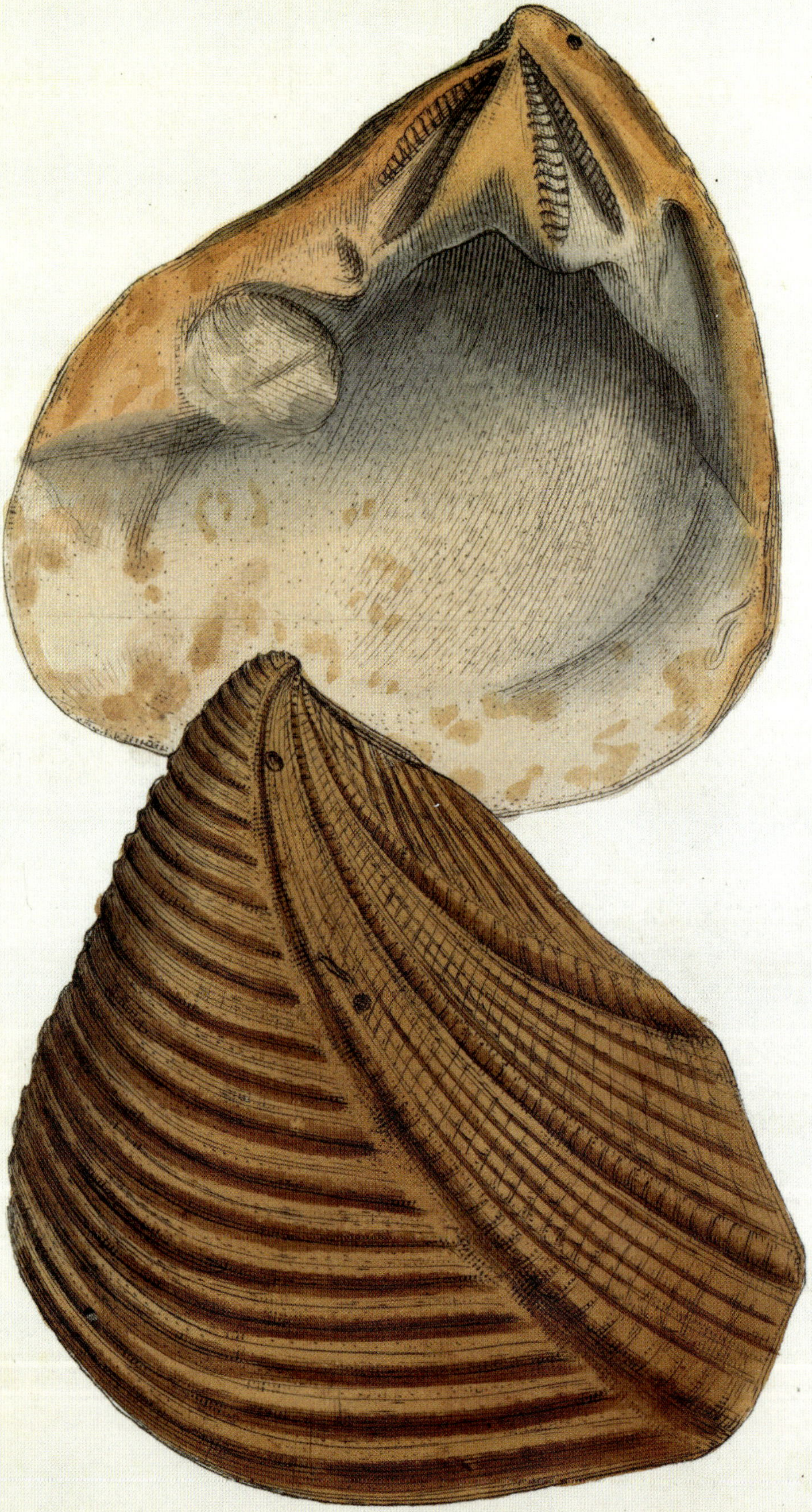

Ap.l 1815.

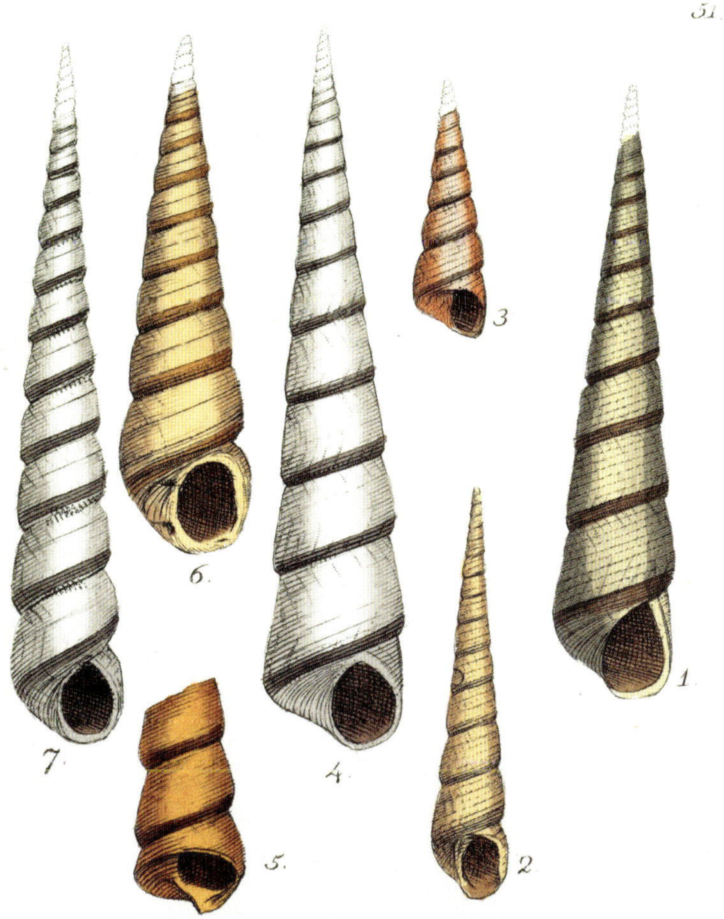

12/4 (opposite) Fossil bivalve shell (*Trigonia costata*). Sowerby discussed the taxonomic use of the one muscular impression and the structure of the hinge. *Mineral Conchology.*

12/5 (above) Specimens of different species of marine gastropods (sea snails) in the genus *Turritella*. Sowerby noted that variations made it difficult to distinguish the species. *Mineral Conchology*.

conditions of past times with the land sometimes under freshwater lakes and at other times under the sea, on the basis of the fossil species they contained.[6] All this was essential grist to Sowerby's mill, relying as he did on his fellow correspondents while he increasingly remained within Mead Place.

As with other Sowerby publications, there was little systematic approach to the sequence of appearance of fossil types. Current interests, new discoveries, pressures of time and the need to consult others were some of the factors determining the appearance of a particular

12/6 Specimens of a rather rare fossil – a helically coiled ammonoid (*Turrilites costatus*) provided from near Lewes by Mantell and from Wiltshire by William Smith.

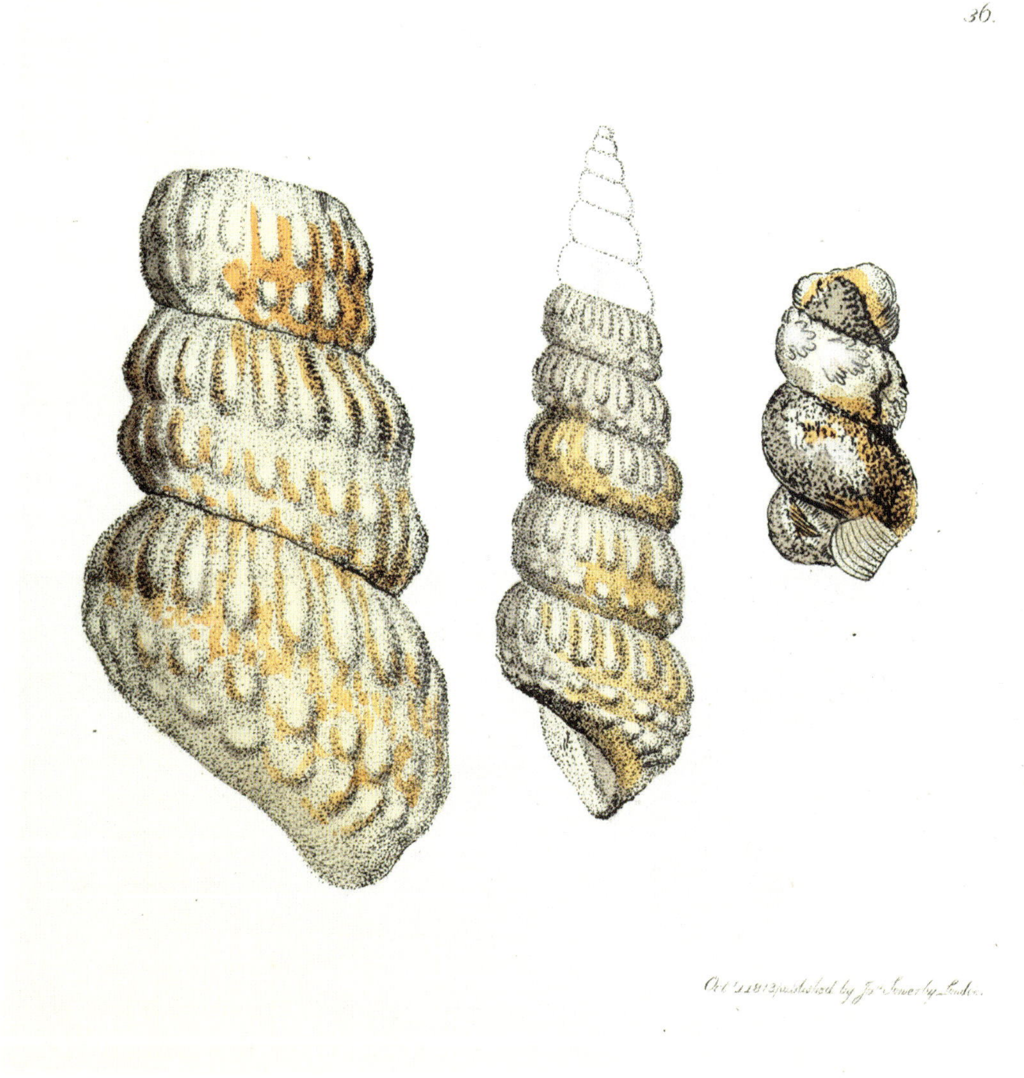

12/7 A fossil clam with spines, found commonly in the chalk of Sussex. Given as *Plagiostoma spinosa* in *Mineral Conchology*.

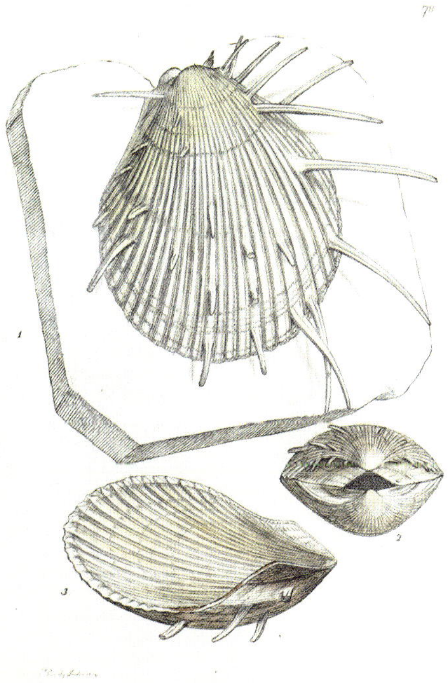

species. Sometimes there was a delay in depicting a particular fossil for fears over the quality of the available specimen. Sowerby wrote to Mantell 'if your friend's Turrilite should expose the septa, so as to make it fit to help my figure – and if there is no chance of a better – it might be borrowed, but if not sufficiently perfect I had better wait', and he expressed his desire to show his readers the true nature of the fossils: 'if any large fragments of the hinge &c of the Inoceramus can be conveyed to give an idea to an ordinary observer of its being sometimes four or five feet in diameter I should be greatly pleased with such'.[7] Users would have wanted the use of good specimens but they were frustrated by the lack of order over their appearance; the idea of having regular updated indices of genera and species was not a characteristic of these publications, the first not appearing until 1835. But an important innovation was a supplementary index, produced by John Farey, giving the stratigraphical position of fossils, this made possible by publication of William Smith's geological map of England and Wales. Farey had been agent to the estates of the duke of Bedford and had also completed an agricultural and mineralogical survey of Derbyshire. His geological excursions in the company of William Smith had given him considerable expertise and he became an enthusiastic and knowledgeable contributor to *Mineral Conchology*, providing it with an enhanced scientific dimension. The supplements appeared in the first four volumes but ceased on his death in 1826.

If Sowerby was not greatly concerned over the sequence of describing the different fossils, he remained, as he always had been, dedicated to accuracy and to recording most carefully the specific characteristics of any species. Of *Mineral Conchology* one professional palaeontologist recently wrote: 'its minutely accurate coloured illustrations and descriptive comments have ensured its working value, as opposed to historical interest, to the present day.'[8]

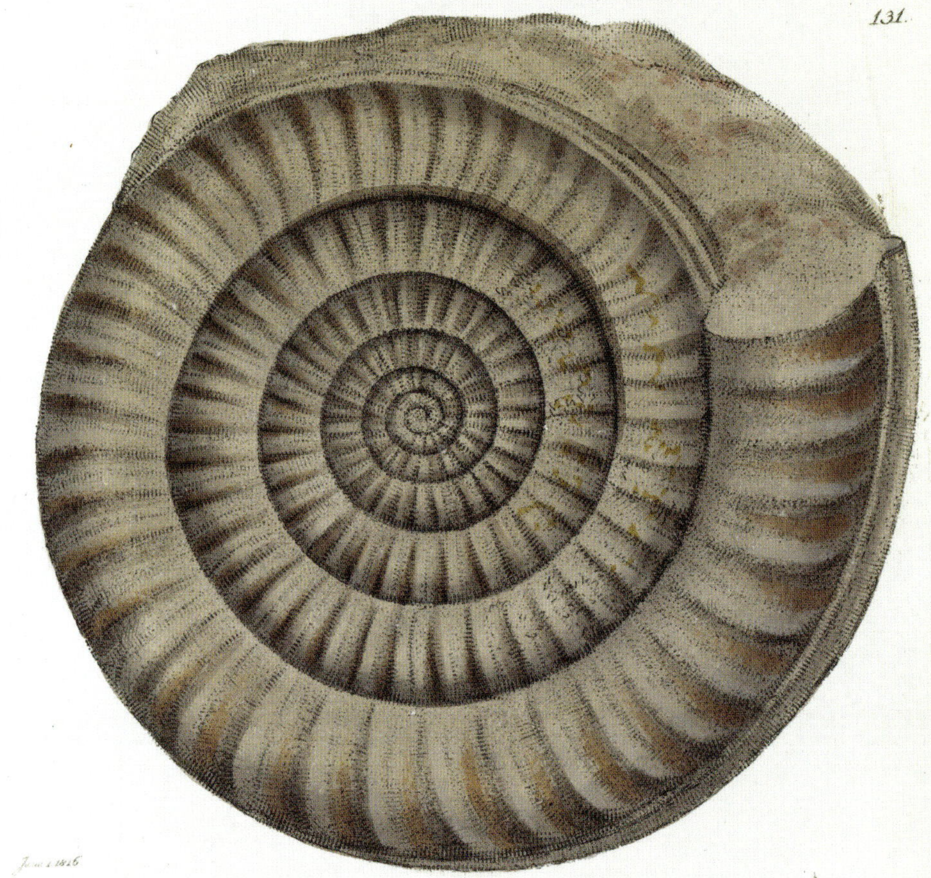
131.

12/8 *Ammonites conybeari.* Engraving, in *Mineral Conchology,* of the ammonite that Sowerby named after William Conybeare who later made major contributions to the study of marine reptile fossils. *A. conybeari* is now named *Metophioceras Conybeari.*

12/9 Specimen of the fossil *Ammonites conybeari* from Sowerby's collection and used for his illustration in *Mineral Conchology.* Credit: The Natural History Museum.

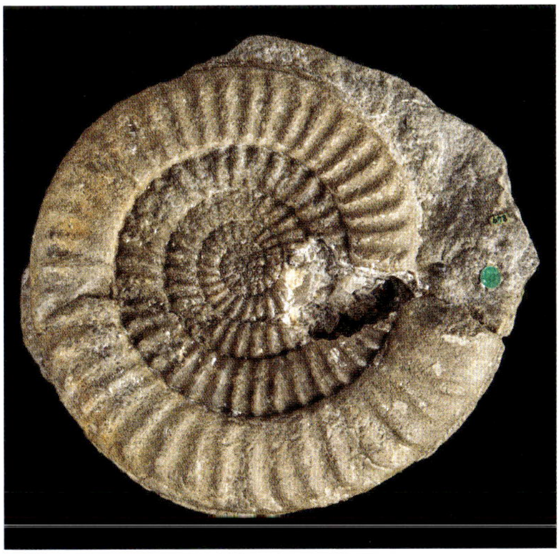

And Sowerby took particular satisfaction in being able, on the rare occasions that a fossil shell was not infilled by hard sediment or mineral deposit, to describe and illustrate the internal, sometimes diagnostically important, structure, and also to make comparisons with living specimens. In all these characteristics his work is that of a true scientist.

New species are abundant in *Mineral Conchology* and Sowerby clearly enjoyed the sense of discovery and novelty, particularly over some ammonites (an extinct group of cephalopods). He took pleasure in giving credit to people by naming species after them, such as *Ammonites conybeari* for the Reverend William D. Conybeare 'whose zeal deserves to be remembered by posterity', and *Ammonites greenoughi* for the President of the Geological Society, George Bellas Greenough, where he opined:

> the attenuated and ramifying sutures of the septa are remarkably striking in the present specimen, and put me in mind of the friendly and attentive Geologist, Greenough, whose genius spreads and ramifies so abundantly, that I could not resist commemorating it with sentiments of friendship, that the suavity of his manners has stamped on my mind.

He also expressed the beauty he saw in some, such as for an ammonite: 'there is peculiar elegance to this species that makes us regret its rarity, and the little chance we consequently have of seeing its exterior… the fine sharpness of the sinuated edges of the septa is beautiful';[9] and he drew attention to other intriguing characteristics, as in the case of a bivalve: 'so flat is this oyster that there is very little room for an animal between the shells'. As in his other publications he praised people for the help and specimens they freely gave, writing, for example, about a previously undescribed limpet (*Emarginula crassa*): 'It was found in the Crag near Ipswich, by Mrs Cobbold, whose zeal for science prompted her to send me the fine specimen from which this figure is taken'. Many donors of specimens or information were to become famous in their own right, such as the geologists

12/10 A fossil gastropod shell (*Murex striatus* in *Mineral Conchology*) from Ipswich, given by Mrs Cobbold.

Charles Lyell and William Buckland. Others were from other countries, particularly France, where collectors and researchers corresponded and exchanged specimens with Sowerby: these included the enthusiastic naturalist Charles De Gerville in the Manche Region of Normandy, after whom Sowerby named species;[10] J. L. De France in Sceaux near Paris, and the accomplished scientists Charles Cuvier and Alexandre Brongniart in Paris.

De Gerville was a particularly interesting correspondent. He had amassed a fine collection of fossils from his region and followed the scientific approach of identifying strata on the fossils they contained, whether in France or England. He sent material to Sowerby, some of which appeared in *Mineral Conchology*, and he often encouraged Sowerby to visit him so as to pursue researches together. But Sowerby never made it, despite inducements such as De Gerville telling him that he had saved a special 'Mr Sowerby's ham' for his visit, sending instead his sons George and James De Carle on different occasions, the travel made easier by the end of the Napoleonic wars.[11] The exchange of specimens and correspondence between Sowerby and his French friends was important in that it started to show the differences, as well as the similarities, in the fossil fauna of the English and French strata.

On occasion Sowerby presented his findings to the Linnean Society, such as when he named the genus *Inoceramus* (extinct bivalves, some of which can be several feet across) and described the species *Inoceramus cuvieri*, but most of his research was presented in *Mineral Conchology*.[12] Once the project was well underway it settled down to a regular publication with Sowerby probably doing most, if not all, of the drawing and engraving using copper plates, while he and his sons, especially George, wrote the text. It is not easy to know just how much the sons did; Sowerby was busy with the project until very close to his death in 1822 some ten years after the start. Initially James De Carle was away from home for over two months, staying and working with the banker and naturalist Dawson Turner in Yarmouth. There he helped Dawson with his work, taught drawing and etching, as well as mineralogy, to his daughters and continued to prepare material for *English Botany*. He told his father that they breakfasted very early so that it was not possible to get out in the morning and that he was busy drawing all day, especially 'since there is work enough for months to come'. But he was able to send some fossils he had collected and occasionally some fresh herrings for the meal table. His family was anxious over his safety when travelling by coach, especially in bad weather, and told him that 'a lamp Lighter was blown from his ladder while endeavouring to light his lamp at Blackfriars bridge and was drowned last evening', and so he should come home earlier than he had promised, to stop his mother worrying.[13]

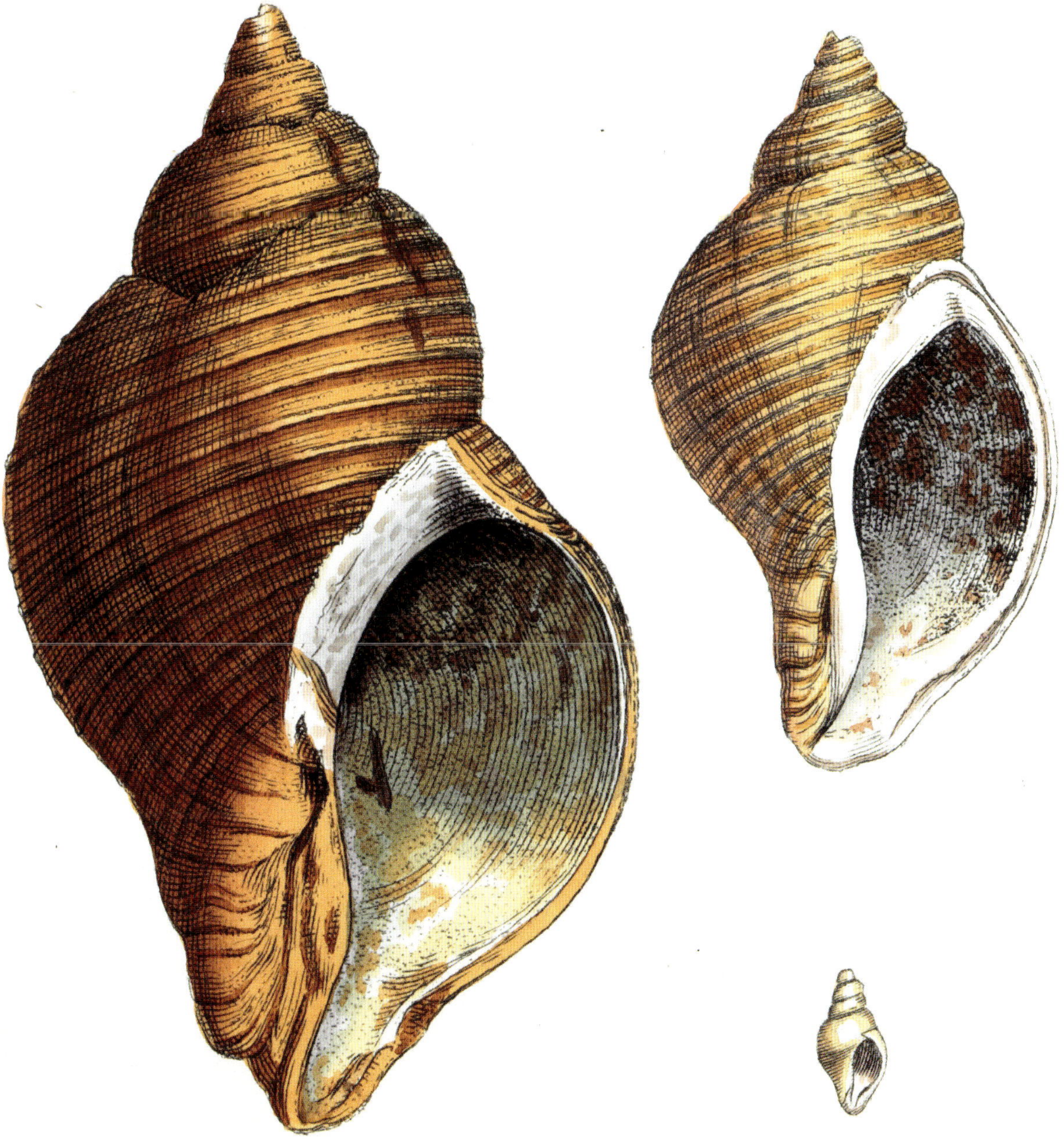

June 1 1813 published by Jas. Sowerby London.

12/11 Specimen of the fossil brachiopod (*Spirifer trigonalis*), with internal spiral appendages, Sowerby used for his description in *Mineral Conchology*. Credit: The Natural History Museum.

12/12 Brachiopod with spiral appendages (*Spirifer trigonalis*), *Mineral Conchology*, 1821. Sowerby took the opportunity to clarify taxonomic issues arising from his earlier paper to the Linnean Society.

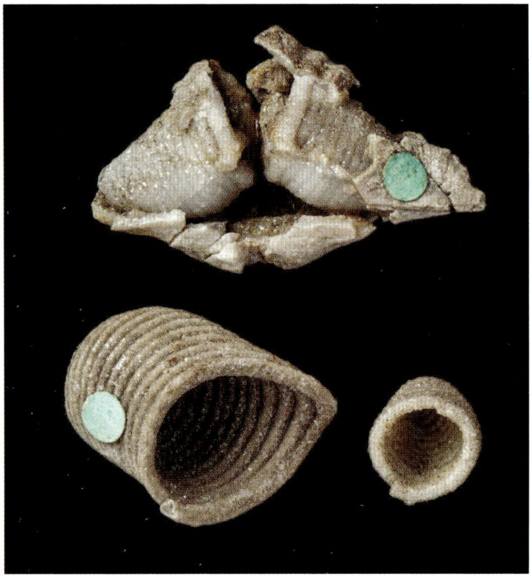

Concern over the possibility of being enrolled in the 'Militia' (for a possible French invasion) was simply dealt with by accepting the risk of being fined for non-attendance and in the hope that Mr Turner would be able to provide some sort of affidavit to excuse him. Any deflection from his work would have been difficult for the Sowerbys' schedules and income. But James De Carle was not thinking much about *Mineral Conchology*: his mind was on many other matters including, no doubt, his forthcoming marriage the following year.

George meanwhile had been married over a year before the start of *Mineral Conchology* but any sense of a new regime must have been disturbed by the arrival in the spring of their first son, also named George Brettingham. The household of Mead Place was thus rather full but it was not until the following year when James De Carle was to be married that a better arrangement came into play. He proposed to his father that he take part of the house as a lodger and that he would sublet three of the eight rooms to George, all at a rental of £50 per annum with George paying half. He assured his father that 'you and Mother will find everything even more comfortable than when you were burdened with two houses'.[14] This is what happened. James married in the following year; he, with his wife Mary Ann, and George with his family, lived in number 3 Mead Place from then on, with several children arriving over the following years, until George finally left to establish his own family home and business in 1818.

Sowerby's collections and expertise brought him new associates, an important one being the geologist William 'Strata' Smith, probably through John Farey, an ardent Smith supporter. The year following the publication of his geological map, Smith produced his work on the identification of strata by their constituent fossils which, in turn, were illustrated by Sowerby's plates and, in a novel way, printed on coloured paper representing the hue of each particular stratum whether oatmeal, murky green or a light blue-green. Suitable references to *Mineral Conchology* were given although the illustrations of brachiopods, molluscs, ammonites and others were distinct

265.

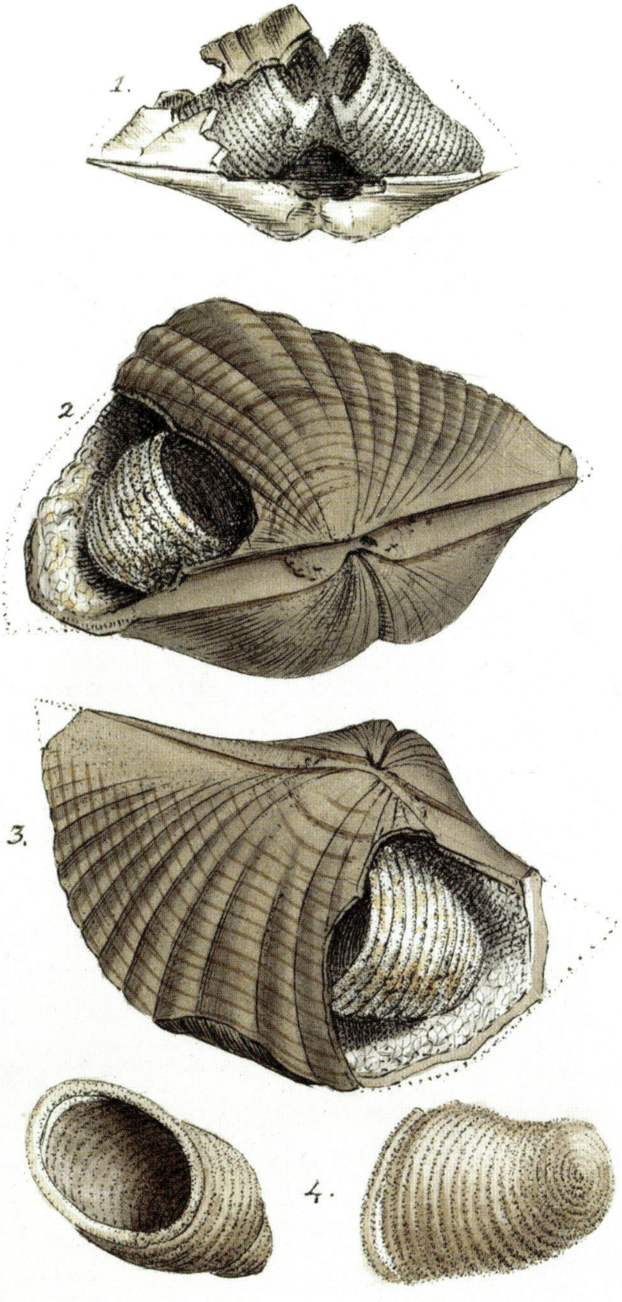

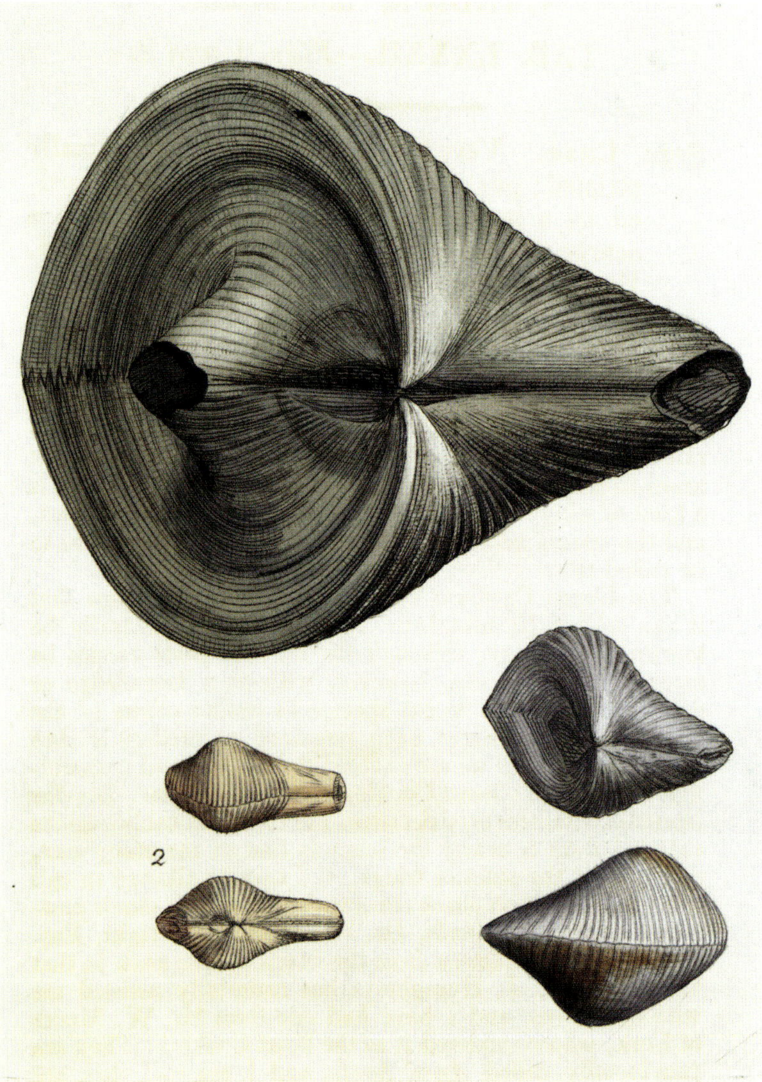

12/13 (left) An unusual fossil 'cockle' shell (named *Cardium hibernicum* in *Mineral Conchology*) over which Sowerby had some uncertainties with classification, it being unlike others in the genus. Also, because the shells were usually broken he had to use several specimens to construct his illustration.

12/14 (opposite) Large numbers of fossil ammonites in a marly limestone found in Somerset where it occurred 'in moderate masses, occasionally big enough to form tolerable sized sideboards; Dr. Lettsom has a very fine one'. Given as *Ammonites planicosta* in *Mineral Conchology*.

and finely executed. But the venture was of limited success, with too few subscribers and only four of the intended seven issues published. In some contrast, Sowerby was asked to supply the illustrations for a book on the minerals and fossils of Scarborough and its vicinity.[15] This kind of commission was much liked by Sowerby: it was relatively easy to satisfy since he used previously published plates and as it sold well it brought in cash. It also induced him to encourage Mantell to publish on the fossils of the Lewes area for which he would readily use illustrations from *Mineral Conchology*, but Mantell was not yet ready for what he saw as a risky venture. A London-based bookseller and conchologist, William Wood, was however prepared to take some risks. His *General conchology*, for which Sowerby prepared several plates, turned out to be a successful and popular publication. But not all associates at this time dealt with fossils; Thomas Purton, a mycologist

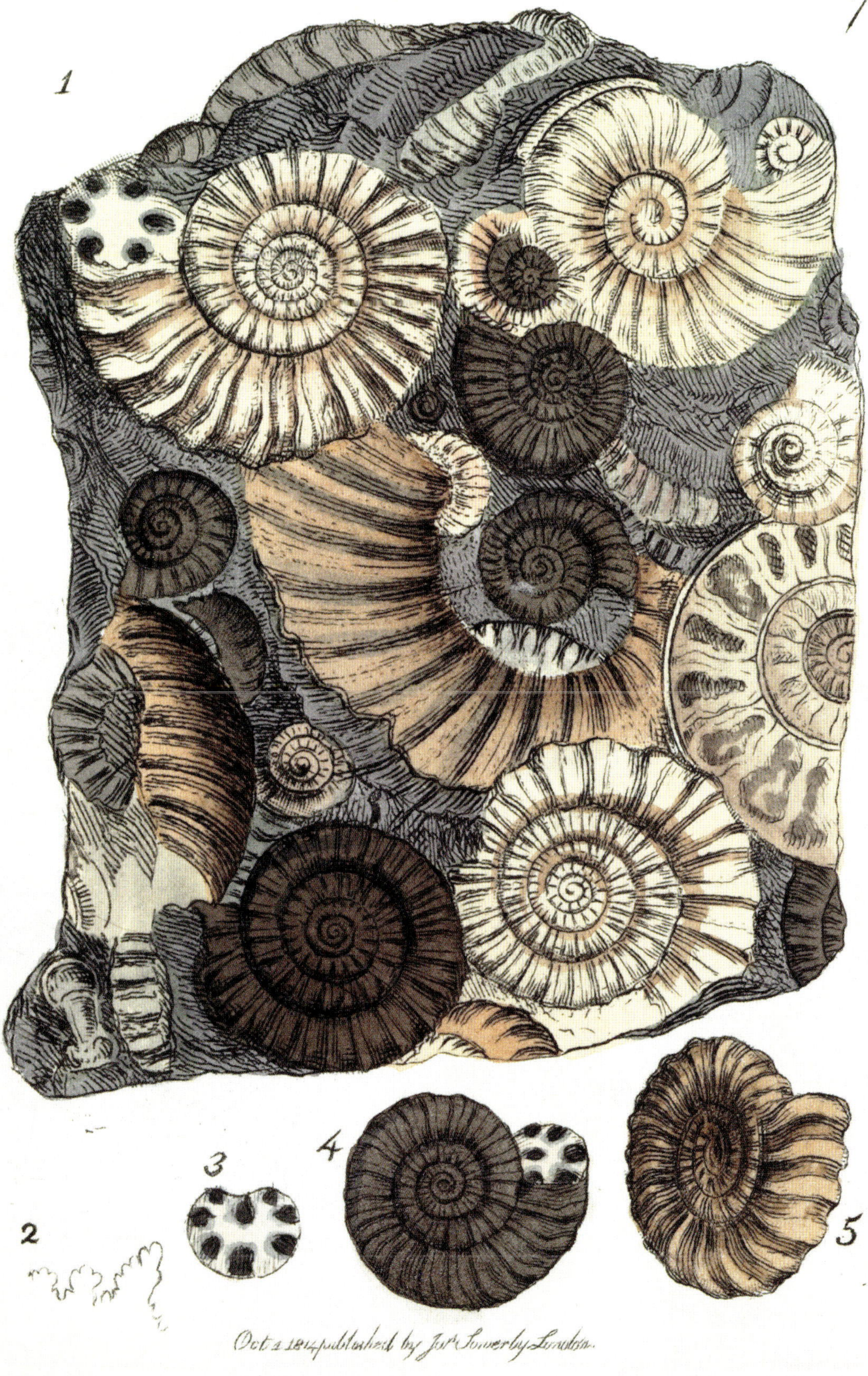

Oct 1 1814 published by Jas Sowerby London.

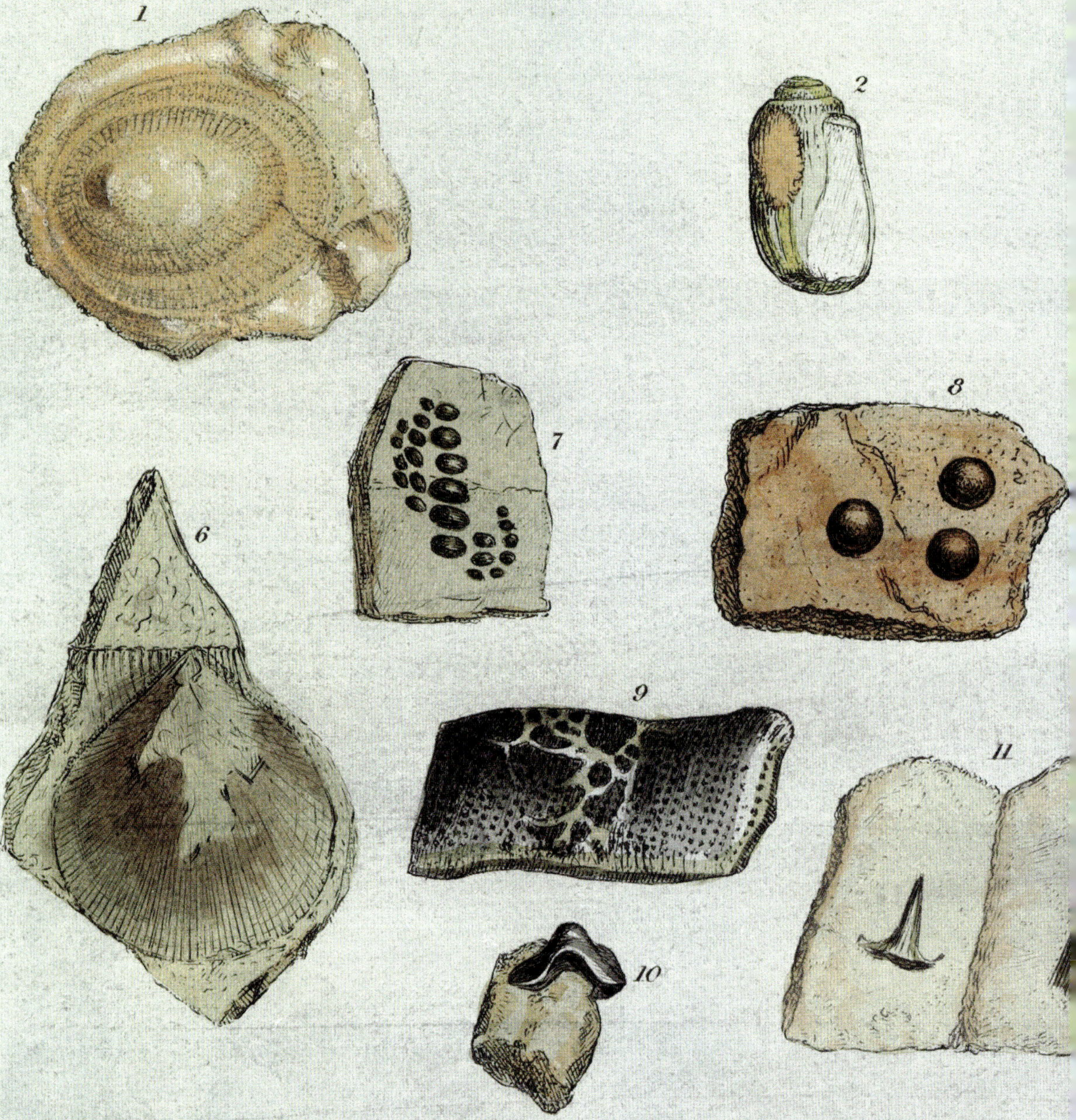

1. *Patella rugosa, M.C.*

2. *Ancilla,*

3. *Rostellaria, imperfect Specimens, Strat. Syst. P. 72.*

4. *Ostrea,*

5. *Pecten,*

6. *Pecten,*

7. *Oval Bufon*

8. *Round Bufo*

IRBLE.

3

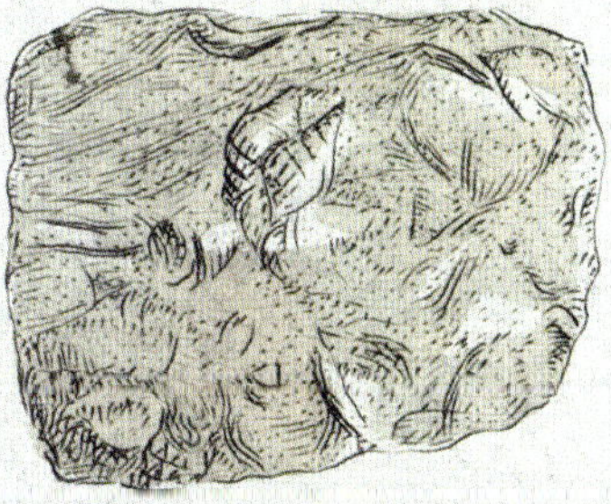

4

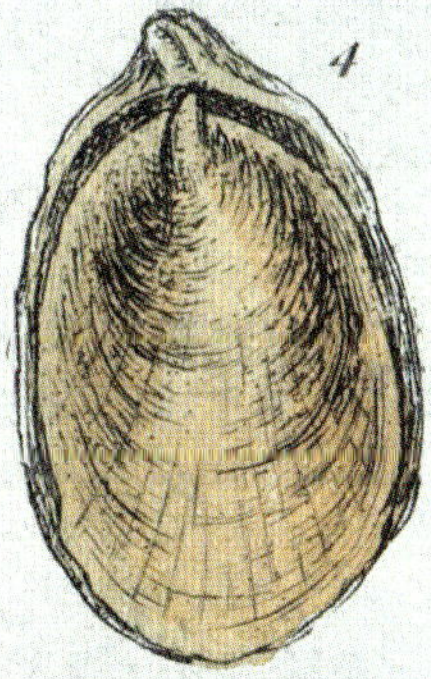

5

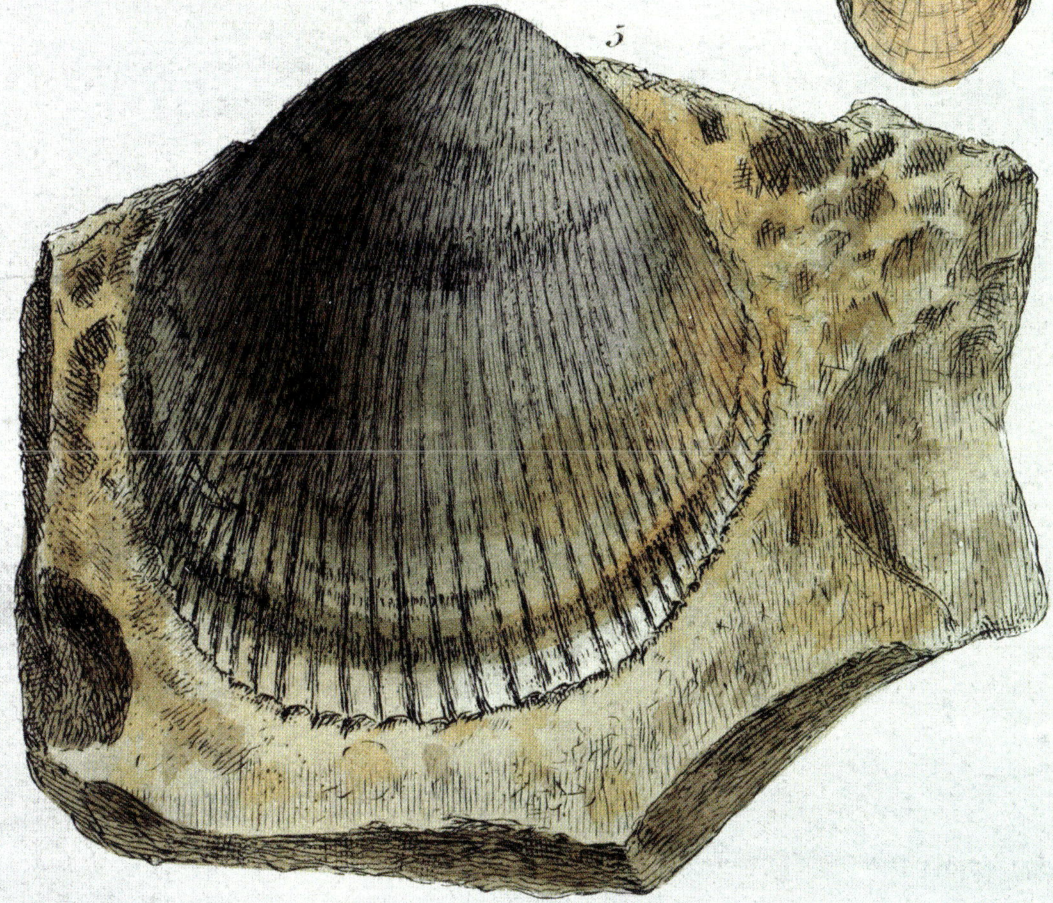

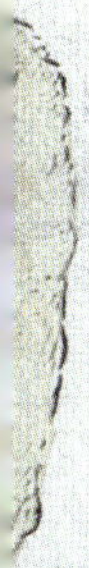

9. *Palate of a Fish,*

10. D°.

or Teeth,

11. *Teeth of Fishes.*

12/15 (previous spread) Fossils in the Forest Marble. Plate from William Smith's *Strata identified by organised fossils*. It includes molluscs, brachiopods, and sharks teeth, and is printed on a pale blue-green paper to represent the colour of the stratum. [Fourth issue, 1819]

and surgeon in Warwickshire, wrote a botanical description of the plants in the 'Midland Counties' using plates from *English Botany*. It gives an interesting glimpse into the views of some botanists for he claimed:

> this flora is particularly adapted to the study of the ladies and to the learners of Botany in general; of course it is divested as much as possible of every allusion to the sexes of plants, which from my own knowledge has been the chief bar to the female sex making it a part of their education… My great aim is to introduce the Study of Botany into female seminarys, in order to give a more religious, moral and useful impression to the youthful mind, which by degrees may damp and in time do away the rage for novel reading.[16]

These novels of which he was thinking might well have included those by Jane Austen. But then William Withering, some forty years earlier in his *Botanical Arrangement of all the Vegetables*, had also dropped the sexual distinctions in the titles to the Classes and Orders 'so that ladies could be involved in botanical study'. He used the word 'chive' instead of stamen and 'pointal' instead of pistil.

Another zoologist, William Leach, collaborated with Sowerby in his serial work on crabs, lobsters and prawns, and which came in the usual blue wrapper until 1820 and showed Sowerby's continuing skill and eye for detail. But the work was not initially always well received. James Brodie, Sowerby's long-standing friend and correspondent, complained that it was published 'in the most immethodical, & I might say, disgusting manner possible, without any Introduction or information as to the plan or expectation of the work', but added that the illustrations 'are beautifully done'.[17] Sowerby defended his position by saying that Leach was still young and they were dealing with many new genera which 'add to the obscurity till the whole is completed' and furthermore that 'I lose much because I do not go a visiting but I cannot make many drawings etc. if I visit much and have a great propensity for spreading the knowledge of natural history.' Leach was later to become assistant keeper of natural history at the British Museum.

It is difficult to envisage the Sowerby household with so much different work going on together with the bustle of family life. It is no wonder that Sowerby wrote to Mantell to tell him that he was 'so much an Hermit at home that I scarce know what is doing out of doors'[18] and this included the Linnean Society, where he even missed the presentation of some of Mantell's work. And to the fossil collector De France he later revealed that he was also 'up to his ears in dealing with and repairing his museum'[19] but not able to resist offers of unusual specimens such as the skin of an African sheep donated by

12/16 The gooseneck barnacle (*Lepas anatifera*) in William Wood's *General Conchology* (1815). Sowerby made the drawing while the animals (alive in a tub of seawater) 'were extending their plumose tentacula in search of food'. These barnacles attach themselves to floating wood, flotsam and so forth. Their peduncles were too long to have been shown fully in the drawing.

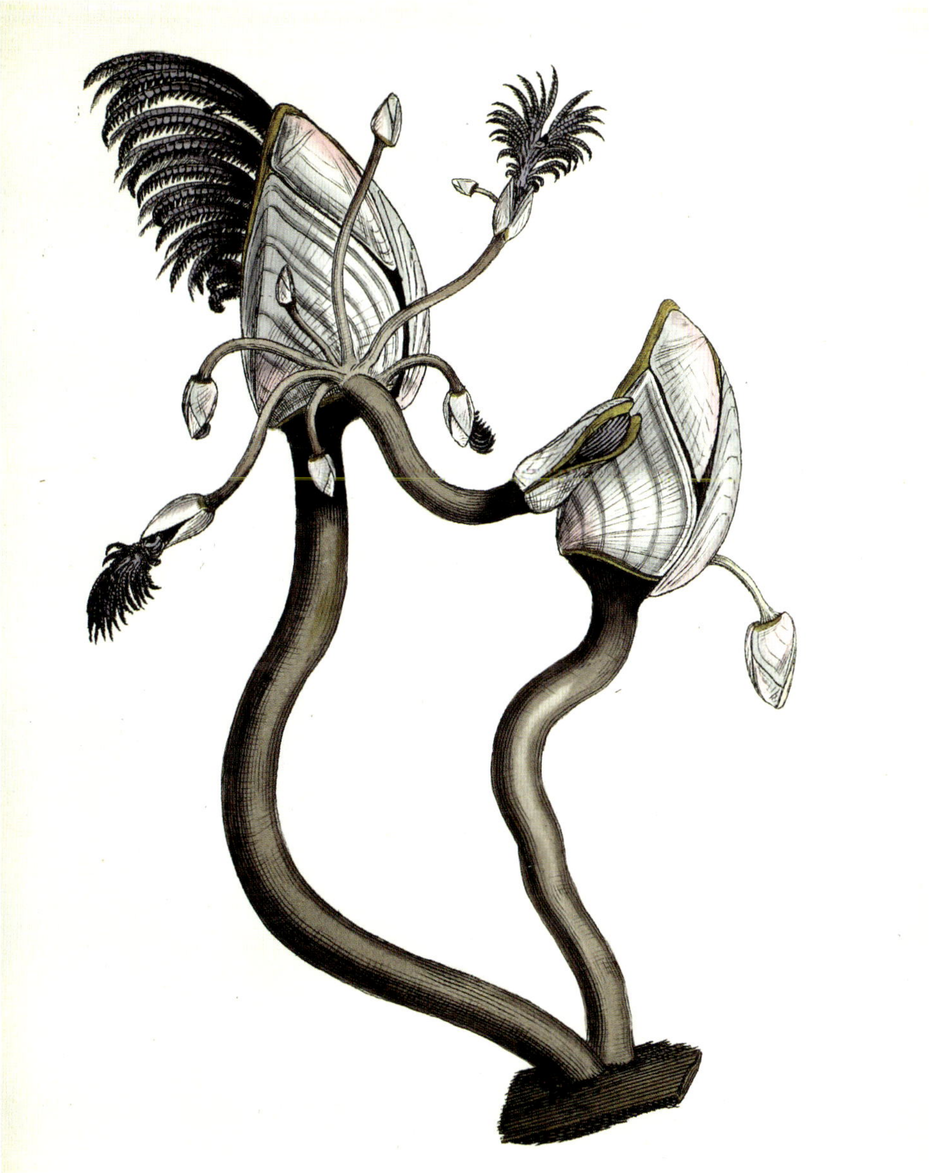

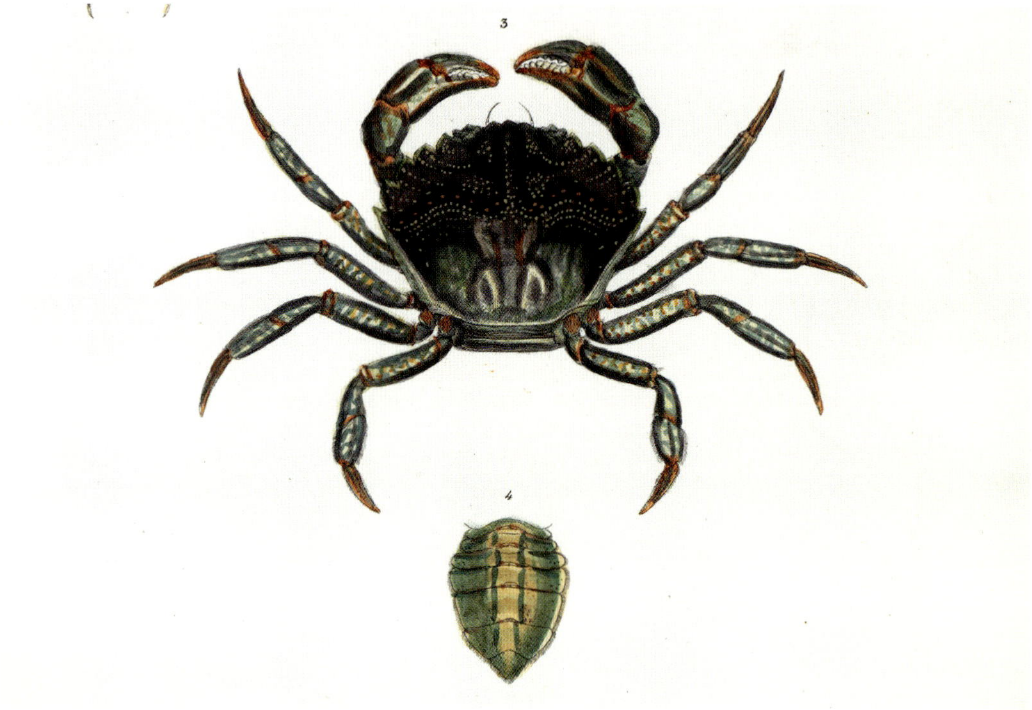
3
4

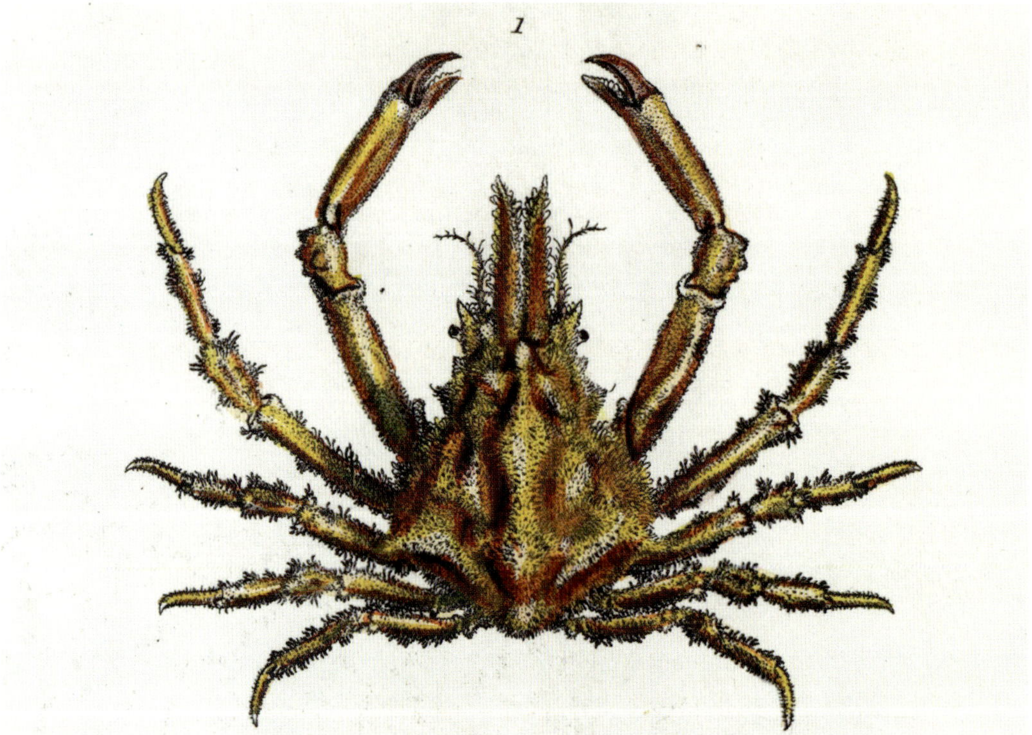
1

12/17 (opposite above)
The crab *Carcinus maenas*, illustrated in Leach's *Malacostraca*. He records it being common along rocky shores of Britain, and 'sent to London in immense quantities and... eaten by the poor, who esteem it a great delicacy'. Nowadays it is considered as an alien invasive species.

12/18 (opposite below)
The crab *Pisa gibbsii* (synonym of *Pisa armata*), found in the eastern Atlantic Ocean, including Devon and Cornwall coasts. Illustrated in Leach's *Malacostraca*.

12/19 (right) The lobster *Galathea strigosa*. Sowerby's engraving based on someone else's drawing, in Leach's *Malacostraca*.

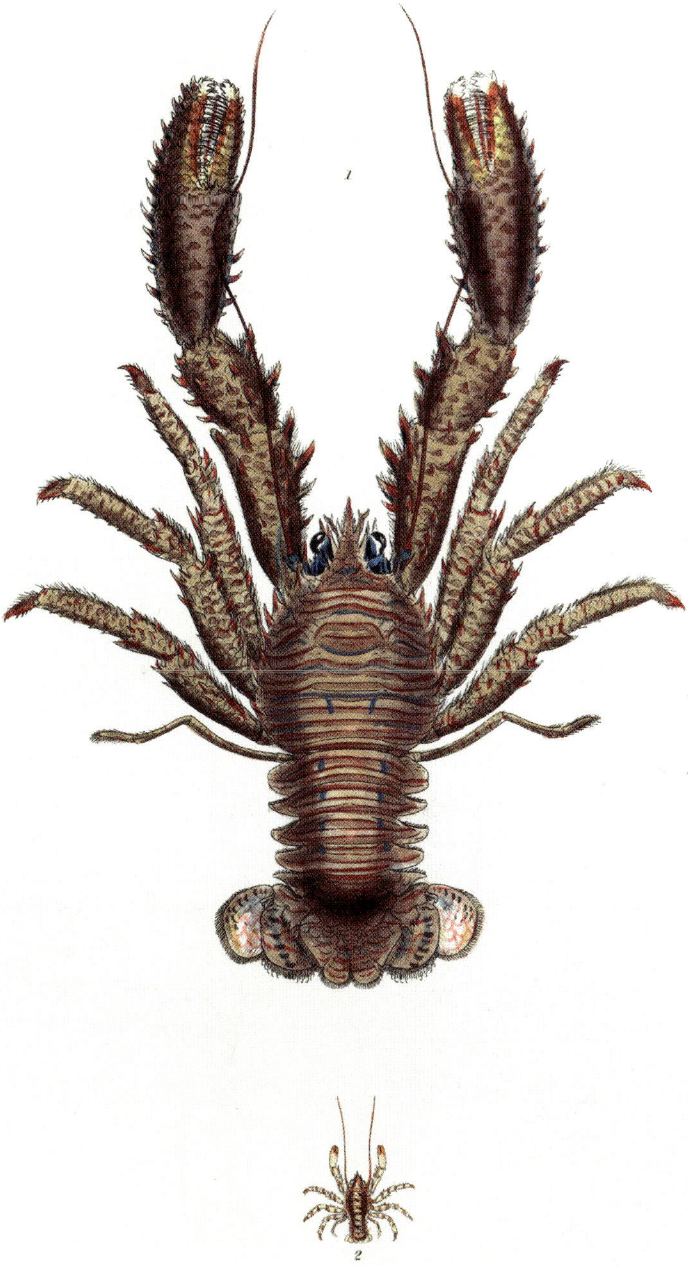

Lord Northampton.[20] Inevitably his increasing reputation in the world of fossil conchology led to more people turning to him for advice over specimen identification and collection, while others still sought help with their mineralogy and botany. Even the geologist Charles Lyell asked him to recommend anatomical as well as mineralogical texts, the latter because he was 'a beginner in mineralogy'.[21] At least Charlotte, now in her teens, was educated at school, which she liked, and not by a home tutor. The high level of activity had not stopped James De Carle sending home a live snake from Yarmouth and then asking in correspondence over the coming months how it was.

Amid this activity and creativity there was one momentary circumstance showing that Sowerby was not totally satisfied with his position and independence. The position of Keeper of Natural History at the British Museum had become vacant on the death of the incumbent George Shaw in July 1813. Sowerby expressed his interest in the job to James Smith, who was naturally someone from whom possible applicants might seek support. Smith assured him of this but warned him that 'it will be no sinecure to any body, but a place of abject drudgery & dependence'.[22] Smith's stark words seem to have banished any such ideas from Sowerby's mind.

Life might never continue smoothly for long in such a creative and active environment but one event in the autumn of 1815, just three years after the start of *Mineral Conchology*, caused a sea change in Sowerby's circumstances and happiness: the unexpected death of his wife, Anne, at the early age of fifty-one, leaving two young daughters in the household. This was a huge shock; he never truly recovered although he was to marry again a few years later. His friend and fellow naturalist the Reverend Hugh Davies attempted to console him on the

> loss of your bosom-friend; such severe trials are permitted by the Supreme Being as proofs of our patience under his gracious dispensations, and how severe so ever the application may seem, the effect may be salutary at the conclusion, if borne with due resignation to the Divine Will.[23]

Sowerby simply replied that he would not be seeing him, as he might have done, at the forthcoming anniversary meeting of the Linnean Society 'as I am not in the humour to stir from home'. James Smith also sent his condolences but in the same letter wrote of his disappointment that Sowerby had not continued with *Exotic Botany* as he would have liked.[24] But Sowerby did continue with energy and application to fulfil his commitments to others as well as to himself, though Mead Place would never be the same again.

12/20 *Nautilus imperialis.* Sowerby's large (290x400 mm) coloured engraving of the fossil shell found in the clay stratum at Highgate. May 1812. Sold as a separate folio plate.

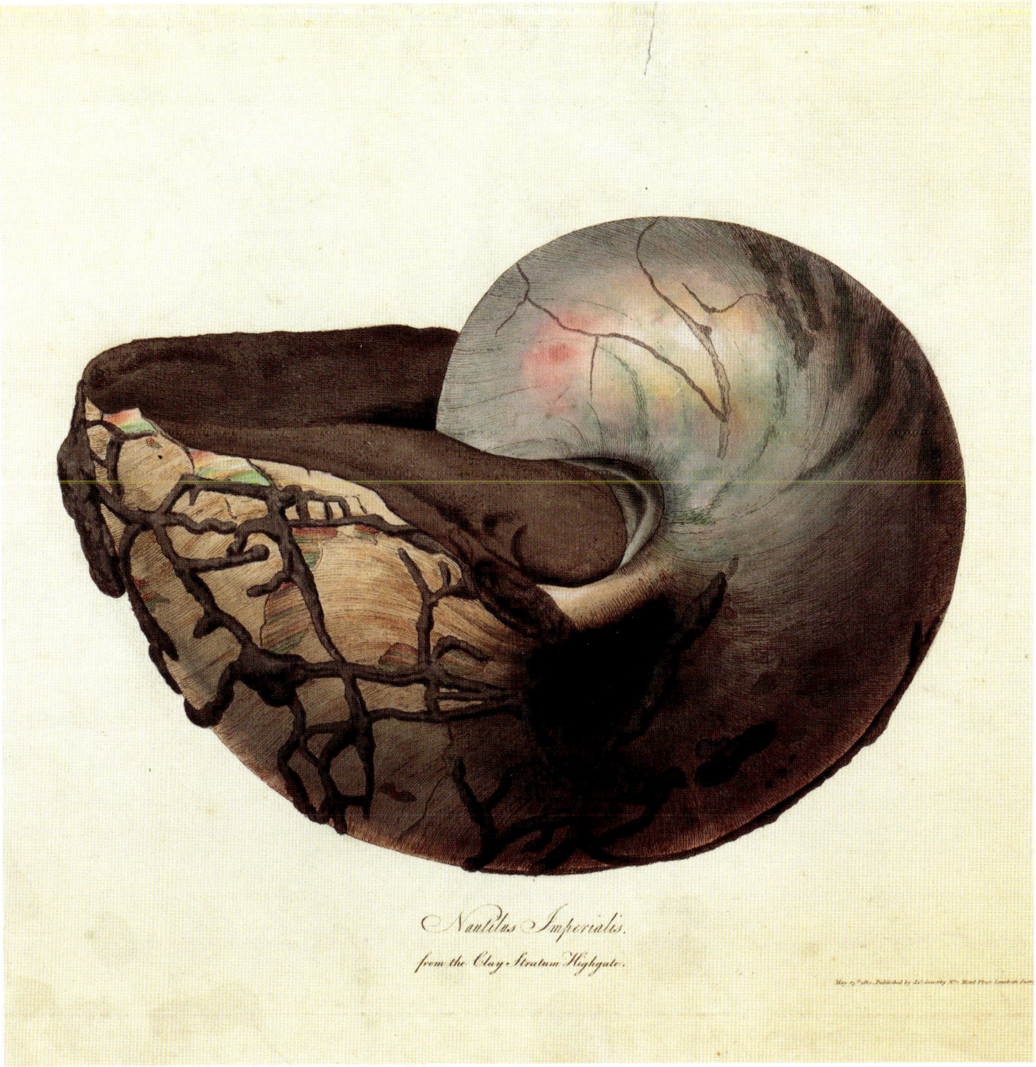

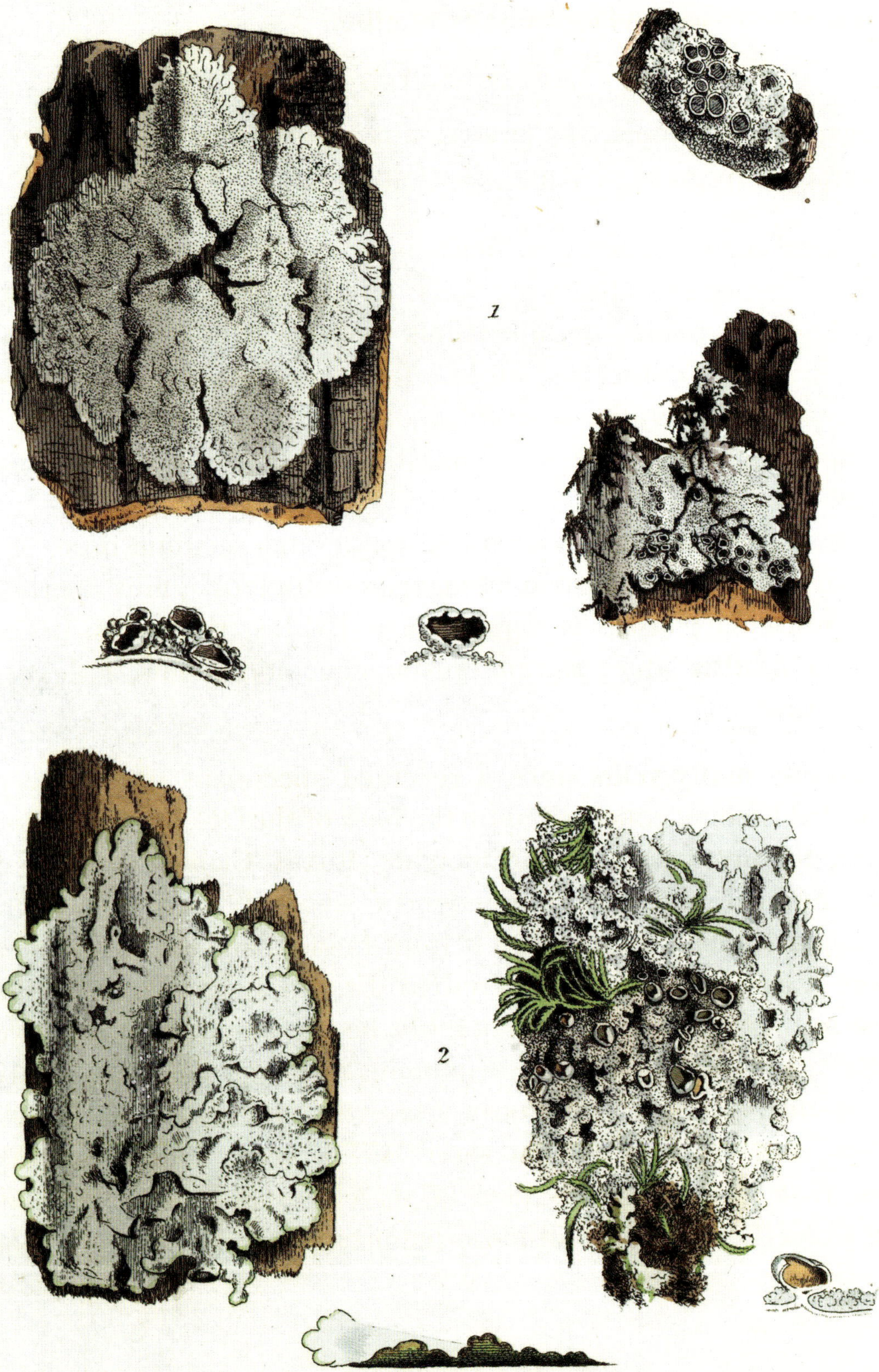
1
2

CHAPTER THIRTEEN

Societies philosophical and chemical

13/1 Lichens. Sowerby's fine engravings of two species of the lichen *Parmelia* for Dawson Turner's paper on new British species for the Linnean Society, 1809.

The formation of societies for those people wishing to come together to discuss philosophical or scientific matters was a quite common phenomenon in Britain during the eighteenth and early nineteenth centuries. Some were small and private and might have had short lives. Others, such as the Geological Society, were to grow significantly and have a sure future.

It was no coincidence that the poet and philosopher Samuel Taylor Coleridge was a member of the same club, the London Philosophical Society, as was Sowerby, nor that the German author and poet Johann Wolfgang von Goethe was a member, along with Sowerby, of the Jena Mineralogical Society. These two societies, although different in their objectives, reflected the desires of intelligent men to foster, through association with others of like mind, a deeper understanding and communication of ideas and knowledge, especially at a time of much discovery and invention. They were fulfilling, or attempting to fulfil, the ideals of the Enlightenment. But Sowerby did not know that he was to be an honorary member of the London Philosophical Society until in the early winter of 1811 he received two letters from the Secretary, Mr T. J. Pettigrew, the first saying he had been elected the previous Friday and the second telling him 'I did the pleasure of proposing you to the Society, I feel extremely happy in the pleasure of informing you of the result. I did it from a conviction that you felt a zeal for the interests and propagation of Science and that superior talents should be honoured.' Sowerby did not rush to meet the other members – he probably did not know who they were – but penned a reply to the effect that he had had no expectation of being elected and that it would be necessary for him to see the rules of the Society before he could consider himself to be a member.[1] He must have been satisfied, for within eighteen months he was named as a member of the Society's council, along with seventeen others, at a time when the president was John Coakley Lettsom, the medical gentleman who had employed him many years before to draw the plants at his house, Grove Hill, in Camberwell.

There is little evidence, despite his position, that Sowerby took much interest in this Society, perhaps because its talks or 'conversations' were on topics that he rarely pondered such as 'Is Man capable of acting from Motives purely disinterested?', and 'Is the Opinion correctly founded which asserts that there is more of Misery than Happiness in the World?' and even 'Is the Education of the Poor a desirable Object, as it affects both their individual happiness and the general Good of Society?' Lectures on natural history were seemingly absent although there were some on chemistry. Meetings were held at Bolt Court off Fleet Street, close to the Sowerby's childhood home, but that did not seem to be a sufficient extra incentive to his attendance. The Society was, however, reasonably successful for several years until an internal squabble over the auditing of its accounts and the flying of false accusations led to the resignation in protest of many of its members, including Sowerby.[2] Coleridge, in disgust over the handling of the procedures, wrote in his long and expressive resignation letter: 'I can not hesitate in withdrawing my name, a mere name I well know, from a philosophical society which has become worse than a mere name – a misnomer, and a false promise.'[3] But the fact remains that many of the members were names of some standing and that Sowerby was included among them, even if he participated only a little. The Society staggered on for a short while but was dissolved in the late summer 1820, not long before Sowerby's death.

Sowerby was, however, delighted to be an Honorary Member of the Jena Mineralogical Society. Goethe, who was an influential member, pursued several scientific interests including botanical, meteorological and mineralogical, and had also written his own *Theory of Colours*. He was actively interested in British science and was well aware of Sowerby's studies including his *Elucidation of Colours* to which he frequently referred. Sowerby seemed a very suitable candidate for the Society especially if, as Goethe intended, it could mean the receipt of specimens and publications such as *Exotic Mineralogy*. Goethe was certainly impressed by the quality of coloured plates in Sowerby's works; he was familiar with Curtis's *Botanical Magazine* and admired the Sowerby plates in Leach's work on crabs. So in the summer of 1816 Sowerby received his diploma – an honour he now added to those on the title pages of his publications. (The German botanist Georg Heinrich Noehden was associated with the Jena Society and arranged for the diploma's delivery.[4] He was a Sowerby correspondent over many years and it is probable that he was also a player in Sowerby's appointment as an Honorary Member of the Physical Society of Göttingen.)

There were other societies Sowerby was invited to join besides the Linnean and Geological. This was an age when societies were increasing at a fast rate and taking the opportunities of increasing urbanisation to improve networking and social interaction around topics of mutual interest including religion. London was a heartland, with as many as three-thousand clubs and societies by the end of the eighteenth century.[5] One, the small *Endeavour Society,* aimed to propagate the principles of the established church of England. It invited Sowerby to join in 1804 and three years later asked him to help form a committee, but Sowerby is unlikely to have been very active.[6] His main contributions were primarily to the Linnean, to a lesser extent the Geological and to one close to home – the Lambeth Chemical Society. This last gives an interesting glimpse into the developing world of chemistry in the early nineteenth century.

The Lambeth Chemical Society gave Sowerby's sons the chance to develop their new interest, sparked by the exciting work of Humphry Davy at the Royal Institution in London and of others such as Antoine Lavoisier in Paris.[7] It formed soon after the demise of the Philosophical and Chemical Society in 1809, which had existed for only a couple of years despite having fifty 'town' members, including Sowerby and his son, James, and many 'country' members. James De Carle had been the treasurer and he and his father were keen supporters of the aim to give members the chance to carry out their own experiments and discuss the findings. Equipment was shared and premises, in Soho, were available for meetings.[8] Exactly why this society had a short life is not known but it was clear that the Sowerbys wanted to continue with the benefits that such a society could bring. The Lambeth one, although strictly outside London, could succeed, especially if it was driven to some extent by the Sowerbys themselves with both sons, James De Carle and George, as enthusiastic members and their father as Treasurer (perhaps a rather surprising position since he was not renowned for keeping his own financial arrangements in good order). The other members, some thirty or so, had varying interests: artists, surgeons, natural historians, and just a few in chemistry. The list included William Leach (author of the work on crabs) and Edmund Davy (cousin of Humphry Davy). The Society met at the Sowerbys' house in Mead Place where there was a workshop and already some facilities for chemical work. Necessary items were purchased – such as evaporating dishes, galvanic batteries, chemicals (including mercury) and, of course, a mahogany balloting box with its black and white balls. It was a wonderful aspiration but undoubtedly fraught with the difficulties over just how tricky chemical experimentation can be.

The Society was well arranged with a hierarchy of electable officers, a rule book, accounts, an annual fee of one guinea, and a programme of experiments. George experimented on 'the combination of sulphur with muriatic acid', Sowerby on phosphorus, James De Carle on 'the combustion of metals in oxygen' and, on behalf of his father, on the 'resinous substances found at Highgate'. Other members performed a range of experiments including on 'hydrogen' and on the 'sources of caloric'. The meetings were intended to be fortnightly during certain months but the regularity was not well maintained.
Each member was given a month's notice by letter from the Secretary of his turn to propose an experiment which he would then perform.[9]

Sowerby had, it seems, given the first introductory talk himself, on the Society's objectives and on the 'reasons for not admitting Ladies every night'. His talk started boldly: 'The Science of Chemistry has of late become so interesting to man, from the extended views it discloses of the works of Divine providence… ', but then continued on the inconveniences for any one person attempting, on their own, to gain an understanding through experiments, and how these could be overcome through 'mutual communication and assistance of friends'. Initially, experiments would be those 'more generally known' before proceeding to 'the more difficult and obscure operations', while possibly to making some 'useful discoveries'.
As to the 'Ladies', whom by custom were excluded from philosophical societies, they should not 'be denied those enjoyments which tend to increase happiness and to enliven the sensibility of her mind, so capable of the highest refinement'.
But occasionally some experiments would likely lead to conditions 'offensive to her natural and admired delicacy which would render her constant attendance uncomfortable'. So it was determined that some meetings could include the ladies, others not, on the decision of the committee and not on the judgement of the ladies themselves. Despite this restraint, the opening of the doors to a wider potential membership was a significant innovation.[10]

The Lambeth Chemical Society started at an already busy time for the Sowerby business, with three of the major publications – *English Botany*, *English Fungi*, and *British Mineralogy* – in progress, plus *Elucidation of Colours* and soon followed by *Exotic Mineralogy and Mineral Conchology*. These, together with the usual business commissions and the family pressures, must have led to circumstances hardly conducive to giving extra time to a society on one's premises. That, and possibly the lack of a sustained interest

13/2 Man using a blowpipe at a lapidary's bench. A small, draft watercolour prepared for John Mawe's book on the use of the blowpipe; a simpler, partially coloured version was finally used. (Drawing by either James Sowerby or his son James De Carle.)

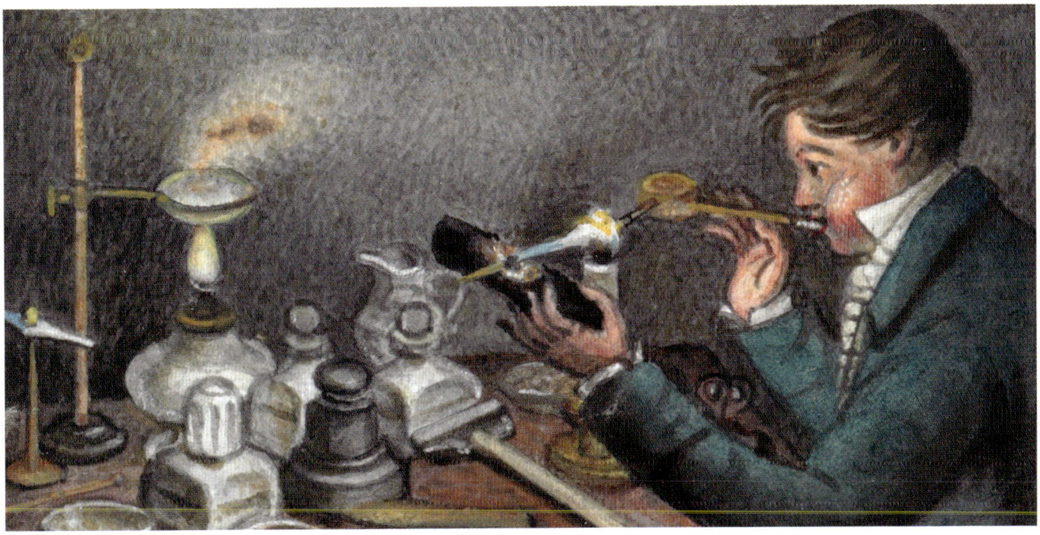

or capacity of the members, led to the Society's dissolution in 1812. The equipment was sold to interested members, with George Sowerby purchasing several items, and the proceeds suitably dispersed. This venture of about three years was not a great success but it revealed the way that chemistry was developing and, especially, James De Carle's enthusiasm for the subject. At one period he worked with Humphry Davy and he was a friend of Michael Faraday, although to what extent is not known.[11]

Despite the excitement generated by discoveries and separation of elements, experimental chemistry was not ultimately a very suitable focus for a society. People perhaps preferred to deal with collections of minerals or plants which they could show to interested friends, rather than involve themselves with obnoxious fumes or awkward apparatus. Some would have had the resources to create their own domestic laboratory, especially if it helped in mineral identification. For this they had the advice and services of John Mawe, a well-known mineral and gem dealer in London with experience of the minerals in Derbyshire, of extended travels into the mining districts of Brazil, and of polishing and analysing minerals of all kinds. His little book *Familiar Lessons on Mineralogy and Geology* included guidance on polishing and on analysis using the blowpipe.[12] Mawe, also, sought Sowerby's help on his book illustrations.

In some contrast to his contributions to chemical and philosophical societies, Sowerby worked loyally and long for the Linnean Society,

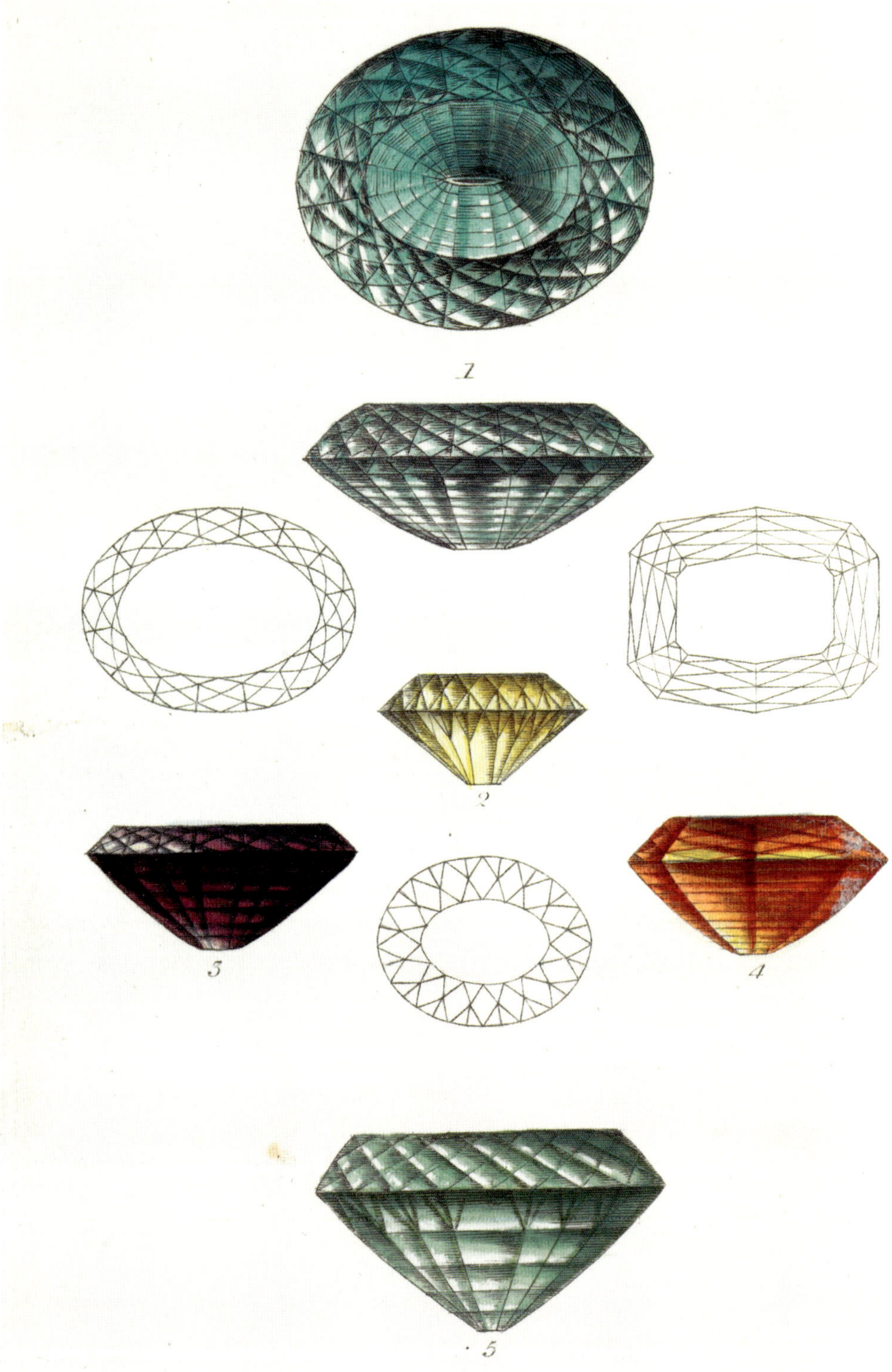
1
2
3
4
5

13/3 Cut gem stones. Frontispiece in John Mawe's *A Treatise on Diamonds*.

13/4 (overleaf left) Fossil brachiopod shells, with an internal spiral tube. Sowerby's drawing for his paper on the structure of these organisms given to, and published by, the Linnean Society, 1814–15.

13/5 (overleaf right) A wheat-destroying insect. Sowerby's engraving for Walford's paper for the Linnean Society (1808), showing the insect at natural size and enlarged, together with the insect's hole in the husk of the wheat.

as part of his own business activities as well as for the communication of natural history. He published four papers in the *Transactions* – on various shells, including one on fossil brachiopods, and on the passion flower *Passiflora*, with some fine taxonomic illustrations, as shown in chapter 4.[13] Others were given orally, over a wide range of topics such as on a sandstone with a tubular structure so that it could be sounded like a horn, on his whale *Physeter bidens*, on rotten wood, and on some crystal properties of calcite.[14] He also presented specimens of various sorts for discussion. Socially he used the Society to meet his correspondents and fulfil exchanges and commissions, although his attendance at meetings diminished considerably in later years; people then wrote to express their disappointment at not seeing him. He was a sponsor, sometimes a principal one, for the election of more than one hundred and forty fellows over thirty years, almost to the time of his death. Names he supported include the artists Franz Bauer and Sydenham Edwards, the Kew gardener William Aiton, John Lettsom, Gideon Mantell, his Scottish friend James Brodie, and the President of the Geological Society George Greenough. He then often assisted his candidates over arrangements for their payment of the £1.11s.6d. entry fee and the 1 guinea annual fee. Sowerby gave generously of his time.

The *Transactions of the Linnean Society* were popular and Sowerby's involvement was to ensure that enough copies of illustrated plates were available to meet the expected demand from members and the public. Many authors employed the Sowerbys to provide, or help with, their papers, while others wanted bespoke productions such as the colouring of all copies. This Sowerby would arrange through the colourist John Adams who, on one occasion, charged £5.4s.2d. for the colouring of five hundred copies of a plate in volume 5 (i.e. at 2 pence each).[15] Sowerby provided some control over quality and his services were used over many years, including providing the illustrations for some authors' papers, such as for Dawson Turner's on lichens.[16] One author, Thomas Walford, in reporting on a wheat-destroying insect excused himself from giving details 'as the drawing is from the accurate pencil of Mr. Sowerby, no description of the insect is necessary'.[17] Sowerby would also have gained much satisfaction from a paper by James Smith naming a new genus of plants *Sowerbaea* in his honour, the genus containing species of some attractive perennial herbs endemic to Australia.[18]

The Linnean Society was one of the constants in Sowerby's life; so was its President, James Smith. Another was his own museum which was always being attended to, and to which visitors and friends would come. One hoped-for constant, his wife, had most unexpectedly gone and his family was inevitably changing its dynamic and ambitions. A time of different challenges was now at hand.

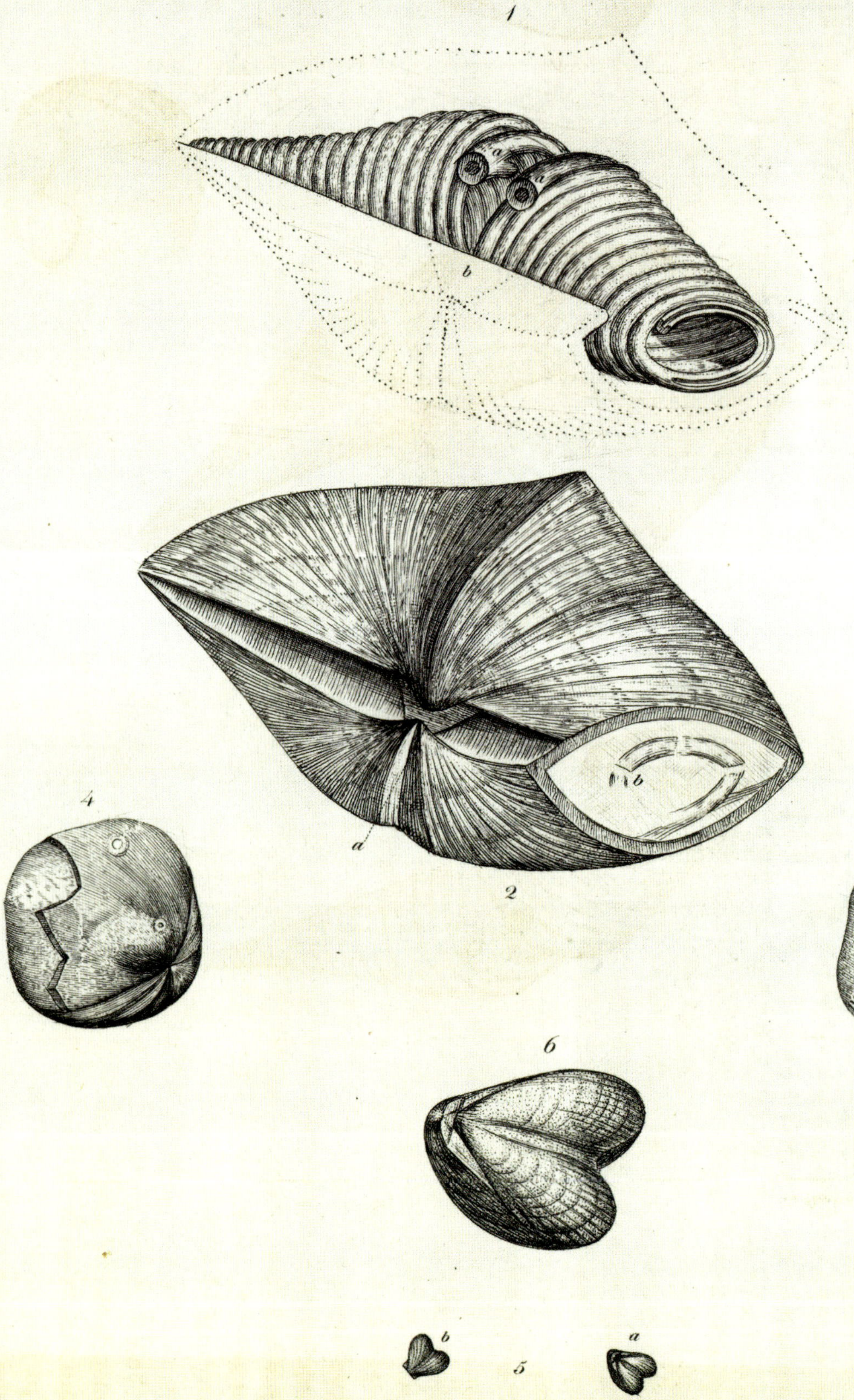
1
a
a
b
2
a
b
3
4
6
5
a
b

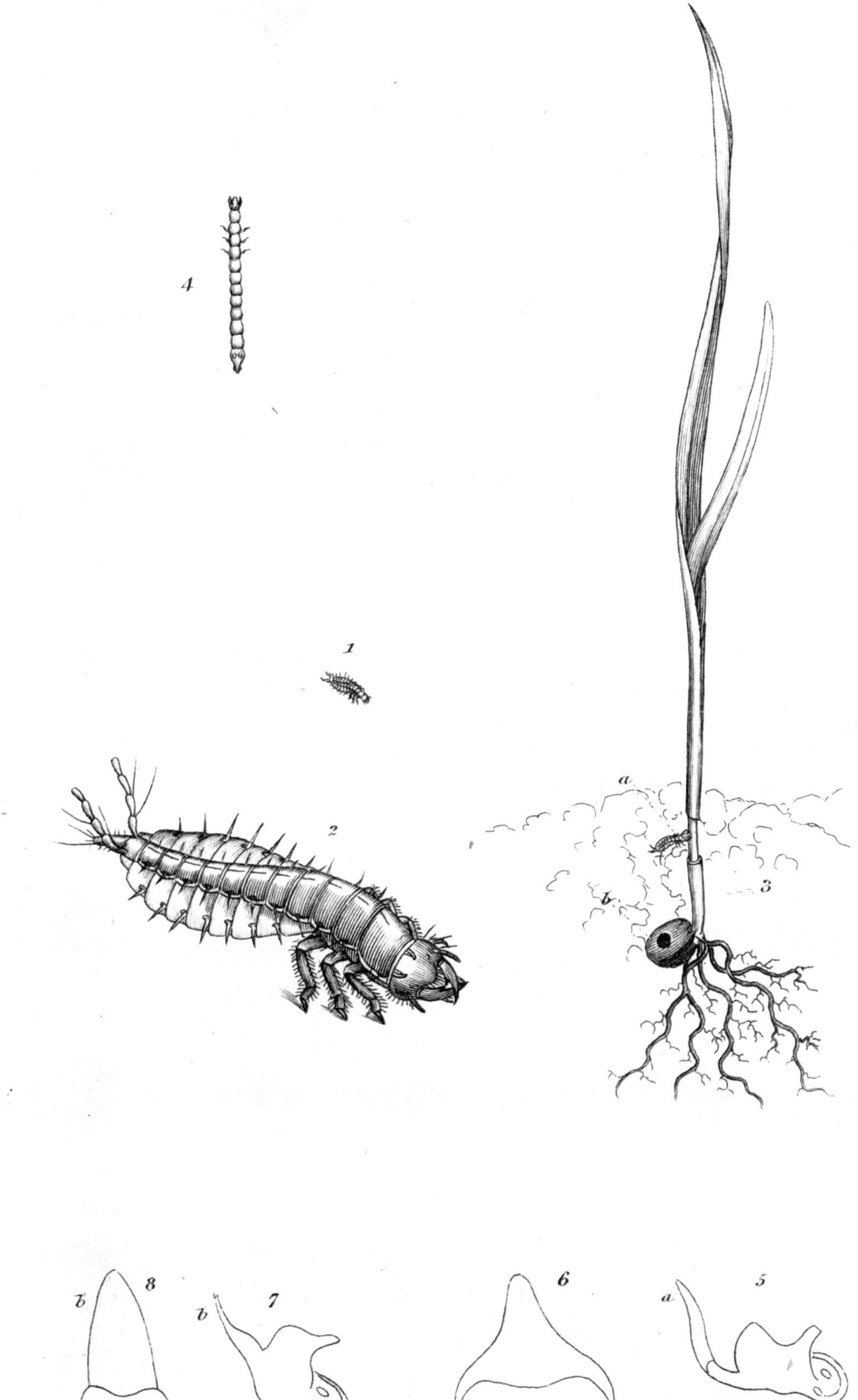
4
1
2
a
b
3
b
8
b
7
6
a
5

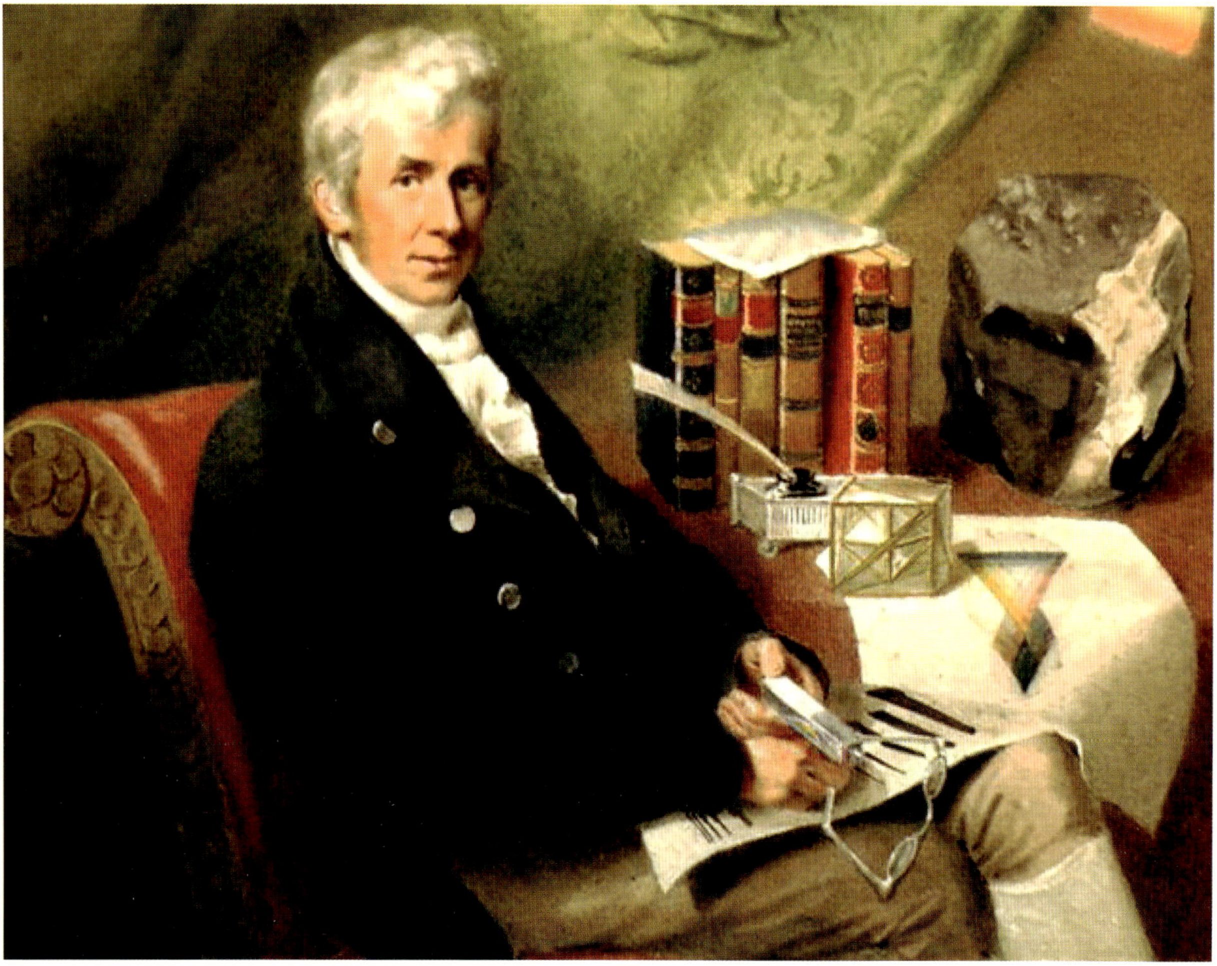

CHAPTER FOURTEEN

Passing the baton

14/1 James Sowerby. Portrait by Thomas Heaphy (1816). On the right is the large specimen of the Wold Cottage meteorite. Credit: R. Cleevely.

In the year following his wife's death, Sowerby decided to have his portrait painted by the artist Thomas Heaphy, showing him smartly dressed, seated in a fine red-leather chair, accompanied by his published books, his chromatometer, a crystal model and the Wold Cottage meteorite.[1] The aim was to display his interests and achievements; it was a kind of acceptance that a major part of his life was passed and this was how he wanted to be remembered. Not that his life wasn't still very full, it most certainly was, but there was a sense of routine with the steady continuation of his many projects. James De Carle expressed his feelings to Dawson Turner: 'we go on in the old plodding way, the vegetables of our country being exhausted, we must live upon rocks, with a few shells for a treat now and then, or anything else that comes in our way'.[2] New work was started, especially one for students, *The Genera of Recent and Fossil Shells*, and a commission from his long-standing friend and co-worker James Smith for a work on the 'grammar' of botany. Some occasional opportunities were passed over: when the botanist William J. Hooker (later to become the first full-time director of Kew Gardens) suggested jointly producing an illustrated work on foreign cryptogamic plants, Sowerby turned him down because 'there are too many obstacles in the way at present'.[3] But he was happy to meet some sudden demands such as drawing plants for the *Botanical Magazine* and for Sir Joseph Banks.[4]

14/2 Engravings of the genus *Conus*, in Sowerby's *Genera of Shells*.

The Genera of Recent and Fossil Shells is a beautiful as well as important scientific publication. Each plate carries fine figures of the external and often internal characteristics of a few species. The artwork is sensitive and accurate so that it can be admired for what it is as well as for its taxonomic value, especially in the coloured versions. The introduction said:

> The Plates of this Work will be all not only original, but executed in his best manner, by a person of generally acknowledged superiority of talent, by one who has already largely contributed by his labours to the advancement of Science in several branches of Natural History, a person whose entire labours for more than thirty years have been devoted to such useful and scientific objects: when we mention the name of JAMES SOWERBY, F.L.S. M.G.S.ETC.

The letterpress for each plate runs to two or more pages and gives the salient information of the genus, followed by its history, broader description, some comparisons among different species and sometimes details of habitat and even specific locations. As the title indicated, both living and fossil species were covered with many new genera and species being described for the first time. This was a serious work, written in very accessible English (other than the short first section on taxonomic characters in Latin) and published during a time of real advances in malacological science. It is not surprising, therefore, that it was well received. As usual it was issued as numbers in blue wrappers, starting at the end of 1821 and continuing approximately monthly until well after Sowerby's death. This was team effort: written by George Sowerby, with the illustrations by his father, and then with James De Carle taking over once the supply of his father's drawings was finished.

The Genera is interesting in another way: its plates could be purchased individually. The Sowerbys were clearly hoping for a wider market than simply that for a taxonomic serial. Initially it had no index, a rather common feature of their publications and one which inevitably aroused some dissatisfaction, despite giving subscribers the freedom to arrange their copies as they wished. But it also showed that such a work was not ultimately dependent on Sowerby the father and that other family members would continue to produce fine natural history works, which indeed was to happen for several generations.

James Smith's *Grammar of Botany* also indicated a change, for it was the first time he gave an explanation of the classification system by the French botanist Antoine Laurent de Jussieu. This system, unlike that of Linnaeus, was based on many characters, not just the number and character of the stamens and pistils. It was difficult for Smith to accept fully that Jussieu's

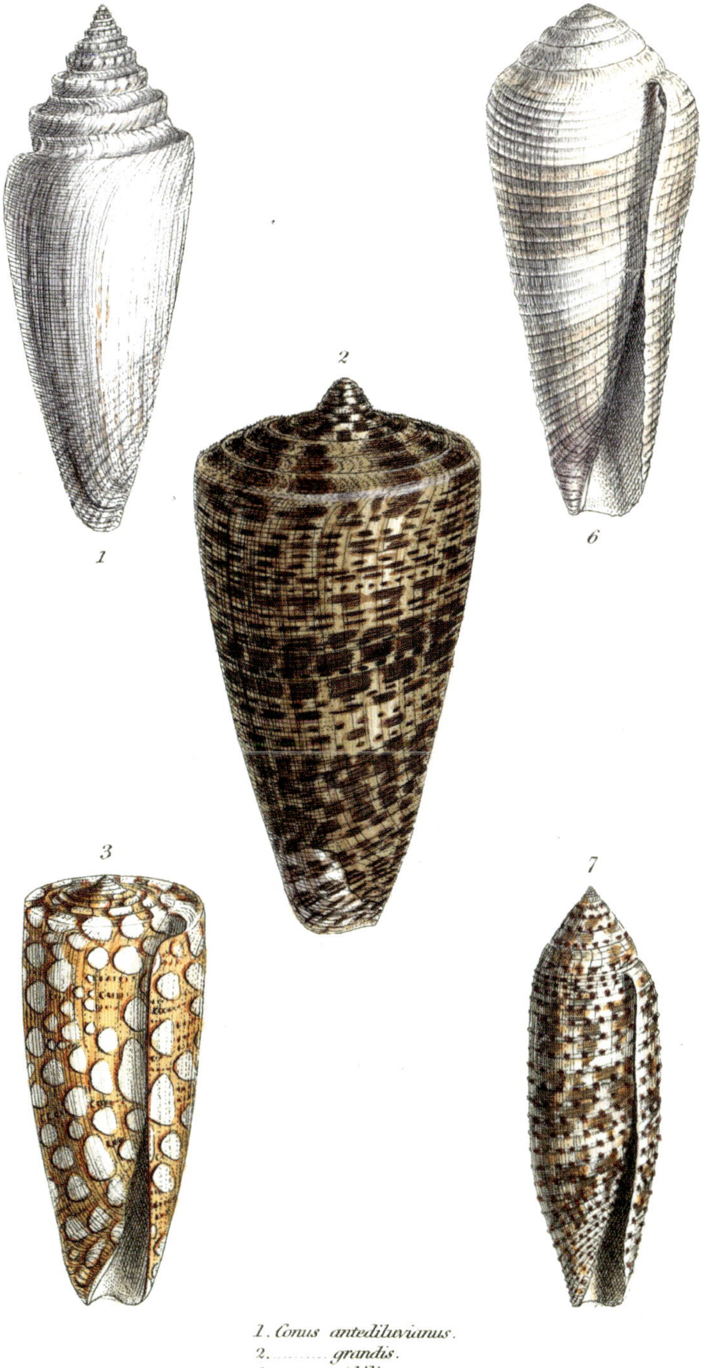

1. *Conus antediluvianus.*
2.......... *grandis.*
3.......... *nobilis.*
6.......... *Terebra.*
7.......... *Nussatella.*

14/3 A bivalve shell of the *Corbis* (now *Fimbria*) genus showing the attention given to the internal structure as well as to the external form. Sowerby's *Genera of Shells*.

14/4 James Smith's *Grammar of Botany*. Engraving showing: upper, the catkins and flowers of the silver poplar tree (*Populus alba*) and, lower, the common fig (*Ficus carica*) and its flowers. The clarity and accuracy of Sowerby's drawings are evident.

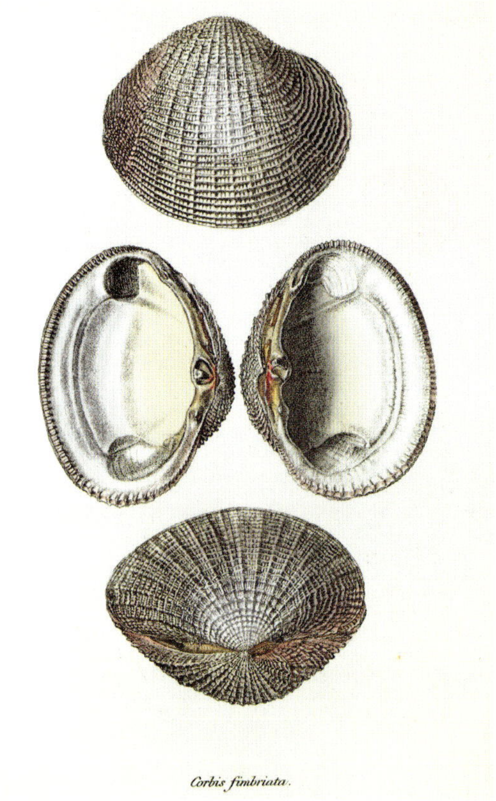

system could be better, for had he not bought Linnaeus's collection and always advocated the strength of his system? But in his preface he said:

> As Professor de Jussieu and his pupils take the lead in the department of natural classification; the botanists of England, who have never been behind their neighbours, in real science, may well desire to know something of the principles or advantages of a system, which deservedly claims so much notice [...] It is evident that no such mode of classification can, at present, serve the purposes of analytical investigation, to make out an unknown plant. That is the exclusive object of the Artificial System of Linnaeus, which, of all the schemes hitherto contrived, is alone, perhaps, universally applicable to the end in question.

The inclusion of the word 'perhaps' indicates that even his faith was being tested.

This small book with twenty-one coloured plates by Sowerby claimed to be an extension of Smith's earlier *Physiological and Systematical Botany* (then in its fourth edition) and was written with the student in mind. The text described the parts of plants and their uses and then dealt with theories of systematic arrangement. Each of Sowerby's attractive plates contained several figures to illustrate the different characters, accurately portrayed even if not always with the same flair of his earlier work. Any student would have been pleased to own a copy. Indeed, Smith also needed to use the plates for his botanical lectures at the Royal Institution in Liverpool, and had chased Sowerby for the proof versions in March 1820 so as to be in good time for his programme in May. But Sowerby's response was typical of his busy life: 'I have been anxiously and industriously employed about

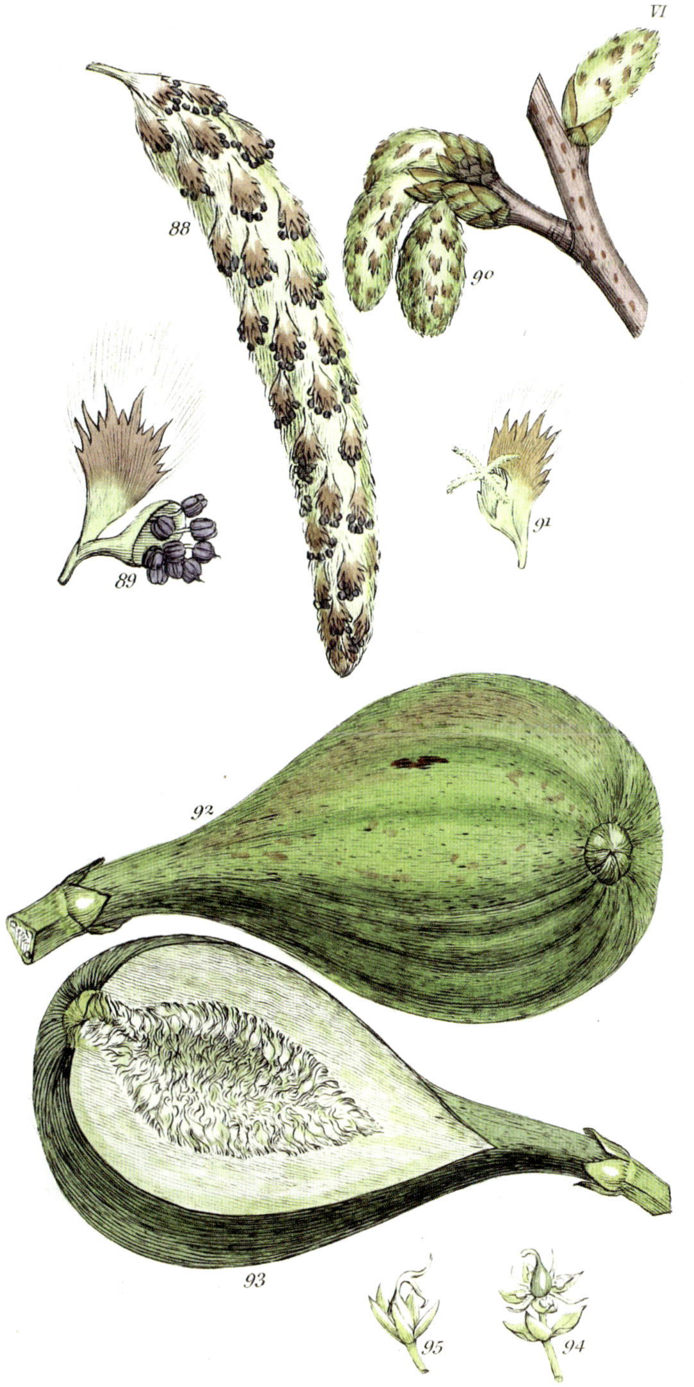
VI
88
90
89
91
92
93
95
94

your Grammar plates ----to be done in about a fortnight.'[5] Perhaps inevitably, the following letters told of further delays and of Sowerby's regret. The book was finally published the following year.

The family business continued, but with a very different atmosphere from when Anne was present. The sons were involved as a team in the works on shells but their family life was divided between numbers 2 and 3 Mead Place. When his daughter Charlotte went to visit relatives in Norwich Sowerby became an even more lonely man. In replying to one of her letters he asks her not to forget to enquire about a particular species but then goes on:

> Felicia came down to me at 6 o'clock this morning without any calling and we had a walk in the park and collected curiosities at Westminster in the way & she desires her love to you and is pleased you particularized her. We are much without other company at meals and she thinks I am too polite and tryes to be a match for me. I am so much hurry with visitors etc. that I grudge the time that I might indulge in going on therefore I leave some in answer to your next.[6]

The relationship with his two daughters was very warm. Charlotte was much excited to see him when he joined her in Norwich a month later, but she must have been anxious over an accident on his way, as later recounted to his son in Mead Place: 'My dear James, First post chaise, horse and postilion fell, but after we had exchanged the horses all recovered except broken knees and the postilion a little harmed... In the evening reached Norwich Charlottes impatience overcome by seeing us sooner than expected.'[7] Such a tone was unlikely in a letter to his son George: their relationship was more formal and detached with George being respectful but wishing for a greater independence than his father was willing to concede; the period of time after Anne's death had not brought any warming in their relationship. Sowerby would write to his son James about family events while in Norwich but not, it seems, to George.

It was no surprise that George finally decided to quit Mead Place and set up business on his own while hoping for some work from his father. He wrote:

> My Dear & Honoured Father,
> As it is your expressed wish to receive this notice and in compliance with the custom upon such occasions, I hereby inform you that I mean to quit the apartments I now occupy in your house No 3 Mead Place, on Lady Day next the 25 of March 1818. I am Your affectionate Son GB Sowerby.[8]

That letter was written on Christmas Day not very long after Sowerby's return from Norwich, and it was followed by an anguished one two months later, showing the tensions between them:

> The difficulty I find in introducing a subject which I fear, from experience, will prove unpleasant in itself and, which I am charged already with having introduced at an unpleasant time, is the reason that I have determined to take this means of asking you for a final determination whether you will employ me or not.
> [...]
> it is desirable that I should know as far as possible whether I can have any and what employment I can have from you, for notwithstanding your expressed determination that upon my leaving this house I should entirely leave your employment, I have some reason for hoping this will not be positively insisted on and therefore to conclude the difficulty is more about how I shall be employed than whether I shall be employed. If I might be allowed to express my wish I should ask you to give me the management of your colouring, for which if you paid, as you undoubtedly would, a fair profit, I should gain enough to settle all my doubts about getting a living.

The letter then continued to raise another difficult topic: the late payment for minerals that George had sold his father nearly a year before. In some despair he was now 'at a time when I am *most* in need of money I shall be *least* able to *command* it'. The letter closed with a final stark sentence: 'You know I can have no objection to communicate with you in any way you think proper but as I have opened the subjects in this manner, you will have an opportunity of considering them and can communicate further as you may choose or not.'[9]

Sowerby wished to be, and indeed was, in control, yet he was about to lose one of the key members of his team, if no longer of 'his family'. George with his wife and three children left Mead Place a month later but on what business terms is not known. He started his own dealership in natural history objects but also continued with his part in *Mineral Conchology* and *The Genera of Recent and Fossil Shells*. Sowerby was far from happy over the outcome. He had lost his wife and now his amanuensis. A year after Anne's death he wrote to James Smith saying how much he was 'confounded' by the loss of his wife, and then he wrote to De Gerville in France explaining that in addition to his usual commitments he had been spending much money and time in his Museum 'which by no means gets done as soon as I expected', and adding 'My son George having left me has lowered my spirits much, and also finding I am not likely for some time, if ever, to see Valonges!'[10]

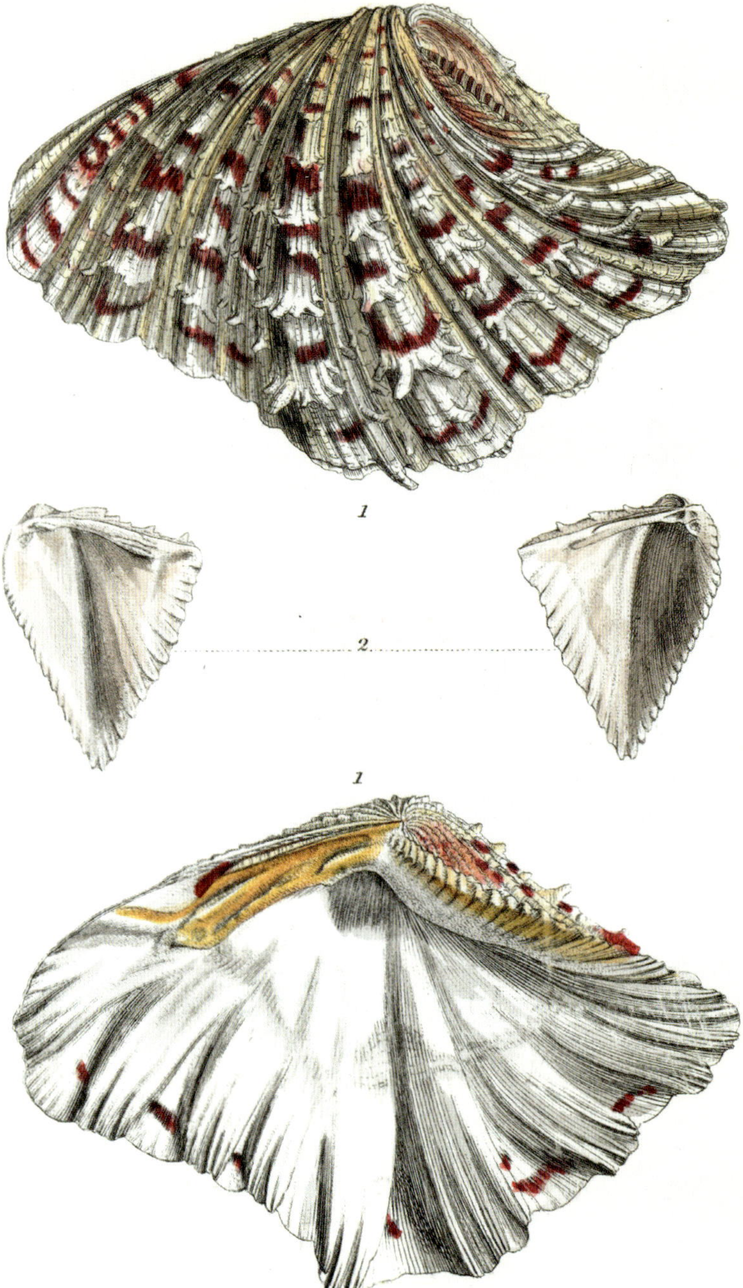

1. *Hippopus maculatus*.
2. ....? *avicularis*.

14/5 Two species of the attractive *Hippopus* genus, in the 'giant clam' family Tridacnidae, specimens of which are often used as an ornament. Sowerby's *Genera of Shells*.

Sowerby was still a driven man but with the loss of his accustomed support he failed to manage his time and affairs smoothly. Letters, especially to Gideon Mantell, reveal his state of mind and a certain degree of disorder in his affairs, although still showing his enthusiasm for his studies and his desire for yet more specimens. He was in a race against time but less able to share his burdens and ambitions. He had, as it were, reluctantly passed a baton to his son George who, with his acquired skills and scholarship, was now working in much the same way, even though no longer at Mead Place. James De Carle similarly had learned a huge amount and had become a practising skilful artist of a similar mould to his father. He was to work with his brother on the *The Genera* and to continue servicing *Flora Graeca*. Indeed, correspondents now sometimes wrote to him rather than his father. Most of the activities of the family business could still continue even if Sowerby were to step aside. But this not being in his nature, and to help resolve his problems experienced over the years since Anne's death, he decided to marry Maria Catharina Reynolds, the daughter of Admiral Reynolds.[11] Such an event was, as might be expected, likely to cause tensions and changes in the working as well as the family environment; the outcome would not always be harmonious, as was soon to be revealed in correspondence between the Sowerby siblings.

In the year following the marriage, the twenty-year-old Charlotte travelled to Paris, accompanied by James De Carle, where she was to stay for several months to learn French and to help build working relationships with some of the French natural historians. She lodged at the home of an eighty-one year-old military man – Monsieur Barzentin, who was fond of history and had three charming daughters – where she had her own room and boarded 'with the family with the liberty of choosing tea or coffee' all at a cost of £5 per month. This home was close to that of Monsieur Dufresne, 'naturaliste-en-chef' at the Muséum d'histoire naturelle, who kept a friendly eye on matters. James De Carle helped her settle in and wrote reassuringly to his father on several matters including that: 'Mdme Dufresne & her daughter will lend their assistance in supplying Charlotte with fashions and making purchases for her so that she may not be imposed upon by the tradesfolk of Paris who charge the English frequently double price.' Once back in England he wrote regularly to her as a dutiful and caring brother, sometimes in French, more often in English, telling her the home news and advising on matters of behaviour and money. But only a few weeks after her arrival he had to write that their father was unhappy with the cost of her stay and refused her money to buy a watch. Later he sent her details of specimens he was preparing for Monsieur Dufresne, which would need her assistance, and gave more news of home, especially about their stepmother:

> Mrs Sowerby Snr is very busy in the shop, she has made up 4 sets and is still making up numbers; I think she will not leave you many to do… Mrs S serves the Booksellers and seems desirous of getting as much of the business into her hands as she can, I know not for what reason, perhaps because she thinks she can manage it better and because she would imitate my late dear Mother in every thing … you will always remember the kindnesses of our late Mother and her merits too well to find an equal in our Stepmother with all her contrivances and patchings.[12]

Perhaps because of the presence of four young and potentially distracting children in 3 Mead Place, Sowerby did not seek a greater family interaction. When James De Carle was in Paris, his wife Mary told him the family news but reported that his father did not seem to want any 'interference', since the door was kept shut between the two houses, although they did come together for daily supper and also for Sunday dinner, which Mary herself cooked.[13] Sowerby, however, missed Charlotte and she missed the family, despite her enjoyable time. Letters from home telling of the birth of a niece, of her father's bad fall which affected his eyesight for a while, and of one of the children having chicken pox only encouraged her desire to return. When her brother George came to Paris to accompany her on her return to London, she finally declared that she would never stay so long away from home again.[14]

With Sowerby not as strong or as active as in the past, James De Carle handled more of the correspondence and projects such as *Flora Graeca*, while Sowerby's new wife dealt with the routines of sales and many enquiries. Sowerby was still very busy with illustrations, modifying his museum and receiving visitors. Furthermore, he still found time to help people in a way he had always loved to do: by teaching the drawing of plants and other objects of natural history. To one request he replied:

> I shall feel pleased to see you and your amiable daughter when you find it convenient. I have a fire in the Museum every Tuesday, but for your amiable daughter to make drawings under my inspection, I will make every convenience I can when you shall find it most agreeable, and if you will make the first appointment, I will attend most punctually to it.[15]

Illness, however, could not be held at bay for much longer. He was suffering from enlargement of the prostate gland which for some months had been troublesome but now, in the summer of 1822, was much more serious. He continued to work, driven as he always had been, but by

21 October he wrote his short will, recognising that there was little time left. Four days later, on the 25 October 1822, he died at Mead Place.

Some notices of his death were written, such as a few short paragraphs in the *Gentleman's Magazine*, but there was no proper obituary.[16] Dawson Turner suggested that one should appear but it came to nothing. James Smith was insufficiently motivated to pen any eulogy; and although he wrote to James De Carle expressing his sadness and sympathy and saying how he had lost a tried and faithful friend ever ready to serve him, he also asked about the business and the continuation of the work for *Flora Graeca*.[17] Indeed, a later attempt by James De Carle to write and publish an account of his father's life and achievements was discouraged by James Smith, although it is not clear why other than that the draft was poorly constructed and written.[18] Even the inscription on his tombstone in the Church of Saint Mary, Lambeth, seemed to have involved the hand of Smith and failed to give any real sense of his achievements. After the preliminary details, it said almost dismissively:

> He was originally a teacher of drawing, but having devoted himself chiefly to the delineating of plants, was noticed by some of our principal botanists, particularly Sir J. E. Smith, the President of the Linnean Society, who employed him to illustrate his works. Thus encouraged, Mr Sowerby attained an extensive knowledge of Natural History, and collected a large museum in the use of which he was very liberal. His publications related chiefly to Botany and Mineralogy, and he occasionally contributed some Papers to theTransactions of the Linnean Society.[19]

If people were shocked and saddened by Sowerby's departure they did not express it much in writing. In one simple sentence, however, from a longer letter to James Smith, the surgeon and botanist Sir Thomas Cullum captured the truth: 'Alas poor Sowerby is no more, what a loss to Science.'[20]

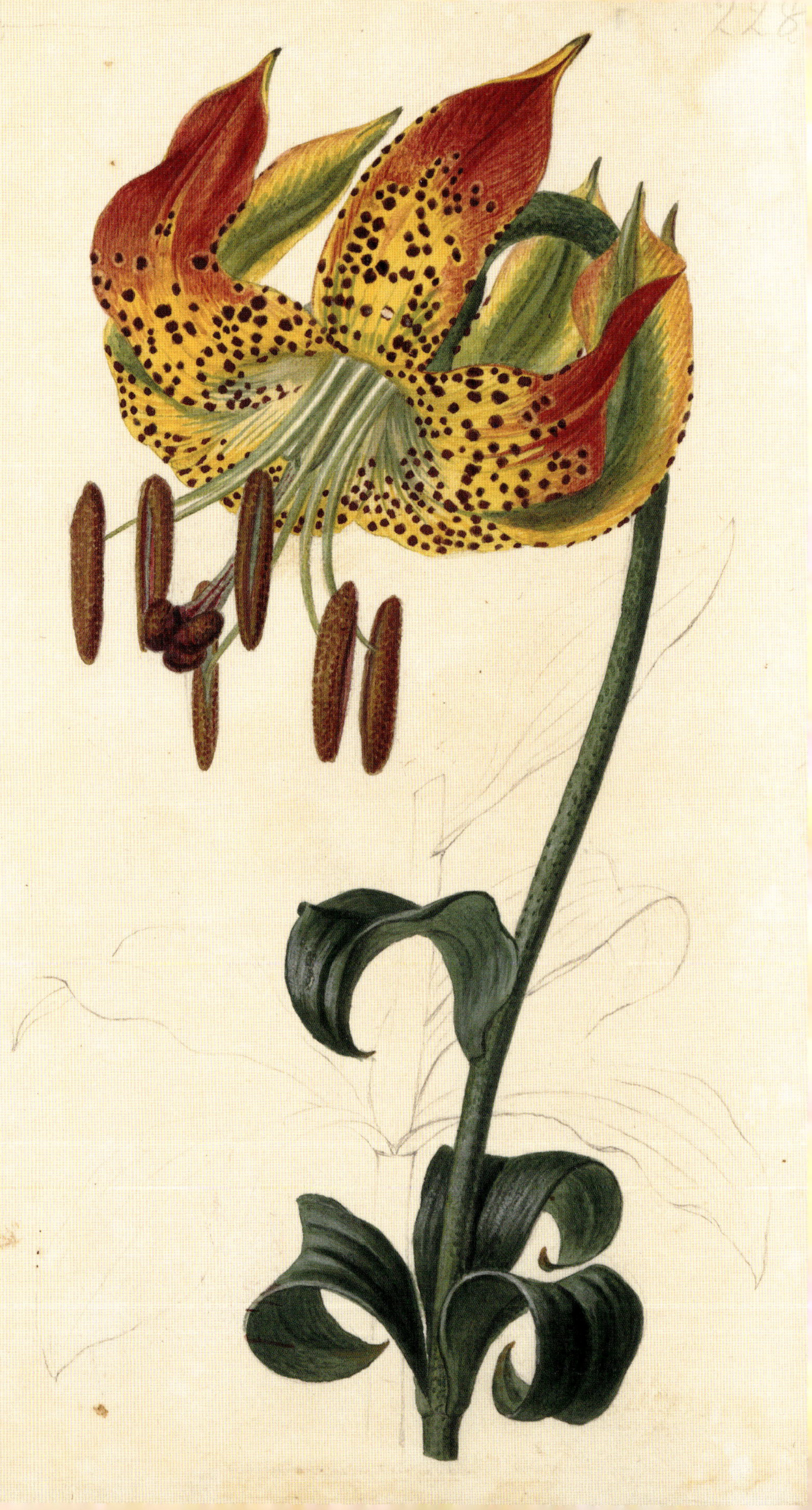
228

Epilogue

In the archives of the Natural History Museum, London, there is a small paper wrapped package containing the fine grey hair of James Sowerby. It is labelled: 'Hair of my Father James Sowerby departed this life October 25, 1822. Aged 65 years.' This touching action was almost certainly done by his eldest son James De Carle, who was closest to him. And it was to this son that Sowerby left the larger portion (four-sevenths) of his property, with the other share (three-sevenths) to his fifth son Charles Edward, on the basis that they would carry on the family business. The will was short and crisp, written just four days before his death.[1] For his second wife, Maria Catharina, it fell to James De Carle and Charles to pay her £50 per annum until her death as well as to arrange that she received back the house in Stock Newington she owned before their marriage. His two surviving daughters, Charlotte and Felicia, were to receive £100 per annum until their marriage, with an additional £50 per annum for seven years to Charlotte after her marriage. George, his second son, on the other hand, was not treated generously receiving the sum of just £100. This small amount might be a reflection of the somewhat strained relationship between father and son as well as the fact that George, who was now running his own business (as a dealer and publisher) and had his own property, could probably look after himself.

The tone of the will was that the business should carry on – and so it did, at least for a while.

E/1. *Lilium carolinianum* (now *Lilium michauxii.*) Original watercolour by James Sowerby in 1815 but not published in a *Botanical Magazine* volume until the year of his death.

E/2. A framed miniature of James Sowerby painted (unknown artist) after his death. Based on the portrait by Heaphy.

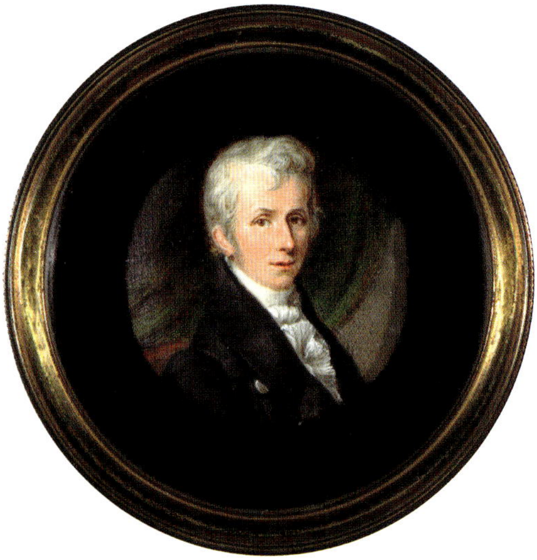

James and Charles went into partnership until 1831 when Charles took over, but without the father the impetus was not as strong and the museum became increasingly a burden rather than a useful resource. James's family grew – all his eight children were born at Mead Place. Charles, by comparison, had only three children. James worked by illustrating material for leading geologists and others; he continued work on the Sowerbys' *Mineral Conchology* (for which he received in 1840 an award from the Geological Society), *The Genera of Recent and Fossil Shells* and with the illustrations for *Flora Graeca*. The two brothers are also likely to have maintained other business activities such as assisting the production of the *Transactions of the Linnean Society*, of which James De Carle became a Fellow the year following his father's death. He, with his cousin Philip Barnes and others, founded the Royal Botanic Society and Gardens, Regent's Park in 1839.

The sons also tried to sustain the museum and to make it available to the public; they opened it on Tuesdays, as in the past, just a few months after their father's death. A few years later it was possible to buy tickets for admission to study in the 'museum and library' for 10s. for three months or £2 for a year. Other subscription systems were tried but were not successful and the bulk of the contents were auctioned in 1831, although how successful that sale was is not recorded.[2] The sale catalogue revealed the numerous treasures that were available and it is fortunate that many items, including part of the herbarium, minerals and fossils, as well as the famous Wold Cottage meteorite, finally found a home in the Natural History Museum, London.[3] These collections are not just historical curiosities; they are relevant to current research in several areas, especially as they contain species which were featured and described by Sowerby for the first time. In that sense alone his contribution has been, and will continue to be, very significant.

If Sowerby's collections were unusual in their diversity, so too were his publications. No other person at that time achieved the breadth and depth of his work, especially in making it accessible to a relatively wide audience. Without self-aggrandisement, Sowerby used his many skills to create a body of work with artistic and scientific integrity that is still valued by scientists, artists and collectors. Yet there is an irony in that those who gave him commissions unwittingly took undue advantage of these skills. It is thus no surprise to see that several botanical works indicate the importance of the plates, yet give little real credit to the illustrator; the author gained prestige for bringing the work to fruition, whereas the artist was merely paid. The main reason Richard Salisbury produced his 1791 book of just ten rare plants in a large folio edition with high quality paper was to show off the illustrations by Sowerby. Salisbury gained recognition and sales would have been much better than for any straightforward botanical description. Other authors did likewise. Even James Smith recognised through his *Gleanings of Botany* (1792) the necessity of good illustrations when he wrote 'the present work is intended to consist of coloured representations of such uncommon vegetables as are not already sufficiently well figured by any author'. It was thus correct of Gill Saunders to say in her work on the history of botanical illustration that 'almost the entire burden of publication may be carried by the pictures.'[4]

Nor is Sowerby's importance simply that he, as author or co-author, produced more than half a dozen major works of over fifty volumes containing more than four-thousand fine illustrations; it is also the success and recognition which he achieved. Sir Joseph Banks wrote to Smith saying: 'How immense has been the improvement in Botany since I attached myself to the study' and how, 'your descriptions & Sowerby's drawings of British Plants would have saved me years of labour had they then existed.'[5] Indeed, *English Botany* achieved its own long-term success with a supplement of three volumes (1831–1843) initiated by the botanist and first full-time director (1841) of the Royal Botanic Gardens, Kew, Sir William Jackson Hooker (with illustrations by James De Carle Sowerby), with further editions involving the Sowerbys to varying degrees until the end of the nineteenth century. This work is undoubtedly the one for which James Sowerby is best known. The *Mineral Conchology* was produced in two 'pirated' editions in French and in German several years after Sowerby's death while the original continues to be held in high esteem by scientists and collectors.[6] As already mentioned, the palaeontologist Graham Elliott noted that

> its minutely accurate coloured illustrations and descriptive comments have ensured its working value, as opposed to historical interest, to the present day. This is in spite of the pioneer nature of much of the contents, and of the uneven geological documentation resulting from the gradual development of stratigraphy (and its terminology) during the years of publication.[7]

Elliott was impressed by the 'amazing accuracy' of the illustrations, so useful to the

E/3. An earth star fungus, illustrated by James Sowerby in his *English Fungi* – the work which stimulated Sacheverell Sitwell to write his verses. Sowerby wrote: 'it differs from the others of its tribe in having several orifices to emit the powder as well as many little pillars or supports to the head above the volva'. In *English Fungi* as *Lycoperdon coliforme*, now *Myriostoma coliforme*.)

taxonomist, while the simple beauty of the plates can also be appreciated by many. *English Fungi* has kept a reputation as a groundbreaking work and even led to the writer, Sir Sacheverell Sitwell, to compose in 1972 a series of verses on some of the depicted species. The last of the seven verses is on the fascinating puffballs:

> But this puff-ball or *tête de mort* like a *Death's Head Moth*
> has its own noise to make,
> Throwing out its white globular part
> 'as far as six or eight inches, no more,
> With a slight but distinct cracking sound';
> Which takes us to those puff-balls
> that from near-to are as little miniature volcanoes,
> With the spores or smoke rising from a vent
> or opening at the top;
> Or odder still the 'Earth Star' or *Geastrum triplex*,
> its crater held up on six lobes like lobster claws;
> And to another *Geastrum* or *Lycoperdon*,
> that besides being raised up on its six lobes or legs
> Has many little pillars or supports to hold up its crater,
> with several orifices to emit the smoke or powder
> And we end on a windy day
> with giant puff-balls *Calvatia gigantea*
> Coming tumbling, rolling towards us,
> till they lodge among the blackberries,
> Under polished green currants of the *'Deadly Nightshade'*.[8]

Sowerby was not, in general, an 'ideas man', preferring instead to observe, record and communicate. But then many natural historians at that time did no more than that, their scholarship resting within descriptive taxonomy. Sowerby did, however, innovate: he took the challenge of *British Mineralogy,* made a meteoritic sword for the Russian emperor, produced an instructive set of fungi models and devised his optical chromatometer, among other things. These initiatives were certainly unusual and had varying degrees of success but however long each lasted they remained as individual features in a changing landscape of activity. Neither did he lack scholarship, for he was well versed in much of the relevant science of the day through his correspondence and reading of papers and texts including many from abroad. He would weigh up the various arguments over particular issues (for example, the opposing viewpoints of the geological Neptunists and Vulcanists) but only rarely committed himself to any one camp, often suggesting that further evidence was still needed. His writing reveals, however, that he marvelled at the wonders of nature and had a questioning mind. Occasionally he would speculate, especially on the origin of minerals or the factors leading to the occurrence of fossils. In one instance, in the year of his death, he wrote

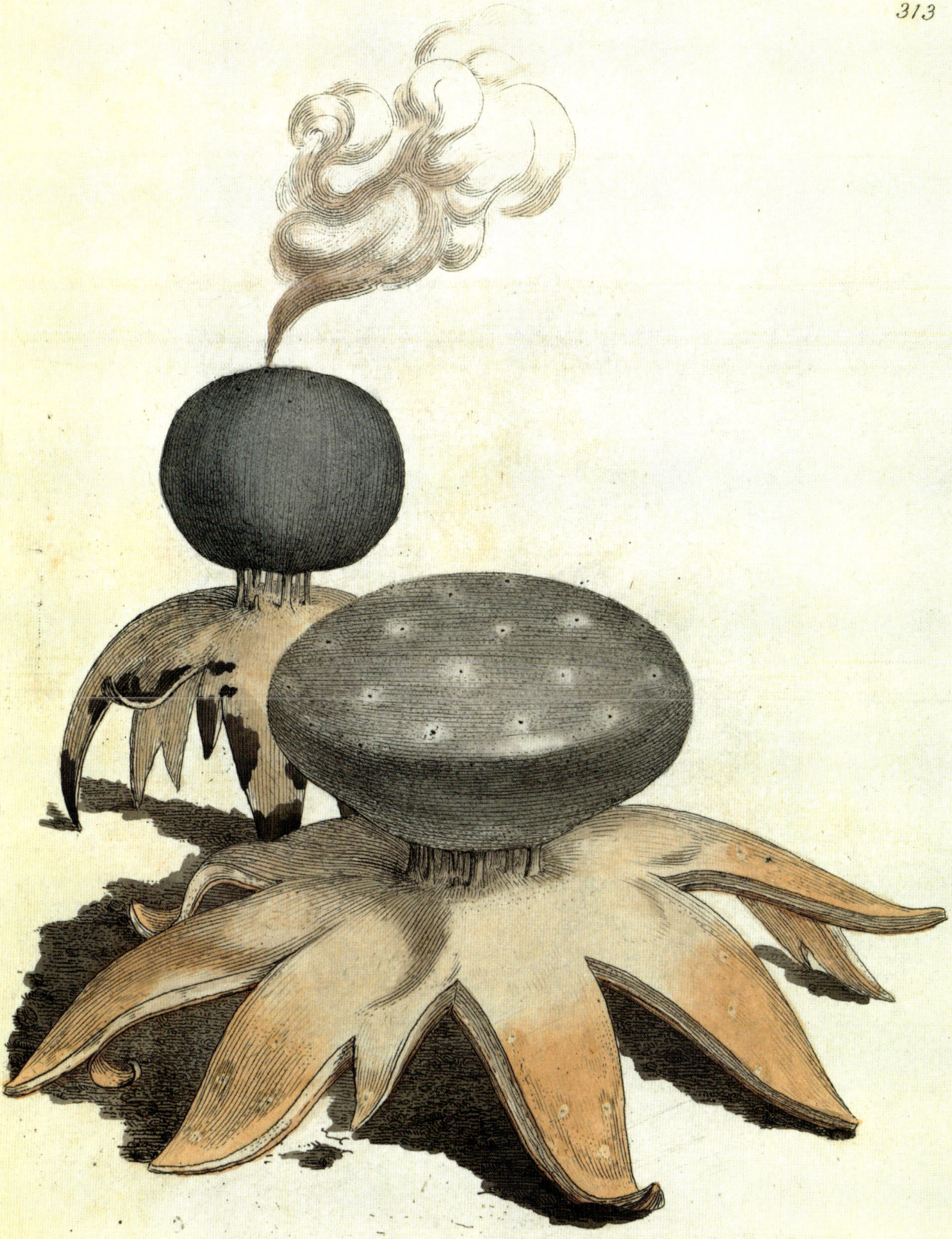

Jan.y 1. 1801. Published by Jas. Sowerby. London.

E/5. *Thalia dealbata*, commonly known as powdery thalia. Large folio engraving by James Sowerby of an aquatic plant, from the United States, that grows to six feet or more. Drawn and engraved in 1794 for John Fraser but probably remained and sold as a separate plate. Sowerby produced numerous plates over his lifetime.

of the potential stratigraphic value of fossils and hinted at possible phenomena related to the then unknown 'continental drift' or plate tectonics:

> May not the several formations that lie in strata, or coats, over the nucleus of the earth, have successive zones replete with the fossil remains of the animals etc. peculiar to them, besides such as are universally distributed through them, which zones might be discovered by a diligent comparison of the fossils of different countries, and indicate the probably position of the poles previously to the destruction of life in those strata, for the order and perfection of many of the remains seem to indicate that they are not far removed from the original sites.[9]

Many members of Sowerby's family were imbued with his mission to show people the world of natural history. Four generations were to continue under his inspiration and he is referred to as the patriarch of a dynasty that lasted into the twentieth century, with fourteen members publishing, writing, or illustrating natural history works between about 1780 and 1954. Sons of Sowerby's sons followed in their fathers' footsteps; George Brettingham (the second) was a fine illustrator of shells and published several works on conchology. James, the first son of James De Carle, published an illustrated work on fungi before his early death in 1834. Many other examples could be given. The family is rightly well renowned for the quality and quantity of its output.

It was one thing to establish a creative family business with sufficient income to support the members. It was another to act as a guide and mentor to young and old alike. The unrecorded acts of assistance, instruction, encouragement and generosity by Sowerby must be legion. And when they are recorded they reveal a man who enjoyed helping others, especially over their attempts to draw or identify specimens. There were times when people would be frustrated by his unwillingness to loan a meteorite or his failure to reply to a letter but in general he was at the hub of an active network that worked very well within a code of mutual cooperation. He was wanted because of his knowledge, his energy, his collections and his skills. It was also because of the person he was, for as one correspondent wrote in a letter thanking Sowerby for his help over an enquiry: 'The urbanity of your manner manifested on the occasion I shall never forget. I gratefully acknowledge it after having often done so to my friends. Such ingenuous courtesy becomes men of science; and in many of them a disposition of this kind appears conspicuous.'[10]

James Sowerby may not have been greatly in the limelight, nor did he achieve during his lifetime the same style of recognition gained by some other natural historians, but he succeeded in handing down a remarkable inheritance which we will celebrate for many years to come.

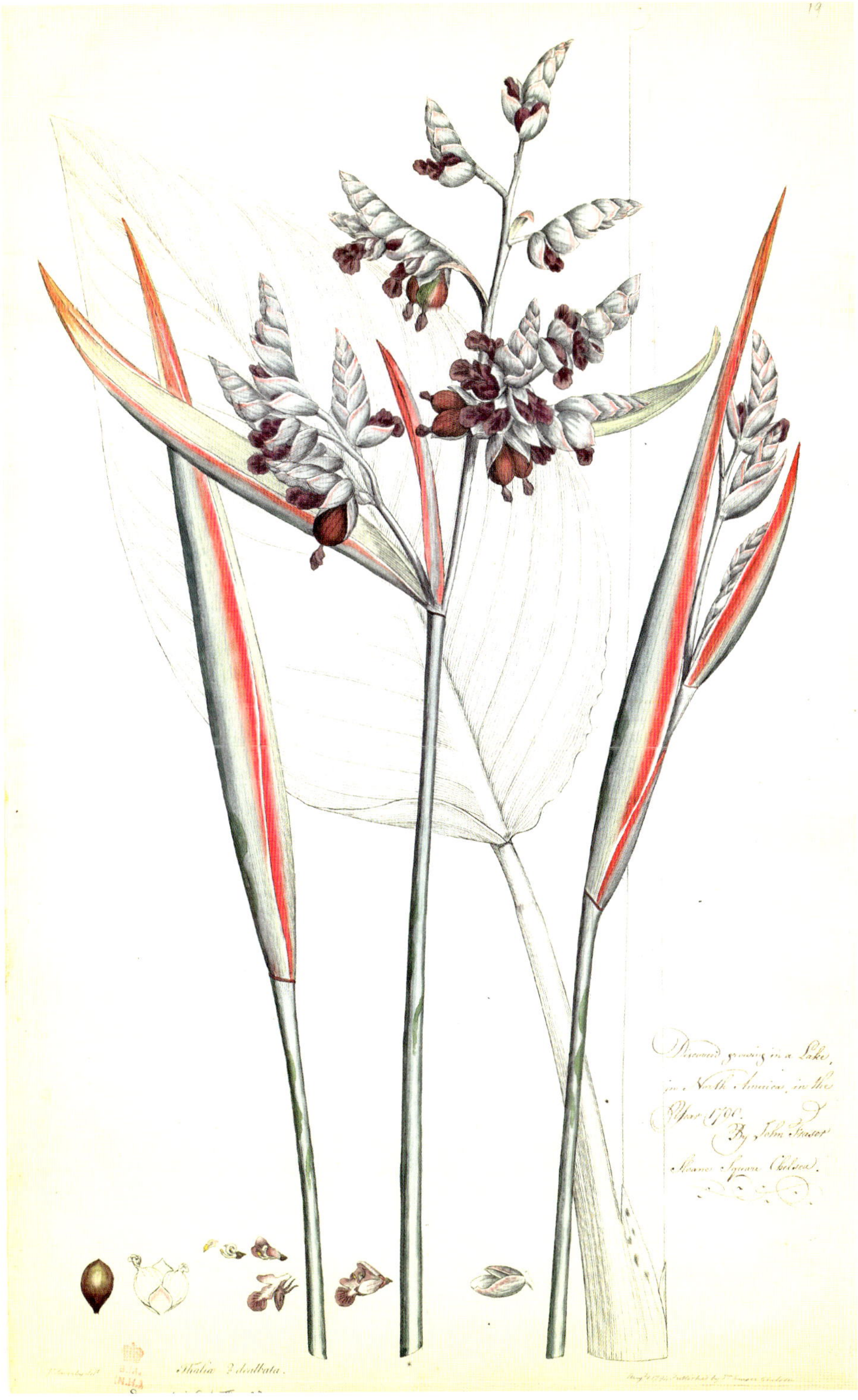
Discovered growing in a Lake, in North America, in the Year 1790. By John Fraser Sloane Square Chelsea.
Thalia ? dealbata.

APPENDIX A

James Sowerby – genealogical tree

John Sowerby = **Arabella Goodreed**
17/11/1718–1766 3/3/1725–10/8/1782

Robert De Carle = **Anne Carr**
8/7/1724–1796 2/1725–1809

James Sowerby = **Anne De Carle**
21/3/1757– 25/10/1822 17/1/1764 – 30/9/1815
(m. 9/2/1786)*

James De Carle
5/6/1787–26/8/1871
m. Mary Ann Edwards
(26/9/1813)

George Brettingham
12/5/1788–26/7/1854
m. Elizabeth Meredith
(16/4/1811)

Markham Valentine
14/2/1791–12/8/1802

John Edward
1793– ?

Mary Anne
1814–1882

James
18/11/1815–1834

Maria
6/1819–1829

Sarah
('Sally')
7/11/1819–1892
m. *John William Salter*
(1846)
7 children

Charles Henry
29/10/1821–~1826

Charlotte

George Brettingham
25/3/1812–26/7/1884

Charles
3/12/1813– ?

* Second marriage, 8/12/1819 to Maria Catharina Reynolds (no children). European date system. Names in italics are known to have been involved in natural history illustration.

Note on sources

Several sources were used to compile this tree. The two main ones were A. De C. Sowerby (1952) and J. De C. Sowerby (1825). Correspondence (1978) from E. de C. Blomfield, a member of the De Carle family, and a genealogical tree constructed (1973) by James Edward Lepper were other useful sources (Natural History Museum, Sowerby MSS, Box 65). Most sources contain some errors and there are inconsistencies between them but, in general, these are small. The version here is considered to be as accurate as possible but must remain as provisional

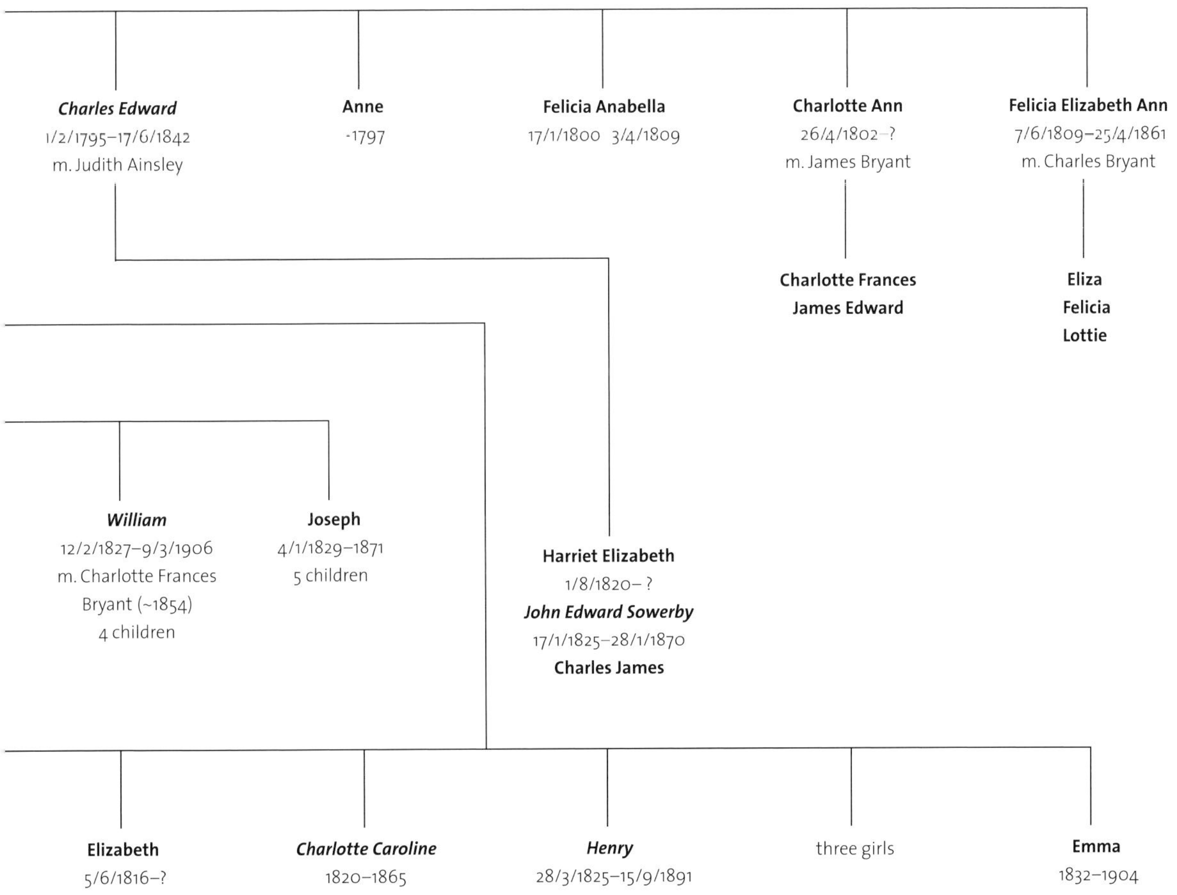

APPENDIX B
Works authored by James Sowerby

The following is a detailed list of works by James Sowerby, either as sole or co-author, in date order. It includes a section on some selected printed plates which were issued separately and paintings exhibited by him. Appendix C lists other publications which contain illustrations by Sowerby.

1788. SOWERBY, James. *An Easy Introduction to Drawing Flowers According to Nature.*
Obl. 4°. Printed by Richard Taylor. Published and sold by Author. London. 6 plates, coloured (4s.) or plain (2s.). Each plate contains several figures to illustrate the component parts of the flowers and states: 'Published by J. Sowerby Nov^{br} 1 1788'. Text after main title on title page: 'Originally designed for the Use of his Pupils. The Author having experienced the want of a Drawing Book sufficiently accurate to enable YOUNG BEGINNERS, who are fond of DELINEATING FLOWERS, to distinguish the different Parts absolutely necessary to characterize each PLANT, he has been induced to offer a few DESIGNS to the PUBLIC, in order to facilitate BOTANICAL STUDIES, and blend Amusement with Improvement.' *Second edition:* **1807**. (Title page with no date). Same printer and publisher as first edition. Four additional coloured plates (VII–X) each with date of publication as 1 August 1807. They cover mosses, algae and fungi. Plate X illustrates the letterpress dealing with opaque colours. Price: 7s.6d. plain, 10s.6d. coloured. A good, complete copy of the second edition in the Wellcome Trust Library, London, has some watermarked pages with dates 1803 and 1806. It is likely that reprinting of this work was common. It was sometimes referred to as *A Botanical Drawing Book*.

1789–1791. SOWERBY, James. *Flora Luxurians; or, The Florist's Delight.*
Folio. 18 plates, 16 pages. Produced as 3 numbers, each with 6 plates, coloured (12s.). Numbers I and II, each entitled *The Florist's Delight*, published 1789 and 10 February 1790, respectively. Number III entitled *Flora Luxurians; or, The Florist's Delight* published 1 May 1791. Plants illustrated include carnations, anemones, polyanthus, hyacinths and tulips.
The Natural History Museum, London, has some plates that are thought to have been prepared for another number of *Flora Luxurians* but which did not appear. (NHM Sowerby MSS, Box 86, folder 40.)

1790–1814. SOWERBY James with SMITH, James E. *English Botany, or, Coloured figures of British Plants, with their essential characters, synonyms, and places of growth.* 36 vols. 8°. 2,592 plates, coloured. Text by J. E. Smith except for plates 16–18 by Dr George Shaw. Issued in numbers of 3 plates each, initially at 1s. General index, 1814, in volume 36.
Volume dates are as follows, with each usually containing seventy-two plates but variations in the plate numbers occur, determined by how the owner chose to bind the original issued numbers. The plate numbers here are consistent with the appendices in each volume. Stafleu & Cowan (1985: 682–4) give information on printers as well as on other plate sequences found in different bound volume sets.
Vols: **1** (1790) pl. 1–72; **2** (1793) pl. 73–144; **3** (1794) pl. 145–216; **4** (1795) pl. 217–88; **5** (1796) pl. 289–360; **6** (1797) pl. 361–432; **7** (1798) pl. 433–504; **8** (1799) pl. 505–76; **9** (1799) pl. 577–648; **10** (1800) pl. 649–720; **11** (1800) pl. 721–92; **12** (1801), pl. 793–864; **13** (1801) pl. 865–936; **14** (1802) pl. 937–1008; **15** (1802) pl. 1009–80; **16** (1803) pl. 1081–1152; **17** (1803) pl. 1153–1224; **18** (1804) pl. 1225–96; **19** (1804) pl. 1297–1368; **20** (1805) pl. 1369–1440; **21** (1805) pl. 1441–1512; **22** (1806) pl. 1513–84; **23** (1806) pl. 1585–1656; **24** (1807) pl. 1657–1728; **25** (1807) pl. 1729–1800; **26** (1808) pl. 1801–1872; **27** (1808) pl. 1873–1944; **28** (1809) pl. 1945–2016; **29** (1809) pl. 2017–88; **30** (1810) pl. 2089–2160; **31** (1810) pl. 2161–2232; **32** (1811) pl. 2233–2304; **33** (1812) pl. 2305–76; **34** (1812) pl. 2377–2448; **35** (1813) pl. 2449–2520; **36** (1814) pl. 2521–92.
J. E. Smith's name appeared on the frontispiece from volume 4 onwards. Reissues of the work included his name also on volume 2; these have watermarks of a later date (1797) than the initial published volume.

1794. SOWERBY, James. Account of the difference of structure in the flowers of six species of *Passiflora*. *Trans. Linn. Soc. London* 2: 19–28. 3 plates. (Paper read to the Society on 5 January 1790.)

(1795) 1796 –1803 and 1809–1815. SOWERBY, James. *Coloured figures of English Fungi or Mushrooms.* 3 vols. Folio (32 x 20 cm). Printed by J. Davis (vols. 1 & 2), R. Wilks (vol. 3), London. Issued as numbers initially with 5 plates each and associated text, initially costing 2s.6d. but later (September 1796) changed to 20 plates at 10s. every four months. 400 plates, coloured. 440 plates, coloured, with supplements. Some early plates dated 1795. The first number January 1796; the last July 1803.
Vol. 1 (of numbers I–XII) 1797, plates 1–120; **vol. 2** (numbers XIII–XVIII) 1799, plates 121–240; **vol. 3** (numbers XIX–XXVII) 1803, plates 241–400 and, in some cases, with plate 401 inserted without text.
Supplement numbers: XXVIII (or No. I of the supplement) 1 February 1809, price 3s., plates 401–6, printer Richard Taylor and Co., London; **XXIX** (or No. II) 1 April 1809, 5s. plates 407–12, printer for this and subsequent numbers R. Meredith, London; **XXX** (or No. III) 1 November 1814, 5s., plates 413–23; **XXXI** (or No. IV) 1 December 1814, 5s., plates 424–9; **XXXII**

(or No. V) 1 January 1815, 5s., plates 430–40. For further details of dates of publication and of plate numbering in the supplements see Ramsbottom (1933). For information about Sowerby's fungus names see Peterson (1983).

1800. TURNER, Dawson and SOWERBY, James. Catalogue of some of the more rare plants observed in a tour through the western counties of England, made in June 1799, by Dawson Turner, Esq. F.L.S. and Mr. James Sowerby, F.L.S. *Trans. Linn. Soc. London* 5: 234–41.

(1802-) 1804–1817. SOWERBY, James. ***British Mineralogy, or, coloured figures intended to elucidate the mineralogy of Great Britain.*** 5 vols. 8°. Richard Taylor, London. 550 plates, coloured. Originally issued in parts. Numbers were published monthly from 1 November 1802 and cost 2s. 6d. each with 4 plates until 1 June 1805 when published bimonthly at 5s. with 8 plates. **Vols: 1** (1804) pl. 1–100; **2** (1806) pl. 101–200; **3** (1809) pl. 201–300; **4** (1811) pl. 301–400; **5** (1817) pl. 401–550. For further details see Conklin (1995).

1804–1805 (–1808). SMITH, James E. and SOWERBY James. ***Exotic Botany consisting of coloured figures, and scientific descriptions, of such new, beautiful, or rare plants as are worthy of cultivation in the gardens of Britain with remarks on their qualities, history, and requisite modes of treatment. The figures by James Sowerby.*** 2 vols. 4° (some numbers issued as 8° at half price). Printed by Richard Taylor & Co., London. Vol. 1 (1804) 118 pages; 60 plates, coloured. Vol. 2 (1805) 122 pages; 60 plates (61–120), coloured. Volume 2 is dated 1805; the date of publication of the plates continues to 1808. This work is sometimes referred to as being authored solely by Smith.

1804–1806. SOWERBY, James. ***The British Miscellany, or, coloured figures of new, rare, or little known animal subjects many not before ascertained to be inhabitants of the British Isles; and chiefly in the possession of the author.*** 2 vols. 8°. Printed by Richard Taylor, London. Published by J. Sowerby, London. 76 plates, coloured. Originally issued monthly in numbers, the first one on 1 December 1804, with blue paper covers, 4 plates, at 2s.6d.; then from 1 June 1805 mostly bimonthly with 8 plates, each, at 5s. Dates of publication of the twelve numbers: **1:** 1 December 1804; **2:** 1 January 1805; **3:** 1 February 1805. **4:** 1 March 1805. **5:** 8 April 1805; **6:** 1 June 1805; **7:** 1 August 1805; **8:** 1 October 1805; **9:** 1 December 1805; **10:** 1 February 1806; **11:** 1 June 1806; **12:** 1 August 1806. Colouring of plates varies in different copies.

1805–1808. SOWERBY, James. ***Description of models to explain crystallography, or, An easy introduction to the understanding of the formation of crystals, so essential to the knowledge of all substances, chemical or mineralogical.*** 2 parts – part 1, 1805 on coal, text of 16 pages and 1 plate; part 2, 1808 on diamond, 15 pages and 1 plate. 12°. Printed by Richard Taylor, London. Each part available with accompanying box of glass models. 10s.6d. each with models; 1s. each without. A printed list (October 1819) of works of natural history for sale by J Sowerby indicates that a third part was planned and partly produced on 'Carbonate of Lime, models showing the molecules produced by the fracture in the long diagonals of Iceland Spar, £1.1s.' One correspondent (M. Gent) wrote on 11 March 1816 to Sowerby that she had the models but wanted a description. This, it appears, was never published. (NHMS)

1807. SOWERBY, James. Additional observations relative to the Mytilus lithophagus. *Trans. Linn. Soc. London* 8: 272–4, 1 plate.

1809. SOWERBY, James. ***A new elucidation of colours, original prismatic, and material, showing their concordance in three primitives, yellow, red and blue; and producing, measuring, and mixing them, with some observations on the accuracy of Sir Isaac Newton.*** 4°. iv, 51 pages. 7 plates (5 coloured); some plates dated 1807. Printed Richard Taylor, London. Price £1.1s.

1811. SOWERBY, James. ***A Short Catalogue of British minerals According to a New Arrangement. Part 1. Combustibles and Earths.*** 12° Printed B. Meredith, London. xvi + 72 pages.

1811–1817 [-1820]. SOWERBY, James. ***Exotic Mineralogy: or, coloured figures of foreign minerals, as a supplement to British Mineralogy.*** 2 vols. 8° Printed by B. Meredith, London. 168 plates, 167 coloured. (Plate numbers are 1 to 169 with no number 73) Originally issued in parts bound in paper covers at 5s. each with 6 plates and letterpress, starting 1 March 1811 and subsequently at various dates to 1 November 1813 (number 12 with plates 67–72), then a gap until 1 March 1817 (with the issue of number 13 with plates 74–79; the plate number 73 for turquoise was omitted) and then bimonthly until number 26 on 1 May 1819. Numbers 27 and 28 were issued in 1820. (The plate for turquoise was issued as plate 93 in number 16 (1 September 1817) but the plate itself still has the engraved number 73 in the first issue.) For further details see Conklin (1995).

1811. SOWERBY, James. Account of a remarkable stone, known by the name of the Blowing-Stone, on the road from Farringdon to Uffington, in Berkshire. Account of a pit about two miles from Farringdon, commonly called the Farringdon Gravel Pit. *Trans. Linn. Soc. London* 10: 405–6. (In: Extracts from the Minute Book.)

1812–1822. [1846] SOWERBY, James. Later: continued by James De C. Sowerby. ***The Mineral Conchology of Great Britain; or coloured figures and descriptions of those remains of testaceous animals or shells, which have been preserved at various times and depths in the earth.*** 7 vols. 8°, 640 coloured plates. Vol. 7 not completed. **Vol. 1** (1812): Printed by Benjamin Meredith, London, 103 plates; **vol. 2** (1818): Printed by Arding & Merrett, London, 102 plates; **vol. 3** (1821): Printed by W. Arding, London, 103 plates; **vol. 4** (1823) 101 plates; **vol. 5** (1825) 96 plates; **vol. 6** (1829) 96 plates; **vol. 7** (1844) 39 plates. Author of vols. 1–4 James Sowerby; vols. 5 on given as 'continued by James D. C. Sowerby'. First four volumes with supplementary indexes by John Farey, including the stratigraphical location of the shells. An alphabetical index to the first six volumes appeared in 1840 and issued with the first part of volume 7. Cleevely (1974b) gives a detailed account of this work.

1814. SOWERBY, James. *General indexes to the thirty-six volumes of English Botany; to which is added an alphabetical index to English fungi, making together, a catalogue of indigenous British plants*. 8°, James Sowerby, London.

1817. SOWERBY, James. Notice on the effect of watering fruit trees early in the spring. *Trans. Hort. Soc. London* 2: 271–2. (Paper read to the Society on 7 January 1817.)

1819. SOWERBY, James. *A list of minerals with Latin and English names and numerous synonyms arranged according to a system founded upon the specific gravities of their component parts: with references to British and Exotic Mineralogy*. 8°. 22 pages. W. Arding, London. This was printed in a form fitted for labels as used in Sowerby's museum, and so was printed on one side of the paper only. The minerals are ordered in Classes: I Combustibles, II Earths, III Metals.

1819. SOWERBY, James. *A list of rocks and strata arranged in the order they generally occur in, placing those which compose the external surfaces of the Globe last; made out chiefly from late authorities with reference to figures in Sowerby's works in mineralogy and mineral conchology*. 8°. 8 pages. W. Arding, London. Printed for use as museum labels.

1819. SOWERBY, James. Some account of the spiral tubes or ligaments in the genus *Terebratula* of Lamarck, as observed in several species of fossil shells. *Trans. Linn. Soc. London* 12: 514–16, 1 plate. (Paper read to the Society 6 November 1814 and 7 February 1815.)

1821–1822 [1834]. SOWERBY, George Brettingham and SOWERBY, James. Later: continued with James De C Sowerby in place of James Sowerby. *The Genera of Recent and Fossil Shells, for the use of students in conchology and geology.* Often as 2 volumes but issued in parts. 8°. 265 plates, of which probably most of those in the first seventeen (of thirty-two) numbers by James Sowerby. Each issue initially 4s. plain, 6s. coloured and printed by E. J. Stirling, London. Petit (2006) gives information on the publication dates and other details. The work was first issued as 'conducted by George Brettingham Sowerby; illustrated with original plates by James Sowerby'. Many library catalogues list the author as James Sowerby but Petit claims it should be George Sowerby. The work was reissued in 1875. See also Petit (2009.)

1822. SOWERBY, James. On a fossil shell of a fibrous nature, the fragments of which occur abundantly in the Chalk strata and in the Flints accompanying it. *Trans. Linn. Soc. London* 13: 453–8, 1 plate (25). The published version of his verbal presentation to the Society on 1 November 1814 dealing with the genus *Inoceramus* and the species *Inoceramus cuvieri*.

Plates

Sowerby published numerous separate plates, some from his other publications so that people could purchase them without any letterpress or need for a subscription. The following were special illustrations, some in a large paper format:

- *Thalia ?dealbata. 'Discovered growing in a Lake in North America in the Year 1790'. By John Fraser, Sloane Square, Chelsea.* Coloured. [Folio]. J Sowerby del. 1 August 1794. Published by Jn Fraser, Chelsea.
- *Meteorolites. These are the only three that have fallen in Great Britain.* Coloured. 1812. (370x495 mm.)
- *Nautilus imperialis from the Clay Stratum Highgate*. Coloured. 25 May 1812. (290x400 mm.)
- *Fossil Shells of France*. 5 plates, coloured. About 1818–1819. (About 300x240 mm.)
- *Productus Scoticus*. Coloured. About 1819. (About 300x240 mm.)
- *Ammonites sedgwicki*. Uncoloured. 1821. (About 300x240 mm.)

Cleevely (1974a) gives further information on plates, some of which were advertised on the wrappers of Sowerby's publications. Thackray (1985) gives further details on some of the above.

Paintings exhibited by James Sowerby

1774 Society of Artists:
[Address: At Mr Wright's, Pimlico][1]
A deceptio visus

1776 Society of Artists (three works):
[Address: 2 Bolt and Tun Passage]
A Landscape; A Ship on Fire; Fish

1779 Royal Academy:
[Address: 2 Bolt and Tun Passage]
A lady; in miniature

1780 Royal Academy:
Portrait of a family

1782 Royal Academy:
[Address: 6, Rolls Buildings, Fetter Lane]
Landscape

1783 Society of Artists:
Flowers and Fruit

1784 Royal Academy:
[Address: 6, Brandon Road, Newington, Surrey]
Portrait of a lady

1788 Royal Academy
[Address: 2 Mead Place, Lambeth]
Flower piece of scarce exotics

1790 Royal Academy:
Passion flowers with fruit

All information from Graves (1907). The location of any of these paintings, if they still exist, is not known.

1 The address is that of Sowerby's residence at the time.

APPENDIX C

Works, by various authors, with illustrations by James Sowerby

List with titles and details of published works, in date order, containing plates drawn and/or engraved by Sowerby. The list does not include many drawings which Sowerby made for other authors' papers published in the *Transactions of the Linnean Society*, principally because in many cases it is not known who the artist was.

1775–1788 CURTIS, William. *Flora Londinensis: or plates and descriptions of such plants as grow wild in the environs of London*. Folio (47x28 cm). 2 vols. 432 plates. Originally issued in 72 numbers of 6 plates each, starting in 1775. Bound into six fascicles of 72 plates each, the first in 1777. Cost in 1777 of each number was 5s. coloured, 2s.6d. plain; of the first fascicle £3.4s. coloured, £1.4s. plain. Cleevely (1974a: 485) has recognised 51 plates as being by Sowerby in fascicles 5 and 6:
Fascicle 5 (1783–1788): 1–4, 7–10, 12, 13, 16, 17, 22–4, 26–9, 31–3, 35, 38, 39, 42, 43, 45–8, 50, 54–9, 60–3, 65–72.
Fascicle 6 (1790? –1798: 20).
In addition the author has noted that plate 53 has Sowerby's signature in some copies (inc. NHM copy), and that plate 21 in fascicle 5 of J. E. Smith's copy in the Linnean Society and the NHM copy each has a faint Sowerby signature and the copy in the Wellcome Library has a clear signature. Collins (1969) gave evidence that plate 48 (*Digitalis purpurea*) of the first fascicle is also by Sowerby. It seems likely in the light of Curtis's comment about employing two artists in 1777 (see chapter 1) and with Sowerby already working for Curtis that some of the plates in the early fascicles are by Sowerby. See also entry for 1785 WITHERING below.
Cleevely (1974a: 485) gives information about later issues and editions of the *Flora* but these did not involve Sowerby in new work.

1777 LIGHTFOOT, John. *Flora Scotia or, a systematic arrangement in the Linnean method of the native plants of Scotland and the Hebrides*. 2 vols. 8° B. White, London. 35 plates, plain. Sowerby draughtsman for plates 27, 29 to 32 and 34, and possibly 33. Others by Moses Griffith. A second edition of 1789 and reissued 1792.

1777 CURTIS, William. *Linnaeus's System of Botany, so far as relates to his classes and orders of plants; illustrated by figures entirely new, with copious explanatory descriptions*. 4° iv + 19 pp. William Curtis, London. 2 plates, each one in a coloured and uncoloured version, drawn by Sowerby, engraved by Sansom.

1784–1785 L'HERITIER DE BRUTELLE, Charles Louis (dated 1784–1785, published 1785–1791). ***Stirpes Novae aut minus cognitae quas Descriptionibus et Iconibus illustravit*.** Folio, 60x30 cm. Paris. Produced as six fascicles, the first and second dated 1784, the rest 1785, but actual publication dates are: fascicle I 1785, II & III 1786, IV 1788, V 1789 and VI 1791.
91 plates (numbered 1–84, some *bis*), some copies coloured, others not. Accompanying text in Latin.
Sowerby drew plate 42 (*Plectranthus punctatus*) in the fourth fascicle, published 1788 (it was engraved by Maleuvre). The rest are by other artists, with most by Pierre-Joseph Redouté.

1785 WITHERING, William. *An Account of the Foxglove, and some of its medical uses: with practical remarks on dropsy and other diseases*. 8° G. G. J. and J. Robinson, London and Birmingham. 1 folio plate. The frontispiece is a coloured plate of *Digitalis purpurea* copied 'under the inspection of Mr Curtis' from Curtis's *Flora Londinensis* (plate 48, fascicle 1) and in some copies signed J. Sowerby del: et sculp. Some authors have suggested that Sowerby was the original artist or re-engraved it for some of the copies of Withering's work. See Cleevely (1974a: 485) for further details.

1785 RELHAN, Richard. *Flora Cantabrigiensis, exhibens plantas agro Cantabrigiensi indigenas, secundum systema sexual digestas*. 8° Cambridge. 7 plates, uncoloured, drawn by James Bolton, engraved by Sowerby. Three are of lichen. Accompanying text mostly in Latin.

1785–1801 DICKSON, James. *Jacobi Dickson fasciculus Plantarum Cryptogamicarum Britanniae*. 4° London. 12 plates, usually plain, those of fascicles II–IV modified from coloured drawings by Sowerby, engraved by D. MacKenzie. The original drawings are held by the Natural History Museum, London. Publication dates: fasc. I 1785; II 1790, III 1793, and IV 1801. Later issues may have had the plates coloured.

1787–1788 L'HERITIER DE BRUTELLE, Charles Louis (dated 1787–1788, published 1791–1792). ***Geraniologia seu Erodii, Pelargonii, Geranii, Monsoniae et Grieli Historia, iconibus illustrata*.** Paris. Folio (~52x34 cm); also a large paper version (~60x45 cm). 44 plates plain; six drawn by Sowerby: 5, 8, 15, 27, 41, and 42 with various engravers. (33 of the plates drawn by Redouté.)
For a discussion of the publication dates see Britten & Woodward (1905).

1787–1799 CURTIS, William. *The Botanical Magazine; or, Flower-Garden Displayed*. 8° London. William Curtis was the editor until his death in 1799, when his friend John Sims took over. Curtis was responsible for volumes up to 13 (1799). Each number of three coloured plates and text cost 1s., 12 numbers form a volume. Sowerby was one of the artists for volumes 1–4. Based on evidence of plate signatures in different copies, original artwork (held at the Royal Horticultural

Society and the Royal Botanic Gardens, Kew) and Cleevely (1974a), Sowerby was the original artist and sometimes engraver for the following:
Vol. 1 (1787) 1, 4, 10–14, 16, 17, 19–21, 24, 26, 28, 32, 33; **vol. 2** (1788) 37, 39, 42, 44, 47, 49–52, 54–6, 58, 61, 67, 70, 72; **vol. 3** (1790) 76, 79, 80, 85, 87–108; **vol. 4** (1791) 109, 111, 113, 115–18, 121; **vol. 42** (1815) 1731, 1762; **vol. 44:** (1817) 1872; **vol. 47** (1820) 2124; **vol. 49** (1822) 2280, 2292, 2305, 2310, 2328, 2354; **vol. 52** (1825) 2568; **vol. 53** (1826) 2654, 2683.
Note that some plates were published years after Sowerby's death. In these cases the artwork had been prepared some years before publication, even by as much as seven years. Plates 18, 23, 25, 27, 29–31, 34–6, 38 and 43 are signed by Sowerby but are by Sydenham Edwards (reported in the note 'To the Subscribers to the Botanical Magazine' by the editor (John Sims) bound after plate 1232 of volume 30 (1809) in the Kew set). This printed note also states that 'the colouring of this work has, from its commencement, been conducted by Mr William Graves, Newington Butts.'). See also Desmond (1987).

1788 DRIVER, A & W. *The Pomona Britanica or fruit-garden displayed: containing scientific descriptions of all such fruit-trees and fruit-bearing plants as are cultivated in Great Britain; with a Botanical Lexicon.* By several hands. Revised by A. and W. Driver Nurserymen, Kent Road. 4° William Darton and Co., London. 8 fully coloured plates, all except numbers 1 and 4 signed Sowerby del. and all dated 1788. (The copy in the Royal Horticultural Society, London, contains 9 plates – the ninth of a plum, unnamed, and without any associated letterpress.) Plates are of fine drawings showing wood, leaves and fruit:
1. *Prunus cerasus* or Harrison's Heart, or Red-Heart Cherry
2. *Rubus idaeus* or Golden Antwerp Raspberry, or Ld. Middleton's Raspberry
3. *Prunus armeniaca* Persica or Peach Apricot
4. *Pyrus malus* Cerasoides or American Crab, or Cherry-like Apple
5. *Pyrus communis* or Swiss Bergamot
6. *Prunus domestica* or Imperial Plum, or Red Bonum Magnum
7. Malus fructo parvo, or Apple
8. *Prunus sativa* or Cultivated Plum.

1786–1787 L'HERITIER DE BRUTELLE, Charles Louis (dated 1786–1787, published 1789–1791). ***Sertum Anglicum sue Plantae Rariores Quae in Hortis juxta Londinum imprimis in horto Regio Kewensi Excoluntur***. Folio (52x34 cm). Paris. Produced as 4 fascicles. Text in Latin. 35 plates (including 15 *bis*), plain, 10 by Sowerby (numbers 1, 3–7, 9, 22, 26 and 29). Carter (1988) gives the date of publication of the first fascicle as January 1789. Further details see Cleevely (1974a: 486).

1789 AITON, William. *Hortus Kewensis, or a catalogue of the plants cultivated in the Royal Botanic Garden at Kew*. 3 volumes, 8°, London. (A second edition 1810–13 by William Townsend Aiton.) Plates plain. Those by Sowerby: vol. 1 plate 1, p. 30, *Calceolaria fothergillii* (engraved by MacKenzie), plate 2, p. 285, *Strelitzia reginae* (engraved by MacKenzie); vol. 3 plate 12, p. 302 *Limodorum tankervilliae* (engraved by MacKenzie), plate 13 (but often unnumbered), p. 496 *Smithia sensitiva*, engraved by Sowerby, (draughtsman ?). Original drawings are in the Natural History Museum, London.

1789 FRASER, John. *A Short History of the Agrostis Cornucopiae or The New American Grass.* Privately published. Folio. London. 2s.6d. for the plain version; 3s.6d. coloured. 7 pages and 1 folio plate by Sowerby (del and sculpt). Plate: 470x270 mm.

1789–1791 SMITH, James Edward. *Plantarum Icones, hactenus ineditae, plerumque ad Plantas in Herbario Linnaeano Conservatas Delineatae*. Demy folio, London, J. Davis. 75 plates, plain, unsigned, mostly by Sowerby. 3 fascicles each of 25 plates. Number I: 1789; II: 1790; III: 1791. Latin text.
Some plates drawn from artwork by Pablo Antonio Garcia (1744–1814), the artist of José Celestino Mutis (1732–1808), a Spanish physician and botanist who sent Linnaeus numerous specimens and drawings, for example plate XXV *Passiflora mixta*, where most of the original is copied in the redrawing and engraving. Fascicles sold initially for one guinea each.

1790–1793 SMITH, James Edward. *Icones Pictae Plantarum Rariorum Descriptionibus et Observationibus Illustratae*. Folio (485x290 mm). London, printed J. Davis. 18 plates, coloured, unsigned, all by Sowerby. 3 fascicles each of 6 plates. Number I: 1790; II: 1792; III: 1793. Each 12s. English version (*Coloured Figures of Rare Plants*) purchasable as an addition (2s.), published at same time. (Some copies on larger paper at 1 guinea, or with both Latin and English letterpress at £1.4s.) Printings were made later, for example sale at Christie's New York auction of a bound volume with the English title-leaf watermarked 1814. (Sale number 8009, 5 June 1997, lot 635.)

1790 SHAW, George. *Speculum Linnaeanum: sive zoologiae Linnaeanae illustration; or Linnean zoology, containing a complete illustration of the zoological part of the Systema Naturae of Linnaeus: with figures elegantly engraved and accurately coloured, representing the most remarkable species of each genus with descriptions at large.* 4° London. 8 coloured plates all by Sowerby.

1790–1794 WOODVILLE, W. *Medical Botany containing systematic and general descriptions with plates of all the medicinal plants, indigenous and exotic, comprehended in the catalogues of the Materia Medica, as published by the Royal College of Physicians of London and Edinburgh: accompanied with a circumstantial detail of their medicinal effects and of the diseases in which they have been most successfully employed.* 3 vols., plus a supplement. 4°. 274 plates in coloured or uncoloured versions, all by Sowerby. **Vol. 1** (1790) plates 1–65; **vol. 2 (1792)** plates 66–135; **vol. 3 (1793)** plates 136–210. **Supplement** (1794) plates 211–74. It dealt also with medicinal plants not in the materia medica.

1791 SALISBURY, Richard Anthony. *Icones stirpium rariorum descriptionibus illustratae.* Folio 60x45 cm, London. 10 coloured plates by Sowerby. (The Linnean Society copy has an uncoloured duplicate of each plate.) The Natural History Museum has some of the original sketches for this work.

1792. SMITH, J.E. *Spicilegium Botanicum or Gleanings of Botany.* Folio (340x210 mm). London. 24 coloured plates by Sowerby. Originally published as two numbers, each with 12 plates and priced 10s.6d. or with Latin and English text 12s.

1793-(1795) SMITH, James Edward. *A Specimen of the Botany of New Holland.* 4° Published by J. Sowerby, London. Originally issued in four parts, 1 & 2: 1793; 3: 1794; 4: 1795. 16 coloured plates by Sowerby based on coloured drawings made in Australia by John White (Surgeon-General to the Colony) and herbarium and actual specimens.

1794 FRASER, John. *Thalia ?dealbata. Discovered growing in a Lake, in North America in 1790.* See Appendix B.

1794 SHAW, George. *Zoology of New Holland.* 4° Published by J. Sowerby, London. 12 coloured plates by Sowerby, several of which are signed as published by him. The copy in the Linnean Society, London, has an additional copy of each of plates I and IV, bound in at the back of the volume, with the imprint 'London. Published June 1 1793, by F. P. Nodder & Co. No 15 Brewer St.' This is erased from the plates within the volume.

1798 ABBOT, Charles. *Flora Bedfordiensis, comprehending such plants as grow wild in the county of Bedford, arranged according to the system of Linnaeus, with occasional remarks.* 8° W. Smith, London. 6 coloured plates, provided by Sowerby. Plate 3 of the marsh violet (*Viola palustris*) is from *English Botany*, plate 444 (volume 7, 1798).

1798 SMITH, James Edward. *Tracts relating to Natural History.* 8vo London. A compilation of twelve articles all, except the last, with the text previously published and including his review of Sowerby's *Introduction to Drawing Flowers* (1788). The last three deal with three new genera of flowers: *Sprengelia, Westringia* and *Boronia* and contain a total of six coloured plates (no signature). Correspondence and draft plates (NHMS Box 49) indicate that Sowerby was the artist for these.

1799–1807 THORNTON, Robert John. *New Illustration of the Sexual System of Carolus von Linnaeus comprehending an elucidation of the several parts of the fructification, a prize dissertation on the sexes of plants, a full explanation of the classes, and orders, of the sexual system, and the temple of flora, or garden of nature, being picturesque, botanical, coloured plates, of select plants, illustrative of the same, with descriptions.* Folio, London. 121(?) plates, coloured or uncoloured (despite title) by various artists. In the section: New Illustration of the Sexual System Of Linnaeus is one plate drawn and signed by Sowerby (engraved by Wetton) of *Ricinus communis*, or common Palma Christi. Plate numbering and sequence vary in different copies.

1802 MONTAGU, George. *Ornithological dictionary, or, Alphabetical synopsis of British birds.* 2 vols, 8°. T. Bensley, London. Sowerby engraved the frontispiece in vol. 1 of a cirl bunting.

1802 KIRBY, William. *Monographia Apum Angliae, or, An attempt to divide into their natural genera and families, such species of the Linnean Genus Apis as have been discovered in England: with descriptions and observations.* 2 vols. 8°. J. Raw, Ipswich. 18 plates. 14 monochrome plates, by Kirby, in volume 1; 4 coloured plates (unsigned) by Sowerby in volume 2. Doubts (Cleevely, 1974a) have been expressed that Sowerby was involved but correspondence from Kirby to Sowerby (NHMS) shows that he was commissioned for volume 2.

1803–1825 LAMBERT, Aylmer Bourke. *A description of the genus Pinus, illustrated with figures, directions relative to the cultivation and remarks on the uses of several species.* Folio. 2 vols. London. About 55 plates by Ferdinand and Franz Bauer, G. D. Ehret, S. Parkinson, J. Lindley and J. Sowerby; engraved by Barlow and others (not Sowerby). Mostly uncoloured but different copies show variations. Vol. 1: 4 plates drawn by Sowerby, engraved mostly by Warner; vol. 2: 6 plates drawn by Sowerby, engraved mostly by Weddell. Different editions were produced and show variation in the plates. Different copies of the same edition also vary (Cleevely 1974a: 489 & 1976: 357).

1804 TURNER, Dawson. *Muscologiae Hibernicae Spicilegium.* 8° London. 16 coloured plates. This work is included here, although the drawing of the plates is ascribed to Sowerby's son, James De Carle, because correspondence indicates that Sowerby was also involved in the plate production, mostly by engraving but possibly also with printing and colouring. (The correspondence is held in the NHM, Sowerby Collection.)

1806–1840 SIBTHORP, John. *Flora Graeca: sive plantarum rariorum histori, quas in provinciis aut insulis Graeciae legit investigavit, et depingi curavit, Johannes Sibthorp, M.D.* 10 vols. Folio (33x46 cm). Printed Richard Taylor, London. Sold J. White (Fleet St.). 966 plates coloured. Engraving by James Sowerby and probably others. After Sowerby's death in 1822, his son James De Carle Sowerby undertook some of the engraving. The Sowerby publishing business was responsible for engraving and colouring the plates and dealing with the printer. Volumes 1–7 edited by J. E. Smith; 8–10 by J. Lindley. Each volume consists of two fascicles with 100 plates, except for volume 10 of 66 plates. Volume 1, fascicle 1 with the Prodromus published 1806, fascicle 2, 1808. Stearn (1967) gives dates for all volumes.

1807 THOMPSON, John Vaughan. *A catalogue of plants growing in the vicinity of Berwick on Tweed.* 8° London. 132pp. Illustration for frontispiece and 1 plate, drawn by author, engraved by Sowerby.

1807 SMITH, James E. *An introduction to physiological and systematical botany*. 8° Printed: Longman, Hurst, Rees & Orme, London. 15 plates, usually coloured, all by Sowerby. Each plate comprises several figures giving a total of 214 illustrations of component parts of flowers. A second edition with the same plates 1809.

1812 MAWE, John. *Travels in the interior of Brazil, particularly in the gold and diamond districts of that country by authority of the Prince Regent of Portugal, including a voyage to the Rio de la Plata, and an historical sketch of the revolution of Buenos Ayres.* 4° London: Longman, Hurst, Rees, Orme, and Brown, Paternoster-Row. 9 plates, one (coloured) by Sowerby of six minerals including one of cubic-shaped gold.

1815 WOOD, William. *General Conchology, or, a description of shells, arranged according to the Linnean System, and illustrated with plates, drawn and coloured from nature*. 8° John Booth, London. 60 plates, coloured. 6 plates by Sowerby. Reissued in 1835.

1815–1820 + 1875. LEACH, William Elford. *Malacostraca podophthalmata Britanniae, or, Descriptions of such British species of the Linnean genus Cancer as have their eyes elevated on footstalks, illustrated with Figures of all the Species by James Sowerby FLS, GS, & WS* 4° Printed by B. Meredith, Published by James Sowerby, London. 47 plates by Sowerby. Issued in parts with initial price. 7s.6d. coloured, 6s. plain. The title on wrappers different from volume title, namely *Malacostraca Podophthalmata Britanniae; or Descriptions of the British species of Crabs, Lobsters, Prawns, and of other Malacostraca with pedunculated eyes*. Parts 1–6, 1815; 7–12, 1816; 13–16, 1817; 17, 1820. A further issue of two parts (numbers 18 and 19) in 1875 with 7 additional plates by G. B. Sowerby (the second) with details of fifteen new species and a generic and specific index for all parts. Published by Bernard Quaritch.

1816–1819 SMITH, William. *Strata identified by organized fossils, containing prints on colored paper of the most characteristic specimens in each stratum*. 4° 32 pages. Printed by W. Arding, James Sowerby, London. 19 coloured plates by Sowerby. Issued in four (of seven planned) numbers: 1: 1 June 1816, 4 plates; 2: 1 October 1816, 5 plates; 3: 1 September 1817, 5 plates; 4: 1819, 5 plates. The plates were printed on coloured paper 'to represent that of the Stratum in which the Specimens were found.' Price 7s.6d. (10s. for 'highly finished copies' for first, second and possibly third number.) For further details see Eyles (1969).

1816 KEITH, Rev. P. *A System of Physiological Botany*. 2 vols. 8°. Baldwin, Cradock, and Joy, London. Nine black and white plates: eight in vol. 1, one in vol. 2, drawn and engraved by Sowerby. Each plate contains several illustrations of parts of plants.

1816 [KENDALL, Frederick]. *A descriptive catalogue of the minerals and fossil organic remains of Scarborough and the vicinity*. 8° Printed by T. Coultas. Scarborough, 316pp. Various versions of this work exist, none with the name of the compiler or author on the title page but some with a following 'Advertisement' signed by Kendall stating the aim of the work. With or without illustrations. One copy in the British Library contains numerous coloured illustrations previously used in some of Sowerby's publications, especially *British Mineralogy* and *Mineral Conchology*, and supplied by Sowerby.

1817 PURTON, T. *A botanical description of British plants in the Midland counties, particularly of those in the neighbourhood of Alcester*. 2 vols. 8°. J. Ward, Stratford-upon-Avon. 8 coloured plates (some copies not coloured) all engraved by Sowerby and including some from *English Botany* and some based on drawings by Purton's daughter. Three are of fungi. Purton produced a third volume (of two appendices) in 1821 with thirty illustrations by various engravers. The extent of Sowerby's involvement, if any, with the third volume is not clear although correspondence (NHM MSS) suggests some.

1820 MAWE, John. *Familiar lessons on mineralogy and geology explaining the easiest methods of discriminating minerals, and the earthy substances, generally called rocks, which compose the primitive, secondary, floetz or flat, and alluvial formations: to which is added a description of the lapidaries' apparatus.* 16°. Second edition. London. Coloured plate with 22 mineral illustrations. Sowerby contributed to the production of the plate for the second edition (and probably for others) but it is not clear to what extent.

1820 MAWE, John. *Instructions for the management of the blow-pipe, chemical tests etc.* London. 47 pages. 18° Frontispiece (partly coloured) by Sowerby (or possibly by James De Carle Sowerby).

1821 SMITH, James E. *A Grammar of Botany, illustrative of artificial as well as natural classification, with an explanation of Jussieu's system*. 8°. Longman, Hurst, Rees, Orme and Brown, London. 242 pages. 21 plates, with numbered total of 277 figures, usually coloured, by Sowerby.

1825 WOOD, W. *Index Testaceologicus, or, A catalogue of shells, British and foreign, arranged according to the Linnean System*. 8°. Richard Taylor, London. 37 plates, coloured or uncoloured. First six by James Sowerby, others by the author. This work first appeared in 1818 without illustrations.

Notes and citations

Abbreviations:
LS: Linnean Society
LSS: Linnean Society collection of Smith MSS.
NHMS: Natural History Museum London Sowerby collection of letters, plates, illustrations etc.
OUS: Oxford University Sherard MSS collection
UBEC: University of Bristol. Special Collection, Eyles collection

Prologue

1. John Hawkins to James Edward Smith. 3 February 1805. OUS.
2. Secord (1994).
3. Harris (2007); Lack with Mabberley (1999).
4. Note 3 ibid; Blunt (1950); Sitwell & Blunt (1956).
5. Porter (1978, 1991); Secord (1994).
6. Linnaeus (1753). See also chapters 1 and 2.
7. The writer and naturalist Thomas Pennant stands out as an example. He acquired an early taste for mineral collecting when in 1747, still an undergraduate at Oxford, he toured Cornwall where he met the Reverend William Borlase who 'gave me a strong passion for minerals and fossils'. Pennant's collection was to become a significant one, part of which is in the Natural History Museum, London. He had a similar strong passion for other areas of natural history and gained considerable renown from his zoological works. Pennant's *Literary Life* cited by Withers (2004).
8. Examples of authors included Arthur Aikin, William Babington (e.g. 1795) and Richard Kirwan (e.g. 1784).

1. Precarious start

1. Sourced mainly from Arthur De Carle Sowerby (1952) and J. De C. Sowerby (1825). There are a few small discrepancies between Arthur De Carle Sowerby's 1952 account and his 1950 genealogical tree (NHMS, Box 54). The 1952 dates are used here and in the family tree in Appendix A. John Sowerby's children were: John (25/2/1744–7/3/1744), John (27/2/1746–?), Arabella (15/2/1748–1814), Charles (16/2/1750–?), Thomas (12/6/1753–?), James (21/3/1757–25/10/1822), William (6/1/1760–?), Elizabeth (4/5/1762–?). John Sowerby's father was Richard Sowerby, of Fetter Lane, London, married to Elizabeth Glover. A genealogical tree (1973; NHMS, Box 65) by James Edward Lepper suggests that Richard Sowerby's father was Charles Sowerby of Keswick.
J. De C Sowerby's short unpublished biography (1825) of his father, to which some additions and changes were made by Sir James Smith, contains some inconsistencies and inaccuracies. See also Simpkins (1974a).
2. NHMS, Box 32, folder B15.
3. Harben (1918); Horwood (1813).
4. Thornbury (1878); Altick (1978).
5. Brotheridge, T. History of Lead Crystal. Website of Brotheridge Chandeliers. www.brotheridgechandeliers.co.uk Last accessed 22 March 2015.
6. Hoare's Bank website: www.hoaresbank.co.uk/history.htm Last accessed 22 March 2015.
7. Bell (1912).
8. Pennant (1791).
9. Bell (1912).
10. J De C Sowerby (1825).
11. Ibid.
12. Personal communication, Alex Kidson, National Museums Liverpool, November 2010; Monks (2008); Nicholson (2004).
13. J De C Sowerby (1825); Monks (2008).
14. Graves (1907).
15. Note 12 ibid; Graves (1907).
16. The evidence that it was this particular William Hodges to whom Sowerby was apprenticed is not complete but it is highly likely. See also J De C Sowerby (1825); Cust (2004).
17. J De C Sowerby (1825); Foskett (1987).
18. Rudé (1971).
19. Blunt (1950); Hunt (2004).
20. Lightfoot (1777).
21. Bowden (2004).
22. W. H. Curtis (1941: 28).
23. Ibid, page 19.
24. Inserted page at end of Curtis's 1777 work on Linnaeus's System:
Just published by the same Author, Price 3£ 4s in Boards, coloured, and 1£. 4s plain,
The First Fasciculus of the Flora Londinensis: containing Seventy-two Folio Plates, with Descriptions.
This work, which is intended first to treat of the Plants which Grow Wild in the Environs of London, and afterwards to be extended to the remaining Plants of Great Britain, is now publishing periodically, in Numbers of Five Shillings each, Coloured, and Two Shillings and Six-pence Plain, and is advanced as far as No XV: each Number contains Six Plates, with Descriptions of each Plant in Latin and English, and an account of its uses in Medicine, Agriculture and other Arts.
25. W. H. Curtis (1941).
26. Hunt (2004).
27. Personal communication, Royal Academy Library, 21 April 2009.

2. Rising reputation

1. Desmond (1995); Schweizer (2004).
2. W. H. Curtis (1941). It was the Honourable Daines Barrington Temple who, having seen Curtis's request, advised him appropriately, in a letter of 26 March 1781.
3. S. Curtis (1805).

4. Desmond (1995).
5. Carter (1988: 121). Joseph Banks (1743–1820) was heir to a landed fortune and from an early age showed a marked enthusiasm for botany. He was a member (with the botanist Daniel Solander) of the *Endeavour* voyage (1768–1771) to Tahiti, New Zealand and Australia. He was introduced to King George III on his return with whom he established a relationship, becoming effectively (in 1773) the director of the Royal Botanic Gardens at Kew. He became President of the Royal Society in 1778, a position he held until his death and which, together with his wealth, gave him huge degrees of influence. He was awarded a baronetcy in 1781.
6. Graves (1907).
7. Personal communication, Royal Academy Library, 22 May 2014.
8. Roscoe et al. (2009). The Coade artificial stone factory, run by Eleanor Coade, was in Lambeth by the River Thames from 1769 to 1840, on the site where the South Bank Centre now stands. It produced a range of decorative moulded 'stones' for architectural, interior and garden ornament purposes. Commissions were received from Robert Adam, John Soane and John Nash among other renowned architects. Robert Brettingham de Carle worked there from the mid-1780s and was responsible for several high-quality pieces.
9. Personal communication, Royal Academy, 29 October 2010. Robert Brettingham De Carle exhibited three portraits in 1785 at the Royal Academy, one of a child and two, in wax, of an unknown lady.
10. J De C Sowerby (1825).
11. Based on drawings in the NHMS, Box 86. One sketch is of three women, an older one holding a baby. The paper has the watermark 1797. The baby might be their daughter Anne, born in that year.
12. B. B. Woodward (2004).
13. T. Woodward to James Sowerby, December 1784, and 29 August 1785 – the latter proposing the publication. He referred to Jacob Christian Schäffer (1718–1790) who was a distinguished German professor of natural history. NHMS, Box 24, folder 71.
14. Walker (1988); Boulger (2004d).
15. There are several ways of determining the approximate value of money over time. Here is used that based on a 'historic standard of living' which can be defined as the purchasing power of a bundle of goods or services. See www.measuringworth.com/ukcompare. (site last accessed: 22 March 2015). From 1790 to 2010 the very approximate multiple is 85.
16. Walker (1988); Boulger (2004d).
17. McConnell (2004).
18. Fascicle XV (1798), number 24. *Lichen rimosus* on crystallised rock specimen. This is in Joseph Banks's copy held in the Natural History Museum, London, and which is in surprisingly good condition.
19. Walker (1988); Geldart (1914).
20. J. C. Lettsom to James Sowerby, 6 Rolls Buildings, Fetter Lane. 9 June 1785. NHMS, Box 11, folder 32.
21. Anon (1804). This pamphlet also states: 'the library contains about six thousand volumes of books, cabinets of shells, insects, and various subjects of natural history; the specimens of woods and of ambers are extensive and valuable... the hortus siccus, or collection of dried plants, occupy sixty volumes'.
22. The paintings (142) comprise two volumes, entitled *Hortus sylva montis,* and held in the Natural History Museum, London. They are considered to have been completed around 1787.
23. Letters over the period 30 May 1785 to 6 January 1786. NHMS, Box 9.
24. James Bolton authored his own books. Two are: *Filices Britannicae; an history of the British proper ferns* (1790), an attractive work with straightforward descriptions of the ferns and containing 31 coloured plates, each drawn and engraved by him; *An history of fungusses growing about Halifax* (1788), a three-volume work with 182 coloured plates also drawn and engraved by him, several of which are of high quality.
25. W. H. Curtis (1941).
26. Collins (1973) suggested that Sowerby was the author but Cleevely (1974a: 486) disagrees. I consider it likely that Sowerby was a contributor but only in part (i.e. one of the 'several hands') on the evidence of the few notes on the drawings in the NHMS, which appear to be written by Sowerby.

3. The distant suburb

1. Horwood's map in Laxton & Wisdom (1985); NHMS, Box 32, folder B20.
2. Royal Society, London. MS/709/52 (1817).
3. T. Allen (1827); Tanswell (1858).
4. T. Allen (1827).
5. Ibid, page 336.
6. Macdonald (1974).
7. Pennant (1791).
8. This statement was printed on the blue paper wrapper covering the plates and text of the first number.
9. Edmund Davall, an English botanist living in Switzerland, was preparing a Swiss flora with illustrations. See chapter 4 for further details. There may be other lesser cases as indicated by general correspondence in NHMS.
10. See chapter 2. McConnell (2004).
11. This is now recognised as an ascomycete species: *Thamnolia vermicularis.*
12. The current name is *Didymodon tophaceus* var. *linearis.*
13. Hawkesworth (1976).
14. Carter (1988).
15. Ibid, page 225.
16. Joseph Banks to James Smith, 15 August 1787. LSS.
17. A current example is: http://www.gutenberg.org/files/38937/38937-h/38937-h.htm Last accessed 22 March 2015.
18. Based on an introductory essay by William Blunt 'The Illustrators of Sertum Anglicum', in the facsimile of L'Héritier's *Sertum Anglicum* published by the Hunt Botanical Library, Pittsburgh, Pennsylvania, 1963.
19. Curtis: *Botanical Magazine,* vol. 3: 103 *Pelargonium acetosum.*

20. Carter (1988: 252).
21. Desmond (1995: 106). Daniel Solander (1733–1782) had been trained under Linnaeus before coming to England in 1760. He was librarian to Joseph Banks as well as assistant librarian at the British Museum (1765–1788). The Swedish botanist Jonas Carlsson Dryander (1748–1810) succeeded Solander as Banks's librarian in 1782.
22. Brunton (2004).
23. Volume 3, page 476.
24. D. E. Allen (2004).
25. Salisbury to James Sowerby, 6 May 1791. NHMS, Box 17.
26. Boulger (2004b).
27. See chapter 2 and note 22 for that chapter.
28. For Lady Rockingham to James Sowerby, 5 June 1790. NHMS, Box 16.
29. Thornton (1803); W. H. Curtis (1941).
30. Roscoe et al. (2009). Quotation of Press accounts in entry for Robert Brettingham De Carle.
31. The other two named members of the Sowerby family are John (presumably James's brother) and Isabella, but it is not known who she was.
32. Victoria and Albert Museum, London. Robert Brettingham De Carle: 'Jug' 1781, Museum number C.90-1985.

4. The two Jameses

1. Walker (1988).
2. Example is a letter, 21 August 1789, from James Smith to Sowerby on two drawings made by him and sending to be engraved for his next fasciculus**.** The title of the work is not mentioned but the date fits with the *Plantarum Icones*. Again, on 9 September 1789, he sent a drawing of *Maratta fraxinea*. NHMS.
3. See entry on *Plantarum Icones* in Appendix C.
4. Letter of 2 July 1789. LSS.
5. Letter from Hillingdon House, 2 July 1789, to James Smith. LSS. The Catesby referred to is the naturalist Mark Catesby (1683–1749) author of *The Natural History of Caroline, Florida and the Bahama Islands* (1729–1747) and *Hortus Britanno-Americanus* (1763). He drew and etched his own illustrations.
6. http://www.illustratedgarden.org/mobot/rarebooks/title.asp?relation=QK98S657 Last accessed 22 March 2015.
7. Letter of 31 January 1873. NHMS, Box 16, folder 46.
8. Christies Sale 4071, 13 April 2006, London, South Kensington Price £3,840 ($6,728).
9. See note 5.
10. Letter of 18 September 1789. The plant is likely to be *Myrtus tomentosa* (now *Rhodomyrtus tomentosa*) as indicated in another letter to Smith of 15 October 1790. LSS.
11. Letter to James Smith of 8 November 1790 asking him to have a copy of the *Icones Pictae* sent to Davall before he might get it from a bookseller. LSS.
12. Boulger (2004a).
13. Walker (1988) and LSS.
14. James Smith letters to Davall 27 April 1790 and 31 March 1791. Dated receipt from Sowerby for £20, via James Smith. LSS.
15. *Analytical Reviews*, August 1790, p515.

5. Sowerby's botany

1. A standard botanical book of the time and one used by Smith and Sowerby was that by the physician and botanist William Withering (1776) entitled: *A botanical arrangement of all the vegetables naturally growing in Great Britain. With descriptions of the genera and species, according to the system of the celebrated Linnaeus.* Withering claimed that good plates were too expensive for general use, so his text is without them. He also dropped the sexual distinctions in the Linnean system of classification so that ladies could be involved in botanical study. Appendix B gives publication and other details of *English Botany*.
2. A translation in the Virgil Pocket Book (1907) arranged by S. E. Win Bolt is: 'what produce the fields of their own free-will will bring forth'.
3. James Smith to Thomas Jenkinson Woodward, 29 November 1790. LSS. Woodward was a botanist, and a magistrate and deputy lieutenant in Suffolk. He was elected FLS in 1789 and co-authored, with Samuel Goodenough, *Observations on British fuci* (1797) (B. B. Woodward, 2004).
4. James Smith to Edmund Davall, 17 March 1792. LSS.
5. Thomas Woodward to James Smith, 25 December 1790. LSS.
6. B. B. Woodward (2004).
7. Thomas Woodward to James Smith, 20 March 1791. LSS.
8. James Smith to Thomas Woodward, 2 July 1791. LSS.
9. Thomas Woodward to James Smith, 29 March 1793. LSS.
10. *Gentleman's Magazine*: Letter from RG in volume 63 (1793) p102; James Smith's reply, page 325.
11. James Smith to James Sowerby, 30 March 1797. Sowerby was given as the author on the title pages for volumes 1–4, then as editor for volumes 5 and 6, and then as proprietor for the remaining volumes. NHMS, Box 19.
12. James Smith to James Sowerby, 16 July 1799. NHMS, Box 19.
13. *English Botany* re. Tab 311.
14. Charles Abbot to James Sowerby of 9 November 1798 and 9 December 1798. NHMS, Box 1.
15. *English Botany*, volume 5, 311.
16. *English Botany*, volume 12, 839.
17. *English Botany*, volume 9, 591.
18. *English Botany*, volume 20, 1377. *Conserva aegagropila*, Globe conserva or Moor Balls. 'The specific name alludes to their resembling the hairy balls found in the stomachs of goats.'
19. E.g. Purple rock lichen, *English Botany*, volume 9, 604.
20. One example is in volume 16, number 1149, where the fern genus *Aspidium* (now defunct) is separated from the genus *Polypodium*.
21. E.g. *English Botany*, volume 4, 221, where Linnaeus is considered to have 'wonderfully erred' in misidentifying *Triticum loliaceum* as *T. maritimum*.
22. *English Botany*, volume 12, 857.
23. E.g. *English Botany*, volume 7, 439, regarding Sowerby's observation on the occasional occurrence of double anthers in a *Bryonia* species.

24. E.g. *English Botany*, volume 7, 441, regarding errors over reporting the number of petals in *Euphorbia*.
25. Dawson Turner to James Sowerby, 5 September 1797. NHMS, Box 22.
26. An example is a letter from Reverend Charles Sutton of Norwich on 25 September 1799 to Sowerby asking him to go to the City (Mack Lane) and get £4.12s. from someone who was owing Sutton this money. NHMS, Box 21.
27. Michael Fly, Bicton, Devon, to Sowerby, 2 September 1798. NHMS Box 7, folder A20.
28. Jonathan Salt to James Sowerby, 6 July 1797 and 11 October 1797. NHMS. Box 17, folder A49.
29. Thomas Pennant, Downing Hall, to Sowerby, 20 September 1797. NHMS, Box 15, folder A44.
30. Walker (1988).
31. James Smith to James Sowerby, 14 April 1801. NHMS, Box 18.
32. Letter to James Sowerby, 26 May 1811. NHMS. Cryptogamous plants were included from volume 2 onwards. They comprised most of the plates in many of the later volumes (e.g. 50 of the 72 in volume 34).
33. See chapter 3.
34. Recorded in a letter, 2 September 1806, from James Smith to Samuel Goodenough. Goodenough often took the Chair at Linnean Society meetings when Smith was absent. In this letter Smith also tells of his correspondence about Salisbury with Salisbury's sisters. 'He wrote me also that he had sent my letters, & copies of his, to his sisters, who (he said) were 'positively imperative' with him to publish this pamphlet & he did not believe they would easily think well of me again. On this I wrote to his sisters, & told them my story. They answer that he had never sent them any letters, nor the pamphlet, nor told them anything about the matter except that he was 'likely to wage war with Smith'.' LSS.
35. Salisbury (1806).
36. Ibid note 35. E.g. Number 50 in the pamphlet.
37. Antoine Laurent de Jussieu produced his *Genera Plantarum* in 1789. He used all parts of plants in their classification and so was able to establish better the real relationships between them.
38. See note 12.

6. Models and museum

1. Christian Hendrik Persoon, a mycologist, published his *Synopsis methodica fungorum* in 1801. He had also published a paper, in 1794, on a classification based on the gross form of the fruiting structures of the fungi. Spores had been observed as early as 1588 by Giambattista della Porta but the prevailing view of spontaneous generation of fungi continued for a further two hundred years (Ainsworth, 1976). Even the work of an Italian botanist, Pier Antonio Micheli (1679–1737) who observed that spores could produce the same type of fungi as that from which they came, was treated with scepticism.
2. Boulger (2004b).
3. The methods are reported in *Trans. Linn. Soc. London* 2: 263–266 (1794). A new Method of preserving Fungi etc. Mr Withering MD FRS FLS. (Read to the Society December 4 1792.)
4. The full scale of the project cannot now be properly assessed for many of the models have been destroyed or lost. The British Museum purchased most of them from Sowerby's son, James De Carle, in 1844, and they were later treated by Worthington Smith to deal with the ravages of time, such as dust accumulation and colour fading. Worthington Smith also wrote a guide (1898) to accompany their display at the Natural History Museum in South Kensington. Only about thirty survived bomb damage to their gallery during the Second World War; some of these were renovated in the mid-1990s with the financial help of the British Mycological Society for display at *Fungus 100* in celebration of the Society's centenary in 1996. Blunt and Stearn (1994: 285) state that the models were made from wire, cane, sheet iron, zinc, lead and Coade's artificial stone as well as pipeclay.
5. Thomas Pennant to James Sowerby, 20 September 1797. NHMS.
6. John Hull to James Sowerby, 17 November 1798. NHMS.
7. For example, a notebook records he was collecting specimens in Peckham Wood on 16 November 1794, another record gives Norwich 1783, NHMS, Box 34, folder 52.
8. Francis Humberston Mackenzie (1754–1815) was the first Baron Seaforth. He was a military man and politician who had an interest in the arts and sciences (as did his wife). He was elected FRS in 1794, made a peer in 1797 and was Governor of Barbados from 1800 to 1806 where he reformed slavery. He often provided Sowerby with specimens. See also Chichester (2006).
9. James Brodie, Brodiehouse, Forres to James Sowerby, 9 February 1795 and June 1798. NHMS, Box 4.
10. George Montagu, Kingsbridge, Devon to James Sowerby, 1 July 1801. Montagu (1753–1815) was a keen naturalist and a collector of birds. He wrote the *Ornithological Dictionary; or Alphabetical Synopsis of British Birds*.
11. Payment slips (Day) of 16 and 18 October 1800 and 6 February 1801. UBEC DM1186.
12. Draft letter by James Sowerby of 4 December 1817 to unknown recipient.
13. Hugh Davies to James Sowerby, 23 September 1797, NHMS, Box 6, folder 16.
14. Ibid, Davies to Sowerby, 22 May 1798.
15. Ibid, Davies to Sowerby, 7 August 1799.
16. Ibid, Davies to Sowerby, 15 October 1799.
17. Ibid, Davies to Sowerby, 12 November 1799.
18. Ibid, Davies to Sowerby, 31 January 1800.
19. J. Murray to James Sowerby, 3 June 1802. NHMS, Box 14, folder 42. Murray was in correspondence with Sowerby about producing a work on British Pomona and sent him some fruit and suggestions on how they should be illustrated. Some original illustrations are in the NHM collection.
20. A descriptive catalogue of Rackstrow's

Museum, 197 Fleet Street. 1782. The museum is said to have lasted until the nineteenth century. Wellcome Library, London.

21. *Flora Londinensis,* Fascicle 5, plates 69–72, for example, are of *Agaricus* species drawn by Sowerby.
22. Jacob Schäffer (1718–1790) was a German academic naturalist who published significant works on mycology and entomology, as well as being an inventor. Pierre Bulliard (-1742 1793) was a physician and botanist who published on botany, especially on fungi. His work was well illustrated.
23. *English Fungi*, volume 2: letterpress to plate 221.
24. *English Fungi*, volume 1: letterpress to plate 2.
25. James Smith to James Sowerby, 14 August 1802, in which he says 'I have done your fungi as well as I could'. NHMS, Box 19, folder 56.
26. This he did at the end of 1802 by sending twenty-one handwritten pages of *Remarks on Sowerby's Figures of English Fungi*. B. M. Forster to James Smith, 18 December 1802; LSS. Bound in volume 1 of *English Fungi*. It deals with volumes 1 and 2.
27. Plate 268 (*Lycoperdon aurantiacum*) in volume 3 of *English Fungi* even shows a small illustration of the mixture.
28. Undated handwritten notes in Sowerby's hand. The particular fungus is not specified. NHMS, Box 34, folder 52.
29. Note. NHMS, Box 34, folder 52.
30. *English Fungi*, plate 407 of *Agaricus virosus* – of eight figures as varieties of the species – numbers 2–6 inclusive being varieties of the same, gathered at Mitcham on the spot by Mr Dickson. Plate 408 has a further sixteen figures of varieties.
31. *English Fungi,* letterpress for plate 417, *Agaricus lanatus*, Supplement No III. Mount Edgecumbe is in Plymouth.
32. A bill dated 28 October 1814 from the printer B. Meredith is for 500 sheets of folio foolscap for *English Fungi* at £3.4s., and 500 blue covers for *English Fungi* no. 30 at £1.5s. NHMS, Box 31.
33. Petersen (1983). He lists 256 new species names according to Sowerby; 46 sanctioned by Fries (*Systema Mycologicum)*.
34. They collected in Bristol, Glastonbury Abbey, Oxford, Weymouth, Penzance, Plymouth, Sidmouth, Dunster, Liskeard, Barnstable, Lyme-Regis, Taunton, Lizard Point, Falmouth, Land's End and other locations.
35. Reported to the Linnean Society as: XXIV Catalogue of some of the most rare Plants observed in a Tour through the Western Counties of England made in June 1799 by Dawson Turner, Esq. F.L.S. and Mr James Sowerby F.L.S.
36. Dawson Turner, Yarmouth, to James Sowerby, 24 April 1800. NHMS, Box 22, folder 65.
37. Dawson Turner to James Sowerby, 24 May 1801. In this letter he expresses delight that Sowerby might be disposed to join him in a jaunt to the Isle of Wight. The 'jaunt' occurred but the dates and actions are not known. NHMS, Box 22, folder 66.
38. Anne Sowerby to James Sowerby, 26 June 1799. NHMS, Box 32.

7. Troublesome exotic floras

1. Dawson Turner to James Sowerby, 24 May 1801. See chapter 6, note 37.
2. Chapter 6 where Sowerby expressed his pleasure of the view at Mount Edgcumbe, Cornwall.
3. Arthur De Carle Sowerby (1952).
4. Harris (2007); Lack with Mabberley (1999). Also see paper by Stearn (1967).
5. John Sibthorp (1758–1796) was the son of the Sherardian professor of botany at Oxford University, and he subsequently took this position. He studied botany in Paris and Montpellier and travelled, especially in Greece where he studied the flora and made significant collections.
6. John Hawkins (1761–1841) was a wealthy geologist with special interests and experience in mineralogy and mining. He was born to a family with considerable mining interests in Cornwall. He studied mining under Abraham Werner and mineralogy under Martin Heinrich Klapthorp in Germany. He travelled extensively including with the botanist Dr John Sibthorp, and was one of the executors of his will. He established a fine mineral collection but this was dispersed after his death.
7. Cornwall Record Office, Truro Documents. Ref. RP16/132. Letter from Ferdinand Bauer in Oxford, 31 July (no year) to J. Hawkins. In the letter he writes: 'I think Dr Sibthorp will beging His tour quite in the old style never come to a determination before the last day, then at ones all in haste which is most unblessent.'
8. John Hawkins to James Smith, 11 November 1799. LSS.
9. John Hawkins to James Smith, December 1799. LSS.
10. James Smith to James Sowerby, 9 January 1800. In this Smith writes, 'I am sorry but dont wonder you cannot now undertake the Flora Graeca – However when we meet I shall talk over the plan with you, and if you can help me anyway I know you will.' NHMS, Box 19.
11. James Smith to James Sowerby, 20 October 1800. NHMS, Box 19.
12. John Hawkins to James Smith, 3 November 1801. OUS. 248, folio 48.
13. James Smith to James Sowerby, 28 November 1801. NHMS, Box 19.
14. James Smith to James Sowerby, 22 October 1802. NHMS, Box 19.
15. The property was leased from Mr John Austin, Carpenter of Vauxhall Walk, Lambeth. Thomas G. Knapp at Haberdashers Hall drew up the lease, starting work in May 1803. Sowerby initially sought a lease for 7 or 14 years at his option, but was persuaded by Knapp to take it for 21 years and still with the option of ending the agreement at 7 or 14 years. Sowerby had occupancy before the end of the year and possibly soon after June. Letters Knapp to Sowerby, May 1803. NHMS, Box 11.
16. James Smith, Norwich, to James Sowerby, 20 June 1804. NHMS, Box 19.

17. James Smith to James Sowerby, 23 December 1804. NHMS, Box 19.
18. James Sowerby draft letter to James Smith, 25 December 1804. NHMS, Box 19.
19. John Hawkins to James Smith, 20 January 1805. OUS 248, folio 48.
20. John Hawkins to James Smith, 3 February 1805. OUS 248, folio 48.
21. Seven children, including Anne who was born around 1797 but little is known of her. She may have died very young. The seven excludes Markham who was killed in 1802.
22. See http://www.bodleian.ox.ac.uk/science/eresources/flora_graeca Last accessed 23 March 2015.
23. Harris (2007).
24. The summons is dated 20 July 1803; Sowerby was sworn in as a Special Constable on 4 January 1804 and he was a parish officer in Lambeth during 1803–1804. NHMS, Box 32, folder B10.
25. James Smith to James Sowerby, 5 December 1804. NHMS, Box 19.
26. James Smith to James Sowerby, 23 November 1804. NHMS, Box 19.
27. James Smith to James Sowerby, 8 March 1806. LS. Cranbrook collection of letters.
28. James Smith to James Sowerby, 6 November 1805. He wrote: 'Pray how come the stigma in all the flowers of the Leptospermum ambiguum Exot. Bot. tab 59 not to be capitate, as in the magnified flower? Are they not so? – You will see by Salisbury's letter that if you have made a mistake it will be the most unfortunate one that could be – I am very anxious about it.' NHMS, Box 19, folder 57.
29. Reported in a letter from James Smith to James Sowerby, 16 January 1806, in which he also wrote, 'everybody thinks it is high time yet I did not mean to turn him out'. NHMS, Box 19, folder 57.
30. James Smith to William Roscoe, 25 November 1804. LSS.
31. The Scottish physician Francis Buchanan-Hamilton (1762–1829) was active as a natural historian during the time he lived in India. He made a significant herbarium of plants from Nepal and Mysore, together with nearly two hundred watercolours. He gave these collections to James Smith with the aim that they be published in *Exotic Botany*. Some were indeed used (by redrawing or engraving) by Sowerby, including plates 97–99, 101, 108, 113–117 (LSS. Letter from James Smith to W. Goodenough, 2 September 1806, and personal communication by Mark Watson, January 2012). The original watercolours, by unknown artists, are in the collections of the Linnean Society, London. Comparison of the art work shows that in some cases (e.g. the root and leaves of the plant in plate 108 of *Exotic Botany*) Sowerby's version lacks some of the finesse and detail of the original.
32. James Smith to James Sowerby, 4 February 1807. NHMS, Box 19.
33. James Smith to William Roscoe, 14 August 1808. LSS.
34. James Smith to William Roscoe, 16 October 1806. LSS.

8. Sowerby's whale

1. Waller (2013) gives further information about the description of the whale and some subsequent published discussions and criticisms. The skull of the type specimen is now in the Oxford University Museum of Natural History (specimen number 06998).
2. Dewhurst (1834).
3. *Asterias equestris*, plate 63; *Millepora compressa*, plate 41 – now named *Porella compressa*.
4. *British Miscellany*, plate 68, *Lernaea spratta* (now named *Lernaeenicus sprattae*). Professor Geoff Boxshall of the Natural History Museum, London, who is an authority on these organisms (copepods), advised on this description and commented that Sowerby was an accurate observer and an excellent illustrator. (Personal communication, 27 November 2012.)
5. *British Miscellany*, letterpress to plate 59, *Carabus chrysostomos*; Golden-mouthed Carabus.
6. Francis Humberston Mackenzie (1754–1815) was the first Baron Seaforth. He was elected FRS in 1794 and made a peer in 1797. He pursued a military and political career and with an interest in natural history. He was Governor of Barbados, 1800–1806, where he made some reforms to slavery.
7. *British Miscellany*, plate 61.
8. *British Miscellany*, plate 8.
9. George Montagu published his *Ornithological Dictionary* in 1802. It soon gained recognition as a work of high quality. In 1803 he published *Testacea Britannica*, an important scientific work on British molluscs.
10. George Montagu corresponded with Sowerby on several topics, including about species of fish and seeking guidance and materials for producing plates for his publications. See also Cleevely (1978).
11 Reverend J. Bale, near Tunbridge Wells to James Sowerby, 23 December 1802. NHMS, Box 2, folder 4. The *Strix otus* is the long-eared owl now named *Asio otus*.
12. Letterpress to plate 36 in *British Miscellany*, and letters (5 May to 18 May 1803) from George Sowerby to his parents; NHMS.
13. Moore (2004).
14. W. Kirby to Sowerby, 8 January 1806, in which he sends descriptions of two insects that 'required a good deal of attention'. *Carabus chrysostomos* (plate 59) and *Staphylinus concolor* (plate 54). NHMS.
15. Letterpress to plate 45 (*Stylops melittae*) of *British Miscellany*, plate dated 1 December 1805.
16. W. Kirby to James Sowerby, 26 April 1806. NHMS.
17. W. Kirby to Sowerby, 31 January 1801 asking him to draw some bee specimens for his monograph *Monographia Apum Angliae*, which was published in 1802 (see appendix C). NHMS.
18. George Sowerby to his 'Honoured Father', 8 September 1802. NHMS.
19. Dawson Turner to James Sowerby, 19 October 1803, in which he says he is pleased with the two plates Sowerby has sent and also that people are

'frightened out of their wits about the French & most families who can afford it, are removing their wives and children out of the Town'. Later (30 November 1803), he again chased Sowerby for more plates, saying that he planned to sell three hundred copies of the book. NHMS.

20. Addition by Dawson Turner to letter from James De Carle Sowerby (Yarmouth) to his father, 15 August 1803. NHMS.
21. James De Carle Sowerby (Yarmouth) to his father, 28 August 1803. NHMS.
22. The original watercolour drawings are held in the Natural History Museum, London.
23. Dawson Turner to Sowerby, 3 and 19 October 1803 and 30 November 1803, NHMS.
24. BBC News, Highland and Islands. 18 September 2013. http://www.bbc.co.uk/news/uk-scotland-highlands-islands-24143838. Last accessed 24 March 2015.

9. Minerals, crystals and colours

1. Examples include the Ecole des Mines in Paris from 1783 and the Mining Academy at Freiburg, Saxony from 1765.
2. Johann Gottfried Schmeisser FRS, FLS (1767–1837) was a German apothecary, chemist and mineralogist. He stayed in London during an extended journey to England in 1794 and in the same year published *A System of Mineralogy* (1794–1795) which was based on the work of Cronsted. It gave a systematic listing with properties of minerals, rocks and metals and methods for analysis. Curtis P. Schuh. *Bibliography of Mineralogy*, available on the Mineralogical Record website: www.minrec.org/library.asp. Last accessed 24 March 2015.
3. The Plymouth City Museum and Art Gallery, Devon, England, holds the Sir John St. Aubyn collections of minerals and rocks. For details see the Museum webpage: http://www.plymouth.gov.uk/museumstaubyncollection. Last accessed 26 March 2015.
4. W. Day to P. Rashleigh, 21 July 1794, Cornwall Record Office, Rashleigh papers, DDR5764/9.
5. James Smith to James Sowerby, 22 October 1802, NHMS, Box 19, folder 56.
6. Two sheets had to be provided to avoid the stamp duty imposed on one-sheet publications. Sowerby informed his potential subscribers through notes printed on the blue wrappers:

   > 'The first Numbers are an Example of the Execution of the Figures, and the whole is intended as a Companion to English Botany' and 'The Figures will be executed from actual Specimens and interesting Parts made conspicuous. The Letter-press will be as intelligible as possible. Reference will be made to scientific Authors, especially Linnaeus, as far as can conveniently be done'.

   He also took the early opportunity to ask his friends to send him specimens and to thank them for their assistance. The title page of the volumes includes the words '(With Assistance)' after Sowerby's name, so giving due recognition to the help he received and no doubt needed.
7. For example, in his *Manual of Mineralogy* Aikin wrote

   > that while the thousands and almost the tens of thousands of species which solicit the attention of the Zoologist and Botanist are very capable of proper graduated arrangement... the Mineralogist, who may draw his distinction from external appearance and impressions, from physical properties or chemical phenomena, has hitherto been completely baffled in his attempts to bring under a similar arrangement the two or three hundred species that form the objects of his study. (Aikin, 1814: 50)

   Sowerby cited the problem several times and stressed the need to become acquainted with the different forms, e.g. letterpress for plate 378, stellated tourmaline.
8. Burchard (1994). In his *Manual of Mineralogy*, Aikin (1814) gives an account of the use of the blowpipe and instructs the reader on how to train himself to blow consistently without fatigue – using the tongue and the nostrils and so forth (as one might for certain musical wind instruments).
9. Letterpress to plate 120.
10. Plate 170. Plumbum sulphatum. An amorphous decomposed form of galena.
11. Letterpress to plate 301.
12: Hallam (1989: ch 1) gives an interesting discussion on the issues of the debate, in his book on great geological controversies.
13. Letterpress for plate 280, a copper hydrate.
14. Letterpress for plate 398, Sulphuret of zinc. The text includes:

    > I chose the present specimen as it would elucidate some other subjects if attentively considered. The matrix is chiefly Quartz, which has probably formed over Cubic Fluor, in a very loose and rough manner, as may be seen by the impression; and after the Fluor was gone, Galena or Sulphuret of Lead has formed in a scattered manner about it. This Galena has, however, mostly passed away, and left Brown Pearlspar in thin walls, which, having covered the cubes, are at right angles in the various positions in which the Galena originally lay.
15. Letterpress for plate 151. Echinus-formed carbonate of lime.
16. For example, the letterpress for plate 62 cites Lord Heathfield, the Duke of Athol and Lord Henry Murray.
17. Wilson (1994: 78).
18. Greville's collection was purchased in 1810 by a special vote of £13,727 (W. C. Smith, 1978).
19. Cleevely (2004a).
20 Rashleigh, Menabilly, near Fowey, Cornwall, to Sowerby, 13 June 1804. UEBC, DM 1186.
21. Rashleigh, Menabilly, near Fowey,

Cornwall, to Sowerby. 30 March 1804. NMHS, Box 16, folder 47.
22. Rashleigh, Menabilly, near Fowey, Cornwall, to Sowerby, 21 April 1804, NHMS, Box 16, folder 47.
23. Wilson (1994: 79).
24. Gilbert Laing in Edinburgh often sent ore specimens to Sowerby and reported, in 1806 (letter 18 March, NHM Sowerby collection), on his experiments on basalt. Henry Penneck of Penzance warned Sowerby (letter 3 October 1806, NHMS) that the cost of sending the granite post would be high and asked for clearer instructions as to what he should do. The Sheffield cutler Jonathan Salt (1759–1810) made several contributions to *English Botany*. His correspondence with Sowerby stretched from 1797 to 1808. He was a Fellow of the Linnean Society (1797, proposed by Sowerby). The letter regarding the bloodstone is dated 14 August 1797. NHMS, Box 17.
25. *The Annual Review and History of Literature*. Editor Arthur Aikin. Vol. II for 1803, published 1804, London. Under Chapter XXIV. Mineralogy (page 926) Art. II.
26. Benjamin Forster to James Smith, 18 December 1802. Linnean Society. Bound in volume one of Smith's copy of Sowerby's *English Fungi*.
27. Edward Clarke, Cambridge to James Sowerby, postmark 19 November 1804. UBEC, DM 1186.
28. *British Mineralogy*, letterpress to plate CCCCXVII.
29. Wilson (1994).
30. The letterpress for *British Mineralogy* plate 58, Pit coal, states: 'we were favoured by Mr. E.D. Clarke of Jesus College, Cambridge in February 1804, with specimens of Lynn Coal, presenting pentahedral prisms, which he observed in it for more than a year past'.
31. M. Gent, Devizes to Sowerby, 11 March 1816. NHMS, Box 8, folder 22.
32. In his *New Elucidation of Colours* Sowerby cited work by Thomas Young and by William Wollaston, particularly those published in the *Philosophical Transactions of the Royal Society* in 1801 and 1802.
33. It is relatively easy to follow Sowerby's instructions on viewing the chromatometer in sunlight and to observe the colours in the sequence he described. To test his method a standard six-inch, triangular glass prism was used.
34. William Kirby to James Sowerby, 17 April 1810. NHMS, Box 10.
35. Various advertisements were produced. The one used here was printed on the back cover of his work of crystal models.
36. Draft biography of his father by James De Carle Sowerby. Simpkins (1974a).
37. Letters George Sowerby, the first to his father undated but likely to be late 1808, second to both parents 26 July 1810. NHMS, Box 33, folder B23.
38. G. B. Sowerby, 2 Mead Place to his father, February 6 1811. NHMS, Box 33, folder B23.

10. Meteorites and a 'sword from heaven'

1. Pillinger & Pillinger (1996). This paper gives an informative and amusing account of the events surrounding the Wold Cottage meteorite fall.
2. James Sowerby, *British Mineralogy*, volume 2, pages 1–19. The number or part (XXVI) appeared in December 1804.
3. Copy of letter J. Sowerby to Major Topham, no date. NHMS, Box 32, B17, folder T.
4. One influential account was the privately published paper by Edward King FRS on *Remarks concerning stones said to have fallen from the clouds, both in these day and in ancient times* (1796). King preferred a volcanic origin.
5. Wrapper for number XXVIII, February 1806, *British Mineralogy*.
6. The folio plate is entitled 'Meteorolites' and dated 25 May 1812. It shows three specimens: Wold Cottage (landed 13 December 1795); Possil, Glasgow, Scotland (5 April 1804); and Tipperary, Ireland (August 1810). Five small illustrations depict the components such as globules and pyrite grains. The specimens were all in Sowerby's own collection. Pillinger & Pillinger (1996) record some details of the folio plate at Wold Cottage.
7. Printed card of no date headed *METEOR STONE*, NHMS, Box 37, folder B110.
8. Grossman (2010). In August 1810 the Mooresfort meteorite fell in Co. Tipperary. Sowerby requested a sample on 2 May 1811.
9. Charles Panse to James Sowerby, 9 April 1820. Royal Society. MS/709.
10. The spasmodic correspondence was over several years; Clarke to Sowerby, 16 February 1811; 26 January 1812 and 15 February 1815. In his letter of 1815 Clarke wrote also that 'the Geological Society have assisted me with all Mr Ferguson's Aerolites – so that your single specimen could be of no use, and it would be now too late'. NHMS, Box 5, folder 13.
11. Dawson Turner to James Sowerby, 9 October 1803. NHMS, Box 22, folder 66.
12. Henderson (2013) gives an account of the manufacture and subsequent history of the sword.
13. Letter from Mrs Stanley, wife of Edward Stanley, to her sister-in-law, Lady Maria Stanley, 16 June 1814 (Adeane & Grenfell, 1906: 86).
14. Sowerby received a bill on 4 May 1814 'To Man's Time cutting & polishing a Piece of Iron' for which he was charged 4s.6d., from 'Henry Maudslay & Co. at their Patent Press Steam Engine and Machine Manufactory, Cheltenham Place, Lambeth.' NHMS, Bills, Box 31.
15. NHMS, Box 37, folder B121.
16. Letter from Mrs E Stanley to Lady Maria Stanley. London, Monday June 23 1814 (Adeane & Grenfell, 1906: 93).
17. The description of the contents of the package is based on pencilled draft notes in Sowerby's hand. The text of the letter of 3 July 1814 from James Sowerby to the Emperor is from a copy made by Sowerby's son, George. NHMS, Box 37, folder B120.
18. It was Henderson (2013) who made the search.
19. Kosadawlew to Sowerby. 16/28 May

1819. NHMS, Box 37, folder B120.
20. William Smith (1769–1839). He recognised that geological strata could be identified by the particular fossils they contained. In 1815 he published the first geological map of much of Britain. It was several years before his achievement was given the recognition that it deserved, leading to him being called 'the Father of English Geology'.
21. James Ryan (c.1770–1847) demonstrated his equipment at No 2 Mead Place, 11 April 1808, before Sir Joseph Banks and some members of the Geological Society. He drilled with an 8 inch diameter tool to a total depth of 9 feet 6 inches. Nineteen people were in attendance. A section of strata based on this demonstration was drawn. NHMS, Box 32, folder B125. Torrens (1997, 2004b) gives details of Ryan's life and inventions.
22. Appendix C gives information on these publications.

11. A lot of rot

1. Ainsworth (1976: 91).
2. Letter 15 July 1812, J. Knowles, Navy Office, to Sowerby; letter 17 July from Navy Office to Sowerby. NHMS, Box 36, folder 95.
3. J. Knowles, Navy Office, 17 July 1812 to Sowerby. NHMS, Box 36, folder 95.
4. J. Knowles, Navy Office, to Sowerby, 18 July 1812. NHMS, Box 36, folder 95.
5. Ramsbottom (1937); Morriss (2004). Ramsbottom's paper is a good resource for detailed information on the problem of dry rot in HMS *Queen Charlotte* and of Sowerby's part in dealing with it.
6. Notebook containing travel log written in Sowerby's hand, NHMS, Box 36, folder 96.
7. Names of these species are now known as follows: *Boletus hybridus* now *Fibroporia vaillantii*; *Boletus medulla-panis* now *Perenniporia medulla-panis*; *Auricularia pulverulenta* now *Serpula pulverulenta*; *Boletus lacrymans* now *Serpula lacrymans.*
8. Sowerby first wrote from Plymouth to the Commissioners on 25 July 1812, noting how the rot was 'deeply planted'. His official report to the Commissioners was on 4 August 1812 and of eleven foolscap pages. National Maritime Museum, MSS: ADM/BP/32C. Notes and copies: NHMS, Box 36, folder 95.
9. Ramsbottom (1937) gives the full report.
10. Fungi reported by Sowerby in his report were *Boletus hybridus* (featured in *English Fungi*, plate 289) 'being the young state of *Boletus medulla-panis'* (*EF* 326), *Xylostroma giganteum* (*EF* 358), *Auricularia pulverulenta* (*EF* 214) and some remains of *Boletus Lacrymans* (*EF* 113). Currently, *Auricularia pulverulenta* is a synonym of *Serpula pulverulenta*, a well-known dry rot fungus, and *Boletus hybridus* is a synonym of *Fibroporia vaillantii.* See note 7.
11. Lionel Lukin letter to Sir William Rule, 25 September 1811. In this he recommends removing decayed wood and 'then introduce between each Frame in each Deck, Mundic, in the state of Vapour which in every instance & situation is destructive to all kinds of vegetation from its containing a large portion of Arsenic; and will stop Fermentation from the considerable portion of sulphur it contains'. National Maritime Museum. MSS ADM/BP/31B. Such a recommendation could potentially be extremely hazardous for any men employed in implementing it.
12. Letter from officers of the Plymouth Yard to Navy Office, 8 August 1812. National Maritime Museum MSS. ADM/BP/32C.
13. J. Knowles, Navy Office to Sowerby, 24 August 1812. NHMS, Box 36, folder 95.
14. Navy Office to Sowerby, 28 August 1812. NHMS, Box 36, folder 95.
15. Lionel Lukin to Navy Board, 5 August 1812. National Maritime Museum MSS ADM/BP/32C.
16. Reported in Knowles (1821).
17. James Sowerby to his son James De Carle, in Yarmouth, undated. NHMS, Box 38, folder C15.
18. Sowerby's report to Navy Board, 17 February 1813. NHMS, Box 36, folder 95; see also Ramsbottom (1937).
19. Report James Sowerby to Navy Board, 29 October 1813, NHMS, Box 36, folder 95.
20. Copy of Sowerby's report on his past activities to the Navy Board, December 1813. Natural History Museum Botany library, folder of report and correspondence.
21. Letter from Navy Office to Sowerby, 5 January 1814. NHMS, Box 36, folder 95.
22. An undated rough draft of a letter, in Sowerby's hand, to the Commissioners of His Majesty's Navy on his objecting to the fact that the Board did not attach any merit to him saving the ship. He wrote, 'I really should be failing in duty to my Country & myself not to make an observation on this statement... On examining the ship I saw the misfortune still in an aggravated state & suggested what might be done'. NHMS, Box 36, folder 95.
23. J. Knowles to Sowerby, 1 January 1814 and Sowerby's draft reply. NHMS, Box 36, folder 95.
24. Minute of 24 December 1813. Navy Board minutes. The National Archives, ADM 106/2679.
25. Minute of 31 December 1813. Committee of Surveyors, letters and minutes. The National Archives, ADM 92/1.
26. Minute of 31 December 1813. Navy Board minutes. The National Archives, ADM 106/2679.
27. Letter from James Sowerby to John Knowles, 6 July 1815 regarding sending a bill for the colouring of the hygrometers on 23 January and not having seen nor heard since. He asks if it can be found and paid. NHMS, Box 36, folder 95.
28. Dawson Turner to James Sowerby, 6 June 1813. NHMS, Box 22, folder 66.

12. Mineral conchology

1. James Sowerby, Norwich, to James De Carle Sowerby, London, 8 August 1797 on small piece of paper. NHMS, Box 38, folder C15.

2. For example, in August 1808 Sowerby had written to the Reverend Francis, Vicar of Roughton in Norfolk and a Fellow of the Linnean Society, to ask if there were any fossils in his area and saying 'I have begun drawings of fossils shells for publication'. NHMS, Box 7, folder 20.
3. Etheldred Benett (1776–1845). A short account is given by Laming & Laming (2007).
4. George Cuvier's major work on the strata of the Paris basin was published with Alexandre Brongniart in 1811: *Essais sur la géographie minéralogique des environs de Paris, avec une carte géognostique et des coupes de terrain*. He later gained additional fame from his work on fossil reptiles.
5. Major roadworks were taking place over several years at the Archway, Highgate in north London, involving extensive digging and exposure of the London Clay. By 1815 there were detached houses and gardens along most of Hornsey Lane to the new Archway, which had been built in 1813. Archway Road goes under the arch and was also 'built' in 1813. There was to be a tunnel but this fell in April 1812.
6. An example of a specific letter on these reports would be: Gideon Mantell, Castle Place, Lewes to James Sowerby, 31 December 1816, Mantell Archive, Alexander Turnbull Library, Wellington, NZ.
7. Sowerby to Mantell, 5 May 1813 and 7 June 1815 respectively. Mantell Archive, Alexander Turnbull Library, Wellington, NZ.
8. Elliott (1975).
9. *Ammonites duncani*, page 129, Tab CLVII.
10. Charles Adrien Dehérissier De Gerville (1769–1853) of Valonges, Normandy. Cleevely (2012) gives an account of the collaboration with the Sowerbys and its significance.
11. C De Gerville to Sowerby, 27 July 1818. UBEC, DM 1186.
12. *Inoceramus* is an extinct genus of bivalves, in which some species attain considerable size. Sowerby gave his talk to the Linnean Society in November 1814 but it was not published in the *Transactions* until 1822, probably through delays by Sowerby in preparing the finished article. His article also discusses growth rates and the hinge structure in some detail.
13. Correspondence between James Sowerby (in London) and his son James De Carle (in Yarmouth) was over a two-month period from about 30 August to 2 November 1812. NHMS, Box 38, folder C15.
14. James De Carle Sowerby to his father, 22 June 1813. NHMS, Box 33, folder B22.
15. This was a kind of catalogue with short descriptions and over one hundred coloured illustrations previously used in *British Mineralogy* and *Mineral Conchology*.
16. Thomas Purton, Alcester, to James Sowerby, 3 November 1816. Royal Society, London, MS/709/48.
17. James Brodie to Sowerby, 11 December 1815, with draft reply by Sowerby. NHMS, Box 4, folder 10.
18. Sowerby to Gideon Mantell, 19 September 1814. Mantell Archive, Turnbull Library, Wellington, NZ.
19. Sowerby's draft reply written on letter from De France, 18 March 1817. NHMS, Box 32, folder B19.
20. Lord Northampton, Castle Ashby to Sowerby, 28 April 1817 and his draft reply. NHMS, Box 15, folder 43.
21. Charles Lyell to Sowerby, 1 and 13 March 1811. NHMS, Box 12, folder A36.
22. James Smith to Sowerby, 27 October 1813, LSS vol. 25.
23. Hugh Davies to Sowerby, 3 May 1816 together with Sowerby's draft reply. NHMS, Box 6, folder 16.
24. James Smith to Sowerby, 15 December 1815. NHMS, Box 20, folder 58.

13. Societies philosophical and chemical

1. Printed letter from T. J. Pettigrew to Sowerby on his election as an honorary member on 18 January 1811; handwritten letter from Pettigrew to Sowerby on 22 January 1811 expressing his pleasure and reasons for nominating him; with Sowerby's pencilled draft reply. Other administrative papers. NHMS, Box 12, folder 34 and Box 32, folder B17.
2. Sowerby was a vice-president in 1817 and was re-elected on 21 October 1819, so although he wrote his resignation letter it seem likely that he withdrew it later. NHMS, Box 12, folder 34.
3. Ratchford (1925).
4. Noehden, Albany Piccadilly to Sowerby, 21 August 1816. NHMS, Box 15, folder 43.
5. Clark (2000).
6. NHMS, Box 6, folder 18.
7. For example, Davy had isolated sodium and potassium in 1807.
8. NHMS, Box 43, folder 106; Averley (1986).
9. Averley (1986) gives an interesting and detailed account of the Lambeth Society. Also: NHMS, Box 11, folder 31. MSS in Lambeth Archives; Lambeth Boroughs archives and records at Minet Library, Lambeth, London.
10. NHMS, Box 11, folder 31.
11. Arthur De Carle Sowerby (1952).
12. John Mawe (1766–1829) was a mineralogist and mineral dealer, with shops first in Covent Garden and later in the Strand, London (Torrens 2004a). He produced several publications including *Travels in the Interior of Brazil* (1812, with a plate by Sowerby), *A Treatise on Diamonds* (1813) with two coloured plates, and his successful *Familiar Lessons on Mineralogy and Geology* (1819) which ran to several editions. It included a coloured plate of twenty-two mineral figures. Notes, drawings and proof plates indicate that Sowerby was involved in the production of the coloured plate for at least the second edition. Sowerby also provided coloured plates of some gems, e.g. sapphire, and was probably involved in the later editions of *A Treatise on Diamonds* with two coloured plates, and with Mawe's work on the use of the chemical blowpipe. NHMS, Boxes 29, folders A95 and 96, 32, folder B17 and 35, folder B83. See also Wilson (1994: 79).
13. For details of the four papers see Appendix B.

14. Examples of papers read by Sowerby but not published in the *Transactions* are: 'Note on some objects of wax-like substance called Asterias Arberis and Nidus Licheniformis' (1799); 'A short account of the long diagonal fracture etc. in carbonate of lime' (1812); 'On rot and decaying wood' (1813); 'Notice relating to several gelatinous substances commonly known as Star-shot, Star-slough etc. and supposed to be of Meteoric origin' (1814). His papers were often accompanied by illustrations.
15. NHMS, Box 30. The bill probably relates to the plate in the paper by John Adams on 'Descriptions of some minute British shells'. Also, Jonas Dryander (librarian to Sir Joseph Banks and contributing author to Aiton's *Hortus Kewensis*) commissioned Sowerby to do one of the illustrations in his paper on the Begonia genus.
16. D. Turner (1808). 'Descriptions of eight new British lichens', *Trans. Linn. Soc. London* 9: 137–150.
17. T. Walford (1808). 'Some observations on an insect that destroys the Wheat, supposed to be the Wireworm'. *Trans. Linn. Soc. London* 9: 156–159.
18. J. E. Smith (1798). 'The characters of twenty new genera of plants.' *Trans. Linn. Soc. London* 4: 213–223.

14. Passing the baton

1. The portrait is a watercolour by the fashionable portrait painter Thomas Heaphy (1775–1835). An engraving of it was made by another portrait painter, J. C. Edwards, and printed by Richard Taylor. Simpkins (1974a).
2. Cited in Cleevely (2004c).
3. W. J. Hooker, Halesworth, to Sowerby, 8 January 1817; NHMS, Box 10, folder 28.
4. Dr James Smith, physician and botanist, was an editor of the *Botanical Magazine*. Correspondence (NHMS) suggests that Sowerby was occasionally asked to do some work for the *Botanical Magazine*. For example, 8 March 1817, Sowerby was asked to go to Joseph Banks at Soho Square to make a drawing of a 'new *Solandra*' from Chelsea Physic Garden. NHMS, Box 18, folder 53.
5. Correspondence between James Smith and James Sowerby, 26 March to 24 May 1820; NHMS, Box 20, folder 59.
6. Sowerby, Mead Place, to Charlotte in Norfolk, around 15 July 1817, draft response to her letter from Bury, 15 July 1817. NHMS, Box 38, folder C15.
7. Sowerby, Norwich to James De Carle, London, 12 August 1817. NHMS, Box 38, folder C15.
8. George Sowerby, 3 Mead Place to his father, 2 Mead Place, 25 December 1817. NHMS, Box 33, folder B23.
9. George Sowerby, 3 Mead Place to his father, 2 Mead Place, 24 February 1818. NHMS, Box 33, folder B23.
10. Sowerby to De Gerville, November 1818. Société des Antiquaries de Normandie, Caen. Cited in Cleevely (2012).
11. Arthur De Carle Sowerby (1952). Maria was the daughter of Admiral John Reynolds (1713–1788), one-time Governor of the Colony of Georgia in North America.
12. James De Carle, Paris to his father, 25 June 1820. NHMS, Box 33, folder B22. James De Carle to Charlotte Sowerby in Paris, 1 August 1820 (in French) and 8 September 1820. NHMS, Box 48, folder G1.
13. Mary Sowerby, Mead Place to James De Carle Sowerby (undated). NHMS, Box 39, folder C19.
14. Charlotte, Paris to Sowerby family, Mead Place, 2 October 1820. NHMS, Box 25, folder A75.
15. Sowerby draft reply to letter from Dr A. L. Emerson to Sowerby, 20 January 1821. NHMS, Box 6, folder 18.
16. *Gentleman's Magazine*, 1822, page 568.
17. James Smith, Holkham Hall to James De Carle Sowerby, Mead Place, 31 October 1822. NHMS, Box 20, folder 59.
18. Simpkins (1974a).
19. Arthur De C Sowerby (1952). The tombstone is no longer there, nor is the exact site of his grave known.
20. Sir Thomas Gery Cullum, Bury, to James Smith, Norwich, 11 November 1822. LSS. Cullum published his *Florae Anglicae* in 1774.

Epilogue

1. UK Public Record Office. The National Archive. Prob 11/1664.
2. A copy of the sale catalogue in the British Library shows that the sale was at Mead Place over four days from 23 June 1831, with the auctioneers Messrs. Thomas & Stevens. The catalogue was entitled: *A Catalogue of the Remaining Portion of the celebrated collection of the Late Mr. James Sowerby*. There were nearly eight hundred lots, over five hundred of them being minerals. Other lots dealt with shells, birds, insects, fish, reptiles and books. The herbarium was not included.
3. MacDonald (1974) gives some details of the purchases by the Natural History Museum. She also gives some biographical information on Sowerby's sons, James De Carle and George Brettingham. Sowerby's herbarium was purchased from James De Carle in 1859 for £100, and the collection of fossils in 1861. Details of purchases by the Natural History Museum of Sowerby-related documents are in the Sowerby Collection file in the Museum Archives (DF601).
4. Saunders (1995).
5. Joseph Banks to James Smith, 25 December 1817, LSS.
6. For details of *Mineral Conchology* and a discussion on these copies see Cleevely (1974b).
7. Elliott (1975).
8. From: *Fungi. A Look at Coloured Figures of English Fungi or Mushrooms by James Sowerby F.L.S. London 1797*, by Sir Sacheverell Sitwell (23.iv.1972), who was the sixth Baronet of Renishaw Hall, and younger brother of Dame Edith Sitwell and Sir Osbert Sitwell. Permission from the Sitwell family to quote this extract is gratefully acknowledged.
9. This example is taken from Elliott (1975).
10. Reverend J. H. Dickenson, Staffordshire, to James Sowerby, 27 July 1812. NHMS, Box 6, folder 17.

Bibliography

This list does not include the works by Sowerby or by his first-generation sons. See the appendices for these.

Adeane, J. H. & Grenfell, M. (eds) (1906). *Before and After Waterloo. Letters from Edward Stanley, sometime Bishop of Norwich, (1802; 1814; 1816).* T. Fisher Unwin, London.

Aikin, Arthur. (1814). *A Manual of Mineralogy.* Taylor, London.

Ainsworth, G. C. (1976). *Introduction to the History of Mycology.* Cambridge University Press, Cambridge.

Allen, D. E. (2004). Salisbury, Richard Anthony (1761–1829). *Oxford Dictionary of National Biography.* Oxford University Press, Oxford.

Allen, Thomas (1827). *The History and Antiquities of the Parish of Lambeth, and the Archiepiscopal Palace, in the county of Surrey.* London.

Altick, Richard Daniel (1978). *The Shows of London.* Harvard University Press, Cambridge, Mass & London.

Anon (1804). *Grove-Hill. A rural and horticultural sketch.* London.

Averley, Gwen (1986). The 'social chemists'. English chemical societies in the eighteenth and early nineteenth century. *Ambix* 33: 99–128.

Babington, William (1795). *A Systematic Arrangement of Minerals founded on the joint consideration of their chemical, physical, and external characters: reduced to the form of tables and exhibiting the analysis of such species as have hitherto been made the subject of experiment.* London.

Bell, William George (1912). *Fleet Street in Seven Centuries being a history of the growth of London beyond the walls into the western liberty, and of Fleet Street to our time.* London: Sir Isaac Pitman and Sons Ltd.

Bjelic, Dusan & Lynch, Michael (1992). The work of a (scientific) demonstration. Respecifying Newton's and Goethe's theories of prismatic colour. In: *Text in Context. Contributions to Ethnomethodology*, eds. G. Watson and R. M. Seiler, pp. 52–78. Sage Publications.

Blunt, Wilfrid (1950). *The Art of Botanical Illustration.* Collins, London.

Blunt, Wilfrid & Stearn, William T. (1994). *The Art of Botanical Illustration.* (New edn.) Antique Collectors' Club with Royal Botanic Gardens, Kew.

Boulger, G. S. (2004a). Davall, Edmund (1763–1798). *Oxford Dictionary of National Biography.* Oxford University Press, Oxford.

Boulger, G. S. (2004b). Forster, Benjamin Meggot (1764–1829). *Oxford Dictionary of National Biography.* Oxford University Press, Oxford.

Boulger, G. S. (2004c). Fraser, John (*bap* 1750, *d* 1811). *Oxford Dictionary of National Biography.* Oxford University Press, Oxford.

Boulger, G. S. (2004d). Smith, Sir James Edward (1759–1828). *Oxford Dictionary of National Biography.* Oxford University Press, Oxford.

Bowden, Jean K. (2004). Lightfoot, John (1735–1788). *Oxford Dictionary of National Biography.* Oxford University Press, Oxford.

Britten, J. & Woodward, B. B. (1905). L'Héritier's Botanical Works. *J. Bot.* 43: 266–73, 325–9.

Brook, Anthony (2001a). Unto the Fourth Generation. *Outcrop* 23: 4–6.

Brook, Anthony (2001b). Patriarch and Senior Son. *Outcrop* 24: 24–32.

Brunton, Deborah (2004). Woodville, William (1752–1805). *Oxford Dictionary of National Biography.* Oxford University Press, Oxford.

Bulliard, P. (1791). *Histoire des Champignons de la France ou traité elémentaire renfermant dans un ordre méthodique les descriptions et les figures des champignons qui croissent naturellement en France.* Paris.

Burchard, R. (1994). The history and apparatus of blowpipe analysis. *Mineralogical Record* 25: 252–77.

Carruth, Nick, Cleevely, Ron & Green, David I. (2007). A Scorodite Specimen Figured by James Sowerby and Analysed by William Gregor. *UK Journal of Mines and Minerals* 28: 4–6.

Carter, Harold B. (1988). *Sir Joseph Banks 1743–1820.* British Museum (Natural History), London.

Chichester, H. M. (2006). Mackenzie, Francis Humberston, Baron Seaforth and Mackenzie of Kintail (1754–1815). *Oxford Dictionary of National Biography.* Oxford University Press, online edn, May 2006 [http://www.oxforddnb.com/view/article/14126. Last accessed 27 March 2015]

Clark, Peter (2000). *British Clubs and Societies 1580–1800. The origins of an associational world.* Oxford University Press, New York.

Cleevely, R. J. (1974a). A provisional bibliography of natural history works by the Sowerby family. *J. Soc. Bibliogr. Nat. Hist.* 6: 482–559.

Cleevely, R. J. (1974b). The Sowerbys, the *Mineral Conchology*, and their fossil collection. *J. Soc. Bibliogr. Nat. Hist.* 6: 418–81.

Cleevely, R. J. (1976). The Sowerbys and their publications in the light of the manuscript material in the British Museum (Natural History). *J. Soc. Bibliogr. Nat. Hist.* 7: 343–68.

Cleevely, R. J. (1978). Some background to the life and publications of Colonel George Montagu (1753–1815). *J. Soc. Bibliogr. Nat. Hist.* 8: 445–80.

Cleevely, R. J. (2000). The contributions of a trio of Cornish geologists to the development of 18th century mineralogy. *Transactions of the Royal Geological Society of Cornwall* 22(3): 89–120.

Cleevely, Ron J. (2004a). Rashleigh, Philip (1729–1811). *Oxford Dictionary of National Biography*. Oxford University Press, Oxford.

Cleevely, Ron J. (2004b). Sowerby, James (1757–1822). *Oxford Dictionary of National Biography*. Oxford University Press, Oxford.

Cleevely, Ron J. (2004c). Sowerby, James De Carle (1787–1871). *Oxford Dictionary of National Biography*. Oxford University Press, Oxford.

Cleevely, R. J. (2011). *Collecting the New, Rare and Curious*. Devon and Cornwall Record Society, Exeter.

Cleevely, R. J. (2012). The collaboration of the French naturalist Charles De Gerville with the Sowerby family and its contribution to early nineteenth-century geology. *Arch. Nat. Hist.* 39(1): 77–110.

Cleevely, R. J. & Bristow, C. M. (2003). Revd William Gregor (1761–1817): The Contributions of a Cornish Cleric and Analytical Chemist. *Journal of the Royal Institution of Cornwall*. 85–108.

Coan, E. V., Kabat, A. R. & Petit, R. E. (2009). *2,400 Years of Malacology* (7th edn). American Malacological Society. [Latest edition available online from the society's website http://www.malacological.org/2004_malacology.html. Last accessed 27 March 2015]

Collins, J. (1969). *Supplementa Sowerbiana; or a catalogue of books and manuscripts written or illustrated by members of the Sowerby family*. Issued as a supplement to Bernard Quaritch's Catalogue no. 894. London.

Collins, John (1973). *A note on the history of the Sowerby family archive*. J. S. Seaton & Co., London.

Conklin, Lawrence H. (1995). James Sowerby, his publications and collections. *Mineralogical Record* 26: 85–105.

Cooper, Michael P. (2001). Keeping it in the family: the Humphreys, Forsters and Heulands. *Matrix* 9: 3–31.

Cooper, Michael P. (2006). *Robbing the Sparry Garniture. A 200-year History of British Mineral Dealers, 1750–1950*. Mineralogical Record, Inc., Tucson, Arizona.

Curtis, Samuel (1805). *Lectures on Botany as delivered in the Botanic Garden at Lambeth by the late William Curtis, F.L.S.* London.

Curtis, W. Hugh (1941). *William Curtis 1746–1799. Fellow of the Linnean Society. Botanist and Entomologist. With some notes on his son-in-law Samuel Curtis, F.L.S.*. Winchester.

Curtis, William (1771). *Instructions for Collecting and Preserving Insects. Particularly moths and butterflies*. London.

Curtis, William (1777). *Linnaeus's System of Botany, so far as relates to his classes and orders of plants; illustrated by figures entirely new, with copious explanatory description. Drawn up for the use of his pupils by William Curtis, author of the Flora Londinensis, and teacher of botany in London*. Gracechurch Street, London.

Cust, L. H. (2004). Hodges, William (1744–1797). *Oxford Dictionary of National Biography*. Oxford University Press, Oxford.

Davies, Hugh (1813). *A systematic catalogue of the native plants of the Isle of Anglesey, in Latin, English, and Welsh; with the habitats of the rarer species, and a few observations*. W. Marchant, London.

Desmond, Ray (1987). *A Celebration of Flowers. Two hundred years of Curtis's Botanical Magazine*. Royal Botanic Gardens, Kew, with Collingridge, Kew.

Desmond, Ray (1995). *Kew. The History of the Royal Botanic Gardens*. London: The Harvill Press with the Royal Botanic Gardens, Kew, London.

Dewhurst, Henry William (1834). *The natural history of the order Cetacea: and the oceanic inhabitants of the Arctic regions*. Published by the author.

'Editor' (1974). Note by editor of the series of papers in this journal giving an introduction: *J. Soc. Bibliogr. Nat. Hist.* 6: 373–79. (The editor was Alwyne Wheeler but name not given.)

Elliott, Graham (1975). The non-descriptive palaeontology of the Sowerbys' *Mineral Conchology*. *Bull. Brit. Mus. (Nat. Hist.), Hist.* 4: 387–99.

Emmerling, Ludwig A. (1799, 1802). *Lehrbuch der Mineralogie*.

Eyles, J. M. (1969). William Smith (1769–1839): A bibliography of his published writings, maps and geological sections, printed and lithographed. *J. Soc. Bibliogr.*

Nat. Hist. 5: 87–109.
Foskett, Daphne (1987). *Miniatures Dictionary and Guide*. Antique Collectors' Club, Suffolk.

Garry, F. N. A. (1905). Notes on the drawings for Sowerby's *English Botany. Journal of Botany* 41 supplement (1903): 1–32, 33–120, 121–276.

Gascoigne, John (1994). *Joseph Banks and the English Enlightenment. Useful knowledge and polite culture*. Cambridge University Press, Cambridge.

Geldart, A. M. (1914). Sir James Edward Smith and some of his friends. *Transactions of the Norfolk & Norwich Naturalists' Society* (Presidential Address). 9: 645–92.

Gilmour, John (1946). *British Botanists*. Collins, London.

Graves, Algernon (1907). *The Society of Artists of Great Britain 1760–1791 The Free Society of Artists 1761–1783 A complete dictionary of contributors and their work from the foundation of the societies to 1791*. London.

Green, David I., Cotterell, Tom F., Jones, Ian, Cox, David & Cleevely, Ron. (2007). Wavellite: Its Discovery and Occurrence in the British Isles. *UK Journal of Mines and Minerals* 28: 11–30.

Grossman, Mark I. (2007). Smithson Tennant. Meteorites and the final trip to France. *Notes & Rec. Roy. Soc.* 61: 265–83.

Grossman, Mark I. (2010). William Higgins at the Dublin Society 1810–20. The loss of a professorship and a claim to the atomic theory. *Notes & Rec. Roy. Soc.* 64: 417–34.

Hallam, A. (1989). *Great Geological Controversies* (2nd edn). Oxford University Press, Oxford.

Harben, Henry, A. (1918). *A Dictionary of London: being notes topographical and historical relating to the streets and principal buildings in the City of London*. Herbert Jenkins Ltd, London.

Harris, Stephen (2007). *The Magnificent Flora Graeca. How the Mediterranean came to the English Garden*. Bodleian Library, Oxford.

Hawkesworth, D. L. (1976). Introduction to facsimile edition of J. Dickson *Plantarum Cryptogamicarum Britanniae*. The Richmond Publishing Co. Ltd, Richmond.

Haüy, René-Just (1801) *Traité de Minéralogie*. Paris.

Henderson, Paul (2013). James Sowerby: Meteorites and his meteoritic sword made for the Emperor of Russia, Alexander I, in 1814. *Notes & Rec. Roy. Soc.* 67: 387–401.

Horwood, Richard (1813) A Map of the Cities of London and Westminster. In: *The A–Z of Regency London*. Paul Laxton (introduction) & Joseph Wisdom (1985). Lympne Castle, Harry Margary, Kent.

Hunt, Robert. (2004) Curtis, William (1746–1799). *Oxford Dictionary of National Biography*. Oxford University Press, Oxford.

Jameson, Robert (1804, 1808). *System of Mineralogy*. Edinburgh.

Kirwan, Richard (1784). *Elements of Mineralogy*. London. (2nd edn: vol. 1 1794; vol. 2 1796.)

Knight, David (2009) Chemists get down to Earth. In: *The Making of the Geological Society of London*, ed. C. L. E. Lewis & S. J. Knell, pp. 93–103. Geological Society, London.

Knowles, John (1821). *An Inquiry into the means which have been take to preserve the British Navy from the earliest period to the present time particularly from that species of decay now denominated dry-rot*. Winchester and Farnham.

Lack, Walter H. with Mabberley, David J. (1999). *The Flora Graeca Story. Sibthorp, Bauer and Hawkins in the Levant*. Oxford University Press, Oxford.

Laming, S. & Laming D. (2007). Etheldred Benett (1776–1845): the first woman geologist? In: *The Role of Women in the History of Geology*, ed. C. B. Burek & B. Higgs, pp. 247–249. Geological Society, London.

Laxton, Paul & Wisdom, Joseph. (1985). *The A–Z of Regency London*. Lympne Castle, Harry Margary, Kent.

Lightfoot, John (1777). *Flora Scotia or, a systematic arrangement in the Linnean method of the native plants of Scotland and the Hebrides*. London.

Linnaeus, Carl (1753). *Species Plantarum*. Laurentius Salvius, Stockholm.

Macdonald, Jessie Bell (1974). The Sowerby Collection in the British Museums (Natural History): a brief description of its holdings and a history of its acquisition from 1821–1971. *J. Soc. Bibliogr. Nat. Hist.* 6: 380–401.

McConnell, A. (2004). Dickson, James (1738?–1822). *Oxford Dictionary of National Biography*. Oxford University Press, Oxford.

Monks, Sarah (2008). Fishy Business: Richard Wright's *The fishery* (1764), Marine Painting, and the Limits of Refinement in Eighteenth-Century London. *Eighteenth-Century Studies* 41: 405–21.

Montagu, George (1802). *Ornithological Dictionary; or, Alphabetical Synopsis of British Birds*. London.

Moore, D. T. (2004). Kirby, William (1759–1850). *Oxford Dictionary of National Biography*. Oxford University Press, Oxford.

Morriss, Roger (2004). *Naval Power and British Culture, 1760–1850. Public Trust and Government Ideology*. Ashgate, Aldershot, England.

Nicholson, A. (2004). Wright, Richard (c.1720–c.1775). *Oxford Dictionary of National Biography*. Oxford University Press, Oxford.

Parkinson, James (1804–1811). *Organic remains of a former world. An examination of the mineralized remains of the vegetable and animals of the Antediluvian world; generally termed extraneous fossils*. London.

Payne, J. F. (2004). Lettsom, John Coakley (1744–1815). *Oxford Dictionary of National Biography*. Oxford University Press, Oxford.

Pennant, Thomas (1791). *Some Account of London* (2nd edn). Dublin.

Petit, Richard E. (2006). Notes on Sowerby's *The genera of recent and fossil shells* (1821–1834). *Arch. Nat. Hist.* 33: 71–89.

Petersen, Ronald H. (1983). An annotated index for James Sowerby's 'Coloured Figures of English Fungi'. *Mycotaxon* XVII: 148–212.

Petit, R. E. (2009). George Brettingham Sowerby, I, II, & III: their conchological publications and molluscan taxa. *Zootaxa* 2189.

Pillinger, C. T. & Pillinger, J. M. (1996). The Wold Cottage meteorite: not just any ordinary chondrite. *Meteoritics & Planetary Science* 31: 589–605.

Porter, Roy (1978). Gentlemen and geology. The emergence of a scientific career, 1660–1920. *The Historical Journal* 21: 809–36.

Porter, Roy (1991). *English Society in the Eighteenth Century*. Penguin, London.

Ramsbottom, J. (1933). Dates of Publication of Sowerby's *English Fungi. Transactions of the British Mycological Society* 18: 167–170.

Ramsbottom, J. (1937). Dry rot in ships. *The Essex Naturalist* 24: 231–67. (Presidential address of 20 March 1937.)

Rashleigh, Philip (1797, 1802). *Specimens of British Minerals, selected from the cabinet of Philip Rashleigh, of Menabilly, in the county of Cornwall*. W. Bulmer & Co.

Ratchford, F. E. (1925). S. T. Coleridge and the London Philosophical Society. *Modern Language Review* 20: 76–80.

Roscoe, I, Hardy, E & Sullivan, M. G. (2009). *A Biographical Dictionary of Sculptors in Britain 1660–1851*. Yale University Press, New Haven, Conn & London.

Rudé, George (1971). *Hanoverian London 1714–1808*. Secker & Warburg, London.

Russan, Lilian & Russan, Ashmore (1923). *Historic Streets of London. An Alphabetical Index*. Simpkin Marshall & Co., London.

Salisbury, Richard, Anthony (1806). *The Generic Characters in The English Botany, collated with those of Linné*. W. Bulmer, London.

Saunders, Gill (1995). *Picturing Plants. An Analytical History of Botanical Illustration*. University of California Press, CA.

Schweizer K. W. (2004). Stuart, John, third earl of Bute (1713–1792). *Oxford Dictionary of National Biography*. Oxford University Press, Oxford.

Secord, Anne (1994). Corresponding interests: artisans and gentlemen in nineteenth-century natural history. *British Journal for the History of Science* 27: 383–408.

Simpkins, Diana M. (1974a). Biographical sketch of James Sowerby, written by his son, James de Carle Sowerby 1825. *J. Soc. Bibliogr. Nat. Hist.* 6: 402–15.

Simpkins, Diana M. (1974b). Childhood reminiscences of James Sowerby. *J. Soc. Bibliogr. Nat. Hist.* 6: 416–17.

Sitwell, Sacheverell (1972). *Fungi. A Look at Coloured Figures of English Fungi or Mushrooms by James Sowerby F.L.S. London 1797*. Smart &Co. (Printers) Ltd. Brackley, Northants.

Sitwell, Sacheverell & Blunt, Wilfrid (1956). *Great Flower Books 1700–1900. A Bibliographical Record of two-Centuries of finely-illustrated Flower Books*. Collins, London. [Note: the bibliography was drawn up by W. T. Stearn, Sabine Wilson and Handasyde Buchanan, and edited by Patrick M Synge.]

Smith, Worthington G. (1898). *Guide to Sowerby's models of British fungi in the Department of Botany, British Museum (Natural History)*. London.

Smith, Worthington G. (1905). Sowerby's drawings of fungi. *Journal of Botany* 43: 156–60, 180–6, 209–15, 239–43, 258–66.

Smith, W. Campbell (1978). Early mineralogy in Great Britain and Ireland. *Bull. Brit. Mus. (Nat. Hist.), Hist.* 6:(3): 49–74.

Sowerby, Arthur De Carle, in collaboration with Sowerby, Alice Muriel & Stone, Joan Evelyn (1952). *The Sowerby Saga. Being a brief account of the Sowerby Family and of its history from earliest times down to the present based upon recent research into available extant literature*. Washington DC.

Sowerby, J. De C. (1825). *A short account of the life of Mr. James Sowerby F.L.S.G.S.W.S.* Unpublished MS. [See Simpkins, D. M. (1974a).]

Sowerby, R. R. (1956). *Sowerby of China. Arthur de Carle Sowerby F.R.G.S., F.Z.S.* Titus Wilson & son, Ltd., Kendal.

Stafleu, F. A. & Cowan, R. S. (1985). *Taxonomic Literature. A selective guide to botanical publications and collections with dates, commentaries and types.* Vol. 5. (2nd edn). Bohn, Scheltema & Halkema, Utrecht.

Stearn, William T. (1967). Sibthorp, Smith, the *Flora Graeca* and the *Flora Graeca Prodromus. Taxon* 16: 168–78.

Swann, Tony (1997). *Great Botanical Books. A bookseller's perspective.* [Printed version of talk given By Tony Swann at the Natural History Museum, London, 13 October 1997.]

Tanswell, John (1858). *The history and antiquities of Lambeth.* London.

Taquet, Philippe (2009). Geology beyond the Channel: the beginnings of geohistory in early nineteenth-century France. In: *The Making of the Geological Society of London*, ed. C. L. E. Lewis & S. J. Knell, pp.155–61. The Geological Society, London.

Thackray, John C. (1985). Separately published prints of fossils in nineteenth century Britain. *Arch. Nat. Hist.* 12(2): 175–99.

Thornbury, Walter (1878). *Old and New London.* Vol.1: 32–53. Cited via British History Online website: http://www.british-history.ac.uk (http://www.british-history.ac.uk/report.aspx?compid=45023 Last accessed 27 March 2015.

Thornton, Robert John (1803). *Sketch of the Life and Writings of the Late Mr. William Curtis.* London.

Torrens, Hugh S. (1997). James Ryan (c.1770–1847) and the problems of introducing Irish 'new technology' to British mines in the early nineteenth century. In: *Science and Society in Ireland: The Social Context of Science and Technology in Ireland, 1800–1950*, ed. P. J. Bowler & N. Whyte, pp. 67–83. Queen's University of Belfast, Belfast.

Torrens, H. S. (2004a). Mawe, John (1766–1829). *Oxford Dictionary of National Biography.* Oxford University Press, Oxford.

Torrens, H. S. (2004b). Ryan, James (c.1770–1847). *Oxford Dictionary of National Biography.* Oxford University Press, Oxford.

Torrens, Hugh S. (2009). Dissenting science: the Quakers among the Founding Fathers. In: *The Making of the Geological Society of London* ed. C. L. E. Lewis & S. J. Knell, pp. 129–44. Geological Society, London.

Walker, Margot (1988). *Sir James Edward Smith M.D., F.R.S., P.L.S. 1759–1828 First President of the Linnean Society of London.* Linnean Society, London.

Waller, G. N. H. (2013) Note on James Sowerby and the discovery of Sowerby's beaked whale, *Mesoplodon bidens. Arch. Nat. Hist.* 40: 270–76.

Wheeler, Alwyne (1974). [See Bibliography entry for 'Editor' (1974)].

Wilson, Wendall E. (1994). The history of Mineral collecting 1530–1799. *Mineralogical Record* 25(6): 1–243.

Withering, William (1796). *An Arrangement of British Plants; according to the latest improvements of the Linnean system to which is prefixed an easy introduction to the study of botany.* (3rd. edn.) London.

Withers, Charles, W. J. (2004). Pennant, Thomas (1726–1798). *Oxford Dictionary of National Biography.* Oxford University Press, Oxford.

Woodward, B. B. (2004). Woodward, Thomas Jenkinson (1745–1820). *Oxford Dictionary of National Biography.* Oxford University Press, Oxford.

Woodward, Horace B. (1907). *The History of the Geological Society of London.* Geological Society, London.

Index

Plant names and book titles in *italic*; illustrations indicated by **bold** page number.

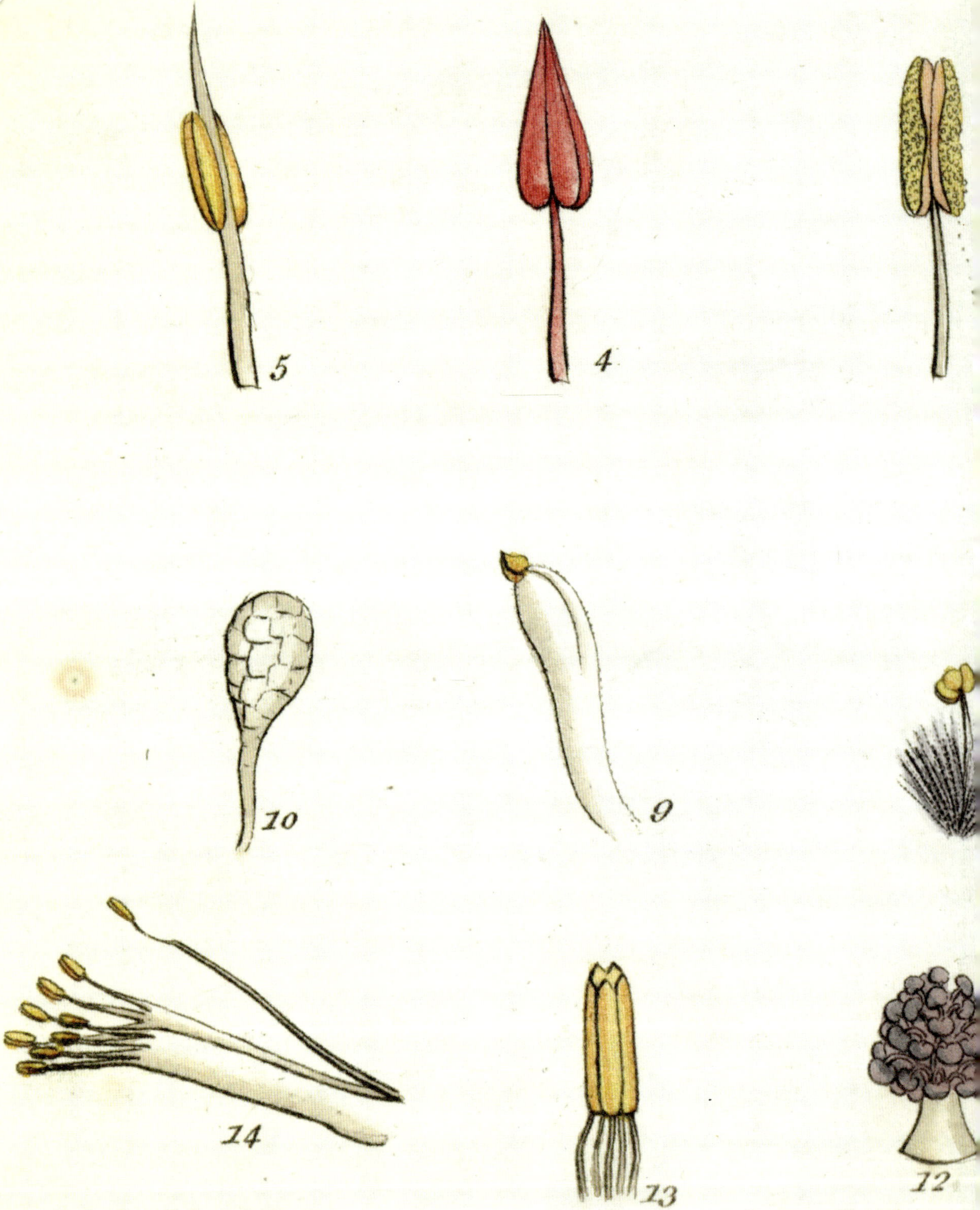

J Sowerby del: et sculp

Publish'd by J. Sow